Mechanical Behavior of Metal-Matrix Composites

Mechanical Behavior of Metal-Matrix Composites

Proceedings of a symposium sponsored by the Composite Materials Committee of The Metallurgical Society of AIME and the Materials Science Division of American Society for Metals held at the 111th AIME Annual Meeting, Dallas, Texas, February 16-18, 1982.

Edited by
John E. Hack
Southwest Research Institute
San Antonio, Texas 78248
and
Maurice F. Amateau
International Harvester Co.
Hinsdale, Illinois 60521

A Publication of The Metallurgical Society of AIME

A Publication of The Metallurgical Society of AIME
420 Commonwealth Drive
Warrendale, Pennsylvania 15086
(412) 776-9000

Printed in the United States of America.
Library of Congress Card Catalogue Number 83-61431
ISBN Number 0-89520-461-4

Foreword

This symposium represents the sixth in the biennial series "Failure Modes in Composites;" approximately ten years have passed since the first symposium was held. Boron and borsic fiber composites, which were the dominant subjects of attention in the earlier years, must now share the limelight with silicon carbide, graphite, and alumina fibers and particles. However, as can be seen from the contents of this volume, the issue of processing effects on mechanical and physical properties, regardless of fiber composition, is still relevant.

The property most often monitored to evaluate the effectiveness of composites is still tensile strength, but fatigue, dimensional stability, and tribological properties are now being cited more often in assessment of composite usefulness.

As the technical community in advanced composites continues to make progress, especially in the metal and ceramic matrix area, this conference series and its proceedings remain important sources of scientific and technical information in the field.

We thank the session chairmen, I. Ahmad, J. C. Hurt, B. A. MacDonald, H. L. Marcus, and R. A. Signorelli, and the contributing authors for their efforts in making this current collection of papers available.

John E. Hack
Southwest Research Institute
San Antonio, Texas

Maurice F. Amateau
International Harvester Co.
Hinsdale, Illinois

February 1983

Table of Contents

FACTORS INFLUENCING THE THERMALLY-INDUCED STRENGTH DEGRADATION OF B/Al COMPOSITES*

J. A. DiCarlo

National Aeronautics and Space Administration

Lewis Research Center

Cleveland, Ohio 44135

Abstract

Literature data related to the thermally-induced strength degradation of B/Al composites were examined in the light of fracture theories based on reaction-controlled fiber weakening. Under the assumption of a parabolic time-dependent growth for the interfacial reaction product, a Griffth-type fracture model was found to yield simple equations whose predictions were in good agreement with data for boron fiber average tensile strength and for unidirectional B/Al axial fracture strain. The only variables in these equations were the time and temperature of the thermal exposure and an empirical factor related to fiber surface preparation prior to aluminum reaction. Such variables as fiber diameter and aluminum alloy composition were found to have little influence. The basic and practical implications of the fracture model equations are discussed.

Introduction

At the high temperatures typically employed for the fabrication of boron/aluminum (B/Al) composites, boron fibers react with the aluminum matrix, forming a weak interfacial reaction product whose growth eventually leads to a loss in fiber and composite strength (1,2). To avoid or minimize this strength degradation problem, it would be of great value to develop a basic understanding of the nature and quantitative contribution of the significant physical factors which influence reaction product formation and its eventual control of fiber fracture. The objective of this paper is to gain such an understanding by carefully examining literature data related to B/Al strength degradation and then analyzing these data in the light of appropriate physical theories concerning interface formation and interface-induced fiber fracture. The results of this study will show that the fracture characteristics of thermally-exposed B/Al composites can be explained well by Griffth fracture theory and the parabolic time-dependent growth of a cracked interfacial reaction product. They will also show that thermally-induced degradation in fiber and composite fracture properties can be empirically described by simple equations involving exposure time and temperature. Aluminum alloy composition and fiber diameter were found to negligibly influence reaction-controlled fracture. However, chemical polishing of the fiber surface prior to aluminum reaction can have a significant beneficial effect.

Discussion

To illustrate typical aluminum reaction effects on boron fiber fracture, the first part of this Discussion section will examine some recent data concerning the thermally-induced strength degradation of aluminum-coated boron fibers. The second part will then present and discuss the assumptions of two theoretical fiber fracture models which have been proposed in the literature to explain the physical influence of the boron-aluminum reaction product on fiber strength. In the third part, the validity of the fracture models will be investigated by comparing their predictions with experimental data concerning the time-temperature dependent fracture of thermally-exposed B/Al composites. Finally, the last part of the Discussion will analyze data which show that chemically polishing boron fibers before subjecting them to aluminum reaction can significantly minimize strength degradation effects.

Strength Degradation of Al-Coated Fibers

In order to obtain a fundamental understanding of fiber-matrix reaction effects in B/Al composites, DiCarlo and Smith (3) recently measured the room temperature tensile and flexure strengths of aluminum-coated 203μm diameter boron fibers which were isothermally exposed for one hour at temperatures typically employed for B/Al fabrication. The pure aluminum coatings were applied at low temperature by ion-plating techniques. Because the coating thicknesses were in the range 2 to 4μm, their load bearing contributions to the fiber fracture stress could be neglected. The results for $\bar{\sigma}_f$ (25mm), the average fiber tensile strength at a 25mm gauge length, are plotted as a function of exposure temperature in the lower curve of Fig. 1. These data show that the fibers retained their original as-produced strengths to 470°C at which point the effects of the boron-aluminum interfacial reaction product

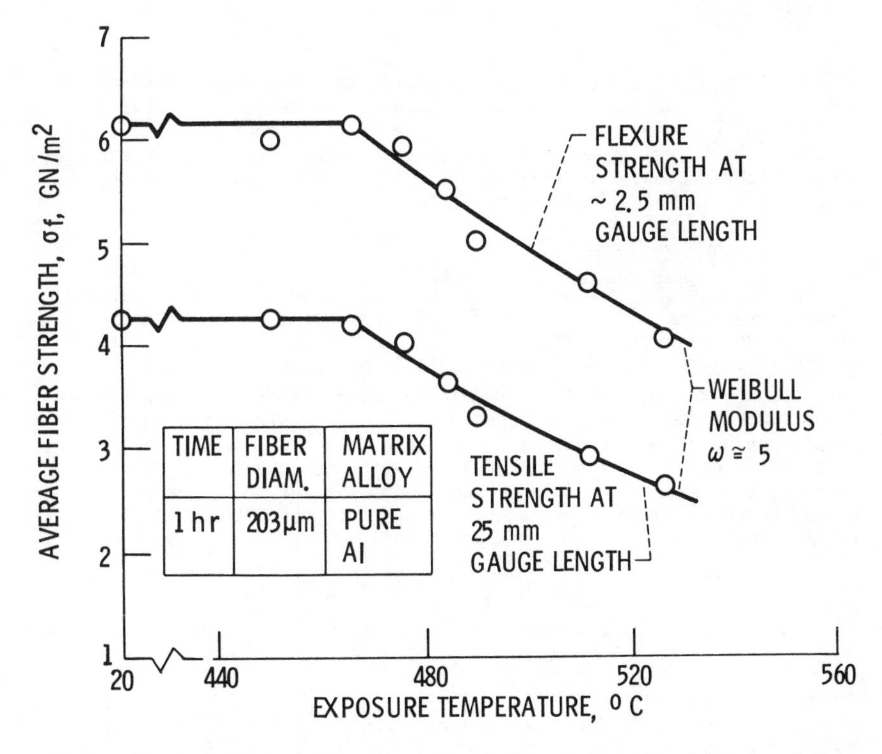

Figure 1 - Tensile and flexure strength degradation for heat-treated aluminum-coated boron fibers (ref. 3)

began to degrade fiber strength. Thus, as temperature increased above 470°C, the average stress required to initiate fiber fracture from reaction-induced flaws decreased to a level below that required for fracture from strength-limiting flaws in the as-produced fibers. Microscopic fracture surface studies revealed that the reaction-induced flaws were located on the fiber surface whereas strength-limiting flaws for the as-produced fibers were located either on the fiber surface or within the fiber's tungsten-boride core.

Because the fiber and composite fracture theories discussed in the next section require knowledge of the Weibull modulus for reaction-controlled fracture, it is of interest here to also examine the average flexure strength results for the aluminum-coated fibers. These data which are plotted in the upper curve of Fig. 1 show a similar threshhold temperature for degradation as the tensile data, but are about a factor of 1.6 larger in magnitude than the tensile data. Since reaction flaws were located only on the fiber surface, DiCarlo and Smith (3) were able to use Weibull statistical theory (4) to show that the higher flexure strengths were due to the fact that the effective gauge length for the flexure test was ~ 2.5mm as compared to 25mm for the tensile test. That is, Weibull theory predicts that the average fiber strength $\bar{\sigma}_f$ should increase with decreasing gauge length L according to:

$$\bar{\sigma}_f(L_1)/\bar{\sigma}_f(L_2) = (L_2/L_1)^{1/\omega} \tag{1}$$

where ω is the Weibull modulus. Thus, from the Fig. 1 data, one obtains $1.6 \simeq (10)^{1/\omega}$ so that the Weibull modulus characterizing reaction-controlled fracture is ~ 5. This ω value is consistent with the scatter in the strength results observed at constant gauge length. For example, strength coefficients of variation from 15 to 20 percent were measured which imply (4) Weibull moduli ranging from 6 to 8.

Reaction-Controlled Fracture Theory

The two primary fracture models that have been proposed to explain aluminum reaction effects on boron fiber fracture assume that the strength decrease with increasing temperature is associated with the growth in thickness of a strength-controlling interfacial reaction product on the fiber surface. Model I proposed by Metcalfe and Klein (1) assumes that due to growth defects within its structure, the reaction product cracks across its thickness, h, at a strain lower than that of the unreacted fiber. Because of good bonding to the fiber surface, the cracked reaction product becomes a surface crack of length h and therefore controls average fiber strength according to Griffth theory; that is:

$$\bar{\sigma}_f = \sigma_f^y\ (L) \qquad \text{for } \sigma_f^y \leq \sigma_f^r \tag{2a}$$

and

$$\bar{\sigma}_f = \sigma_f^r\ (L) = B/[\bar{h}(L)]^{1/2} \qquad \text{for } \sigma_f^y > \sigma_f^r\ . \tag{2b}$$

Here σ_f^y and σ_f^r are the average fiber strengths controlled by as-produced flaws and interfacial cracks, respectively; B is a material constant; and $\bar{h}$ (L) is the average crack size controlling σ_f^r for a test gauge length L. The gauge length dependence was introduced into the thickness h to account for the Fig. 1 results which show that after the same thermal degradation treatment, $\bar{\sigma}_f$ decreased with increasing gauge length, implying by Eq. 2b that the average size $\bar{h}$ of the strength-limiting cracks was increasing with gauge length. Thus, according to Model I, the distribution in fiber strength is explained by a distribution in reaction product thickness.

Model II, a fracture model proposed by Shorshorov et al. (5), assumes that fiber fracture occurs simultaneously with reaction product fracture because the local stress at the newly formed crack tip is greater than the fiber cohesive strength which is assumed equal to 10 percent of the fiber elastic modulus. These authors also assume that the average strength of the interfacial reaction product $\bar{\sigma}_i$ is controlled by Weibull statistics; that is:

$$\bar{\sigma}_i\ (V_1) = \bar{\sigma}_i\ (V_2)\ [V_2/V_1]^{1/\beta}\ . \tag{3}$$

Here V is the reaction product volume and β is the Weibull modulus characterizing the product strength distribution. Presumably β is related to the size and spatial distribution of growth flaws within the reaction product. Thus, according to Model II, average fiber strength under reaction conditions should obey the relation

$$\bar{\sigma}_f = \frac{E_f}{E_i} \bar{\sigma}_i (V_1) = \frac{E_f}{E_i} \bar{\sigma}_i (V_2) \left[\frac{L_2}{L_1}\right]^{1/\beta} \left[\frac{D_2}{D_1}\right]^{1/\beta} \left[\frac{h_2}{h_1}\right]^{1/\beta} . \quad (4)$$

Here E_f and E_i are the elastic moduli of the fiber and interface layer, respectively; and the interface volume V has been replaced by the product of the test gauge length L, the fiber diameter D, and the interface thickness h. It follows then that under this model, if D is constant and h is position-independent, any observed gauge length dependence for average fiber strength can be used to measure the Weibull modulus β. Assuming this to be the case for the Fig. 1 results, these data yield $\beta = \omega \simeq 5$. On the other hand, if h does depend on position as, for example, $(h_2/h_1) = (L_2/L_1)^{1/\eta}$, Eq. 4 and Fig. 1 yield $\beta = \omega(1+1/\eta)$ so that $\beta > 5$ for $\eta > 0$. Thus, if Model II is applicable for the boron-aluminum reaction, Eq. 4 with $\beta \geq 5$ should predict fiber strength degradation as h increases.

To put the fracture model equations into forms suitable for direct comparison with time and temperature dependent fracture data, consideration should be given to the physical mechanisms and kinetics influencing the growth of the interface thickness h. Microscopic studies using thermally-exposed B/Al composites have observed that the boron-aluminum reaction product consists of acicular or needle-type crystals emaninating from the fiber surface (1,2). The shape and structure of these crystals were found to depend on alloying constituents in the aluminum matrix (6). Obviously this type of growth pattern is far from the uniform interface structure implicit in the assumptions of Model I and II. Nevertheless, for the purpose of determining the general applicability of these models, one might as a crude approximation assume that crystal height above the fiber surface is equivalent to interface thickness h. Then, because the growth kinetics of boron-containing interfacial reaction products are typically characterized by a diffusion-limited parabolic time dependence (7,8), one might also assume that the crystal height for the boron-aluminum reaction product increases with time and temperature according to:

$$h = \alpha \, t^{1/2} \exp\left[-Q/2kt\right] . \quad (5)$$

Here α is a normalizing constant, t is exposure time, T is exposure temperature in Kelvin, k is Boltzman's constant, and Q is the activation energy controlling product growth. Since crystal shape is observed to be matrix dependent, the parameters α and Q may depend on alloying constituents in the aluminum.

Support for the parabolic time-dependent growth for the boron-aluminum reaction product can be obtained from scanning electron micrographs of Klein and Metcalfe (2) who studied B/6061-Al composites that were exposed for various times at 504°C. Using the micrographs to measure maximum crystal height and plotting these heights as a function of $(t)^{1/2}$, one obtains the results shown in Fig. 2. Although the scatter is large, the data clearly support a linear relationship between h and $t^{1/2}$.

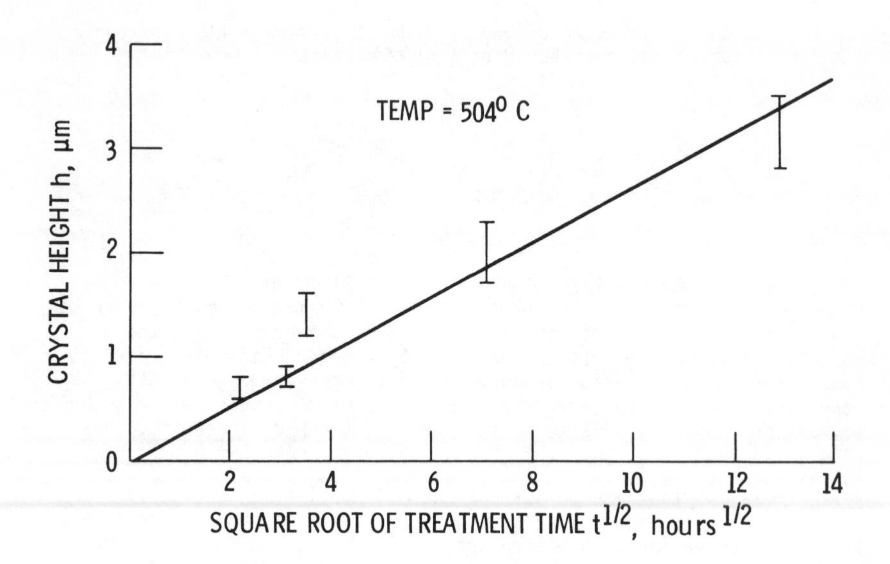

Figure 2 - Maximum height of interfacial crystals as measured from scanning electron micrographs for isothermally-treated B/6061-Al composites (ref. 2)

Assuming the parabolic growth Eq. 5 and the fracture model Eqs. 2b and 4, one then obtains the general result that under reaction-controlled conditions, average fiber strength should depend on time and temperature according to:

$$\sigma_f\ (L) = C\ t^{-1/2m}\ \exp\ [U/T] \qquad (6)$$

where C is an empirically determined constant. For Model I, C = C(L), m=2, and U=Q/4k. For Model II, C=C(L,D), $m=\beta \geq 5$, and $U=Q/2\beta k$.

The degradation in average fiber strength predicted by Eq. 6 can also be used to predict the time and temperature dependent degradation in fracture strain of a unidirectional B/Al composite. Unlike composite tensile strength which contains both a fiber and matrix contribution, composite axial fracture strain ε_c is, to a good approximation, independent of matrix behavior. For this reason, it is possible to directly relate ε_c to the $\sigma_f(L)$ for the reinforcing fibers. In fact, if composite fracture occurs by the cumulative fracture of individual fibers (9), the Appendix shows that:

$$\varepsilon_c\ E_f = \sigma_{bf} = G\ \bar{\sigma}_f\ (L)\ . \qquad (7)$$

Here σ_{bf} is the effective fiber bundle strength within the composite and G is a constant which is independent of reaction conditions. Thus, according to Eqs. 6 and 7, the two fiber fracture models predict that

$$\varepsilon_c = \varepsilon_c^u \qquad \text{for } \varepsilon_c^u \leq \varepsilon_c^r$$

and (8)

$$\varepsilon_c = \varepsilon_c^r = H\ t^{-1/2m} \exp\ [U/T] \text{ for } \varepsilon_c^u > \varepsilon_c^r .$$

Here ε_c^u is the composite fracture strain under conditions in which the fibers maintain their as-produced strengths, ε_c^r is composite fracture strain under reaction-controlled conditions, and $H=CG/E_f$ is a normalizing empirical constant. As discussed in the Appendix, axial fracture strain and thus the H parameter should be independent of composite gauge length for test sections longer than the ineffective length (9) which is typically in the range 2 to 8mm for B/Al composites.

Time and Temperature-Dependent Fracture

Having established theoretical equations for thermally-induced fiber and composite strength degradation, let us now examine their validity by comparing their predictions with time and temperature-dependent fracture data.

Turning first to multifilament composite fracture, the time-dependent axial fracture strain data of Klein and Metcalfe (2) are plotted in Fig. 3 for B/Al composites which were isothermally exposed at 538°C.

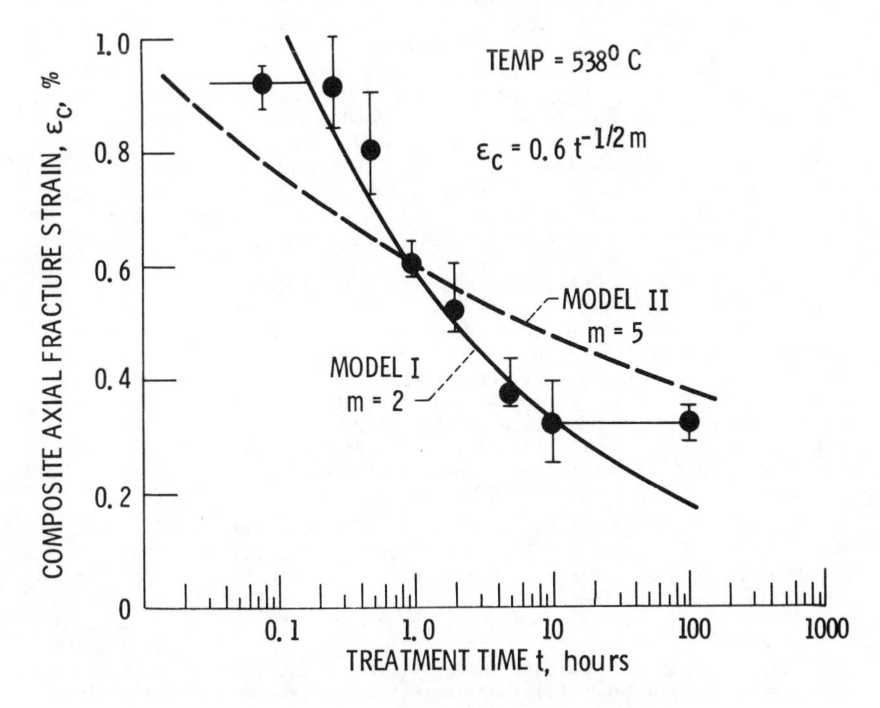

Figure 3 - Comparison of the time-dependent fracture strain predictions of models I and II with axial fracture strain data for isothermally-treated B/6061-Al composites (ref. 2).

The composites consisted of a 6061 aluminum alloy matrix reinforced by 48 volume percent of 142μm diameter fibers. Best fitting Eq. 8 to these data, one obtains the solid and dashed curves in Fig. 3 for fracture Models I and II, respectively. Clearly the $m \geq 5$ values for Model II do not fit the data, indicating that the assumptions of Model II are probably not valid for the boron-aluminum reaction product. On the other hand, the good empirical fit of Eq. 8 using m=2 supports the assumptions involved in Model I fracture, parabolic growth, and cumulative weakening. The Fig. 3 result showing deviation from m=2 behavior at strains below 0.3 percent appears to be related to a change in composite fracture mode since the average fracture strain of fibers extracted from highly degraded composites can fall well below 0.3 percent (10). This implies that Eq. 8 for ε_c degradation should only be compared with ε_c data greater than 0.3 percent whereas Eq. 6 for $\bar{\sigma}_f$ has no such restriction.

To verify whether Eqs. 6 and 8 using Model I parameters could also predict reaction-controlled fracture at other temperatures, the logarithms of fiber strength data were plotted in Fig. 4 as a function of the reciprocal absolute temperature at which the fibers were heat treated. For this plot, two types of strength data were obtained from the literature. The first type were average strengths measured at 25mm gauge length both for aluminum-coated fibers and for fibers which were extracted from heat-treated multifilament B/Al composites. These $\bar{\sigma}_f$ (25mm) data are plotted as open points. The second type were fiber bundle strengths as calculated from $\sigma_{bf} = E_f\varepsilon_c$ where the ε_c are axial fracture strain data for multifilament B/Al composites and E_f is the fiber modulus taken as 400 GN/m^2. Only those experimental results were used in which ε_c data were directly measured or could be easily calculated from fracture stress data and published stress-strain curves. The σ_{bf} data are plotted as closed points. To account for differences in exposure time t (hours), all strength data were normalized to a one-hour exposure by multiplying the experimental data by $(t)^{1/4}$. That is, it was assumed that the time dependence obeyed parabolic growth and the Model I fracture assumptions.

Examination of Fig. 4 shows that over a large temperature range, all reaction-controlled strength data can be fit well to the same straight line. The implications of this result are many. First, it indicates that with $C=3.5\times10^{-4} GN/m^2$, $H=8.8\times10^{-7}$, m=2, and U=7060K, Eqs. 6 and 8 can be used to give good estimates of reaction effects on average fiber tensile strength and on the axial fracture strain of unidirectional B/Al composites. Second, it shows that under reaction-controlled conditions, G=1 so that little difference exists between average fiber strength measured at 25mm gauge length and effective fiber bundle strength in B/Al composites. Third, it indicates that at least empirically, fiber diameter and matrix alloy composition (1100 or 6061) have little effect on fiber fracture as described by Eq. 8. The apparent absence of a diameter dependence is another fact in opposition to Model II fracture theory (cf. Eq. 4). Fourth, the assumptions of Model I fracture, parabolic interfacial growth , and cumulative weakening for B/Al composites appear to conform to reality. Finally, assuming the validity of these assumptions, the best fit U value suggests that Q, the energy controlling interfacial growth, is 4kU=2.4ev (56 kcal/mole).

It should be mentioned that although the effects of time and temperature on composite fracture strain degradation can now be accounted for by Eq. 8, this simple empirical equation should only apply

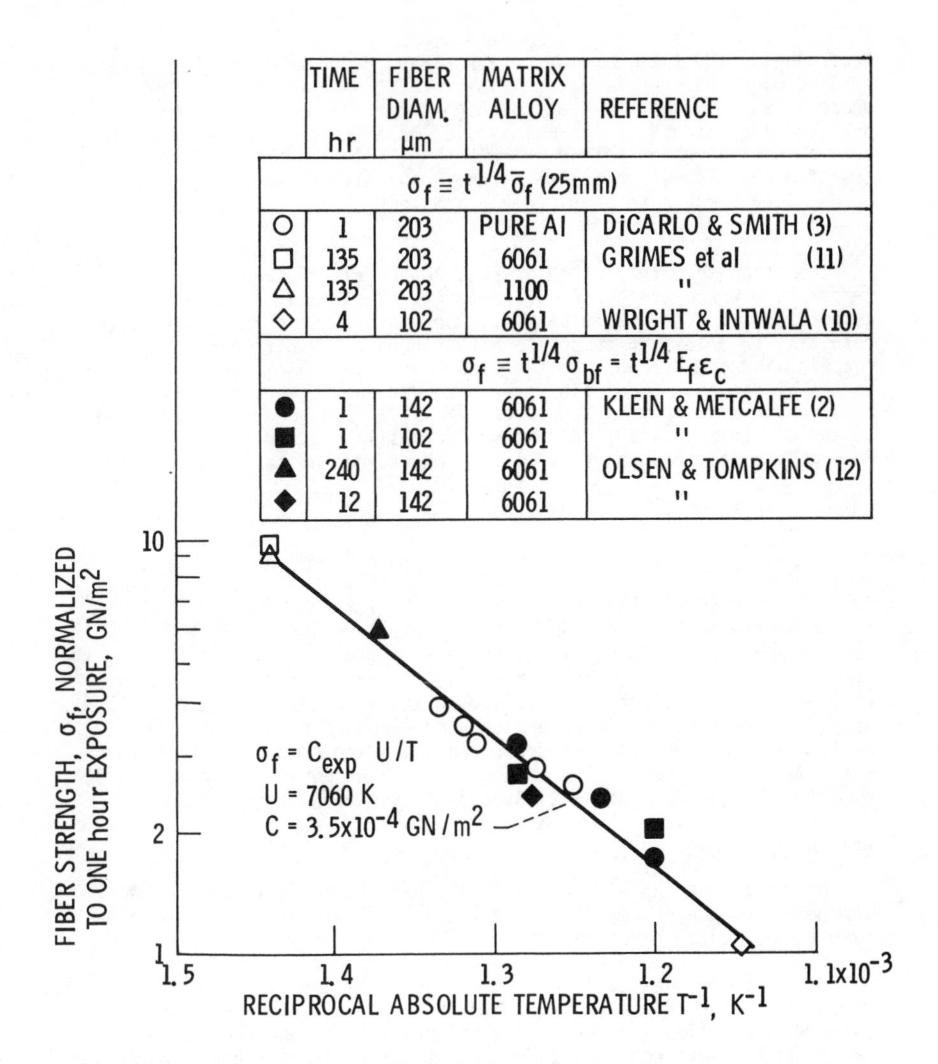

Figure 4 - Comparison of fiber strength data with the time and temperature-dependent strength predictions of model I.

for conditions involving continuous thermal exposure and should not be accurate for cyclic types of thermal exposure. This is due to the fact that after being thermally cycled between a low and high temperature, the strength of B/Al composites have been observed to degrade to a lower level then the strength of composites which were isothermally exposed for the same equivalent time at the high temperature (11, 12). This result has been explained primarily by the cycling-induced breakup of interfacial structure caused by the mechanical fatigue effects associated with the mismatch in fiber and matrix thermal expansion. Thus, under these conditions, the protective nature inherent in the parabolic type of interfacial growth would not completely exist.

Accordingly, the time dependence for strength degradation after thermal cycling might be expected to be closer to $t^{-1/2}$ rather than $t^{-1/4}$ where t is equivalent time at the high temperature. An additional problem associated with thermal cycling is a fatigue-induced debonding between the fiber and matrix which can not only lead to reduced stress transfer and a lower composite strength but also to an exposure of the boron fiber surface to detrimental high temperature reaction with oxygen (13).

The strength data plotted in Fig. 4 were all measured at room temperature where the boron fibers deform elastically. However, if B/Al composites are tested at elevated temperatures, the boron fibers creep, resulting in a loss in composite strength entirely different than the reaction-induced strength loss. DiCarlo studied this creep problem (14) and concluded that under fiber creep conditions, composite axial fracture strain ϵ_c is to a good approximation independent of the time t' and the temperature T' during which tensile loading is applied. However, composite strength will fall off according to:

$$\sigma_c \cong \overset{\circ}{\sigma}_c / A\,(t', T') \tag{9}$$

where $\overset{\circ}{\sigma}_c$ is the room temperature composite strength and A (t', T') is a fiber creep function which increases from a value of unity as the time and temperature of loading increase. It follows then that for high temperature applications, B/Al tensile strength could depend both on the time-temperature conditions involved in the exposure and also on the time-temperature conditions involved in the loading. On the other hand, B/Al fracture strain will depend only on exposure conditions. Thus, even if B/Al composites were under axial loading at boron-aluminum reaction temperatures, Eq. 8 with the empirical constants from Fig. 4 should still yield a good estimate of composite fracture strain.

Chemical Polishing Effects

In their study of aluminum reaction effects, DiCarlo and Smith (3) also measured the thermally-induced strength degradation of aluminum-coated 203μm diameter boron fibers which were chemically polished in nitric acid prior to the coating and thermal exposure. The initial polish treatment yielded fibers with higher average strength and lower strength scatter than the original as-produced fibers. The improved strength properties were caused by the removal of low-strength high-variability flaws from the as-produced fiber surface, thereby leaving fiber fracture to be controlled only by higher-strength lower-variability flaws located within the fiber's tungsten-boride core. After coating the polished fibers with aluminum at low temperature, the fiber strength properties were found to be unchanged. However, after one-hour isothermal exposure at temperatures above 500°C, reaction-related strength degradation effects were observed as shown by the Fig. 5 results for average tensile and flexure strength. These data were measured at room temperature using coated pre-polished fibers with reduced diameters of 195, 180, and 140μm.

Comparing the two data sets of Fig. 5 with the corresponding data sets of Fig. 1, one observes that for the same test and reaction conditions, the average strength of the coated pre-polished fibers was on the average a factor of 1.6 greater than that of the coated as-produced fibers. In addition, the threshhold temperature for tensile strength degradation was ~ 45°C higher for the pre-polished fibers.

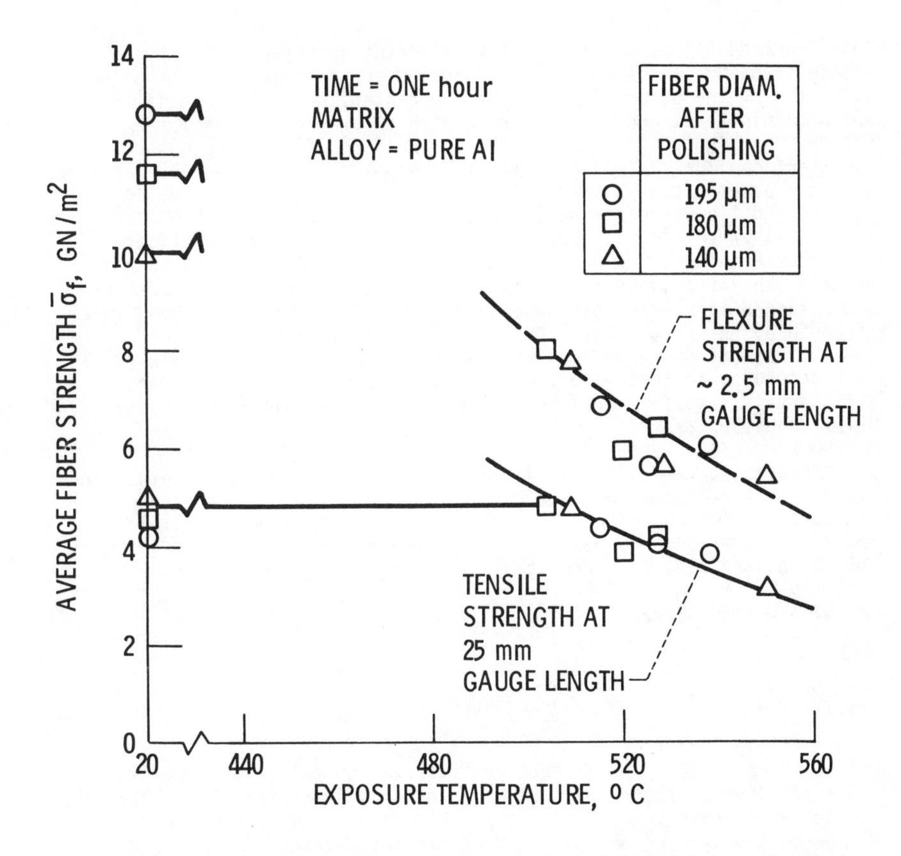

Figure 5 - Tensile and flexure strength degradation of heat-treated aluminum-coated boron fibers which were chemically polished prior to the coating and heat treatments (ref. 3).

Thus, chemical polishing not only improved the strength properties of unreacted fibers, but also significantly increased the stress levels required for reaction-controlled fracture. This in turn shifted the strength degradation curve for pre-polished fibers to higher temperatures. The practical implications of these results for producing stronger B/Al composites at higher fabrication temperatures are discussed in some detail by DiCarlo and Smith (3).

Any fracture model proposed to explain the beneficial effects of chemical polishing must account for the fact that only slight polishing is required to improve the strengths of both unreacted and reacted fibers. Wawner (15) has suggested that the strength improvement for unreacted fibers is due to the smoothing of the crack-like structure associated with growth nodules generally found on the surface of as-produced fibers. Because little change in nodule height was observed with slight polishing, DiCarlo and Smith (3) suggest that for unreacted fibers, the primary strengthening effect of polishing is to significantly increase the average radius of curvature r_0 at the nodule boundary.

This crack blunting concept together with the interfacial fracture mechanism of Model I suggests that under reaction conditions, the lower strengths of the as-produced fibers are controlled by interfacial cracks which terminate at nodule boundaries with an average tip radius r_o, whereas the higher strengths of the pre-polished fibers are controlled by interfacial cracks which terminate randomly on the fiber surface with a larger average tip radius r_i. Presumably r_o is dependent on the as-produced fiber surface structure, whereas r_i is dependent on mechanical properties of the interface alone. Assuming that fibers heat-treated and tested in the same manner have the same interfacial crack depth, it follows then from simple fracture theory (16) that the pre-polished fiber strength should be larger than the as-produced fiber strength by the factor $\phi=(r_i/r_o)^{1/2}$. For the range of reaction and test conditions examined, the results of Figs. 1 and 5 yield ϕ=1.6. Thus according to this model, sharp nodule boundaries not only act as detrimental flaws for unreacted fibers but also act as stress raisers on interfacial cracks.

Summarizing the polishing results of Fig. 5 and the time-temperature results of Fig. 4, one can now express the thermally-induced strength degradation of boron fibers in contact with aluminum by two simple empirical equations for average fiber tensile strength $\bar{\sigma}_f$ (25mm) and for composite axial fracture strain ϵ_c. These equations are

$$\bar{\sigma}_f \text{ (25 mm)} = E_f \, \epsilon_c \tag{10}$$

and

$$\epsilon_c = (8.8\text{x}10^{-7}) \, \phi \, t^{-1/4} \exp\,[7060/T] \quad . \tag{11}$$

Here t is exposure time in hours, T is exposure temperature in Kelvin, E_f is 400 GN/m^2, ϕ=1 for as-produced fibers, and ϕ=1.6 for chemically polished fibers. The data of Figs. 3 and 4 indicate that Eqs. 10 and 11 should be accurate to within ± 10 percent as long as $\sigma_f < \sigma_f^u$ and 0.3 percent $< \epsilon_c < \epsilon_c^u$ where σ_f^u and ϵ_c^u are the average strength and composite fracture strain for the as-produced unreacted fibers. The excellent predictive accuracy of Eq. 10 for chemically polished fibers is shown by the solid line through the $\bar{\sigma}_f$ (25mm) data of Fig. 5.

Concluding Remarks

This study has shown that the thermally-induced strength degradation of B/Al composites can be explained well by a fiber fracture model based on the parabolic time-dependent growth of a cracked interfacial reaction product. Simple analytical equations were derived whose predictions were found to be in good agreement with literature data for the average tensile strength of reacted boron fibers and for the axial fracture strain of isothermally-exposed unidirectional B/Al composites. The only variables in these equations were the time and temperature of the thermal exposure and an empirical factor related to fiber surface preparation prior to composite consolidation. Such factors as fiber diameter and aluminum alloy composition were found to have little influence on reaction-controlled boron fiber fracture.

With the development of the fiber and composite fracture equations, it should now be possible to obtain good estimates of reaction effects for B/Al composites which are subjected to continuous high temperature exposure either during composite fabrication or during structural application. Whereas the composite strain equation could be used to evaluate thermally-induced losses in the axial fracture properties of unidirectional composites, the fiber strength equation could be used to understand general reaction-related fiber weakening for either unidirectional or angle-plied composites. An additional advantage of the fiber strength equation is that it should allow a better understanding of those temperature, time, and fiber surface conditions which must be avoided if the strength properties of the original as-produced fibers are to be retained after composite fabrication.

Appendix

Composite Fracture Theory

Characterizing strength by two-parameter Weibull theory (4), the average tensile strength $\bar{\sigma}_f$ for a group of fibers tested individually at gauge length L is given by:

$$\bar{\sigma}_f = \gamma L^{-1/\omega} \Gamma(1+1/\omega) \quad \text{(A1)}$$

where γ is a scale parameter, ω is the Weibull modulus, and Γ is the gamma function. If these same fibers are tested as a parallel bundle, the fiber bundle strength is given by:

$$\sigma_{bf} = \gamma L^{-1/\omega} (\omega e)^{1/\omega} \, . \quad \text{(A2)}$$

If the fiber bundle is infiltrated with matrix material to form a unidirectional composite, the fiber bundle strength is generally observed to be greater than that measured without the matrix material. This can be attributed to the fact that the matrix localizes the loss of load carrying ability in a broken fiber, thereby allowing a greater number of individual fibers breaks to occur before complete bundle fracture. Rosen (9) analyzed this cumulative mode of composite weakening and concluded that for a large number of fibers, the fiber bundle strength can be predicted by Eq. A2 with L replaced by the "ineffective" length δ. For a ductile matrix like aluminum, δ can be estimated from:

$$\delta = \sigma D/2\tau_{ym} \quad \text{(A3)}$$

where $\sigma=\sigma_{bf}$, D is fiber diameter, and τ_{ym} is the shear yield strength of the matrix. An important consequence of Rosen's theory is that the fiber bundle length for a composite is determined by material properties, so that as long as the composite test length L_c is greater than δ, the fiber bundle strength σ_{bf} should be independent of L_c.

Assuming cumulative weakening and neglecting any residual stresses on the fiber, it follows from Eq. A2 that the axial fracture strain ε_c of a unidirectional composite is given by:

$$\varepsilon_c = \sigma_{bf}(\delta)/E_f = \gamma \delta^{-1/\omega} (\omega e)^{-1/\omega} /E_f \quad \text{(A4)}$$

where E_f is fiber modulus. Alternatively, using Eq. A1 to eliminate γ,

$$\varepsilon_c = \bar{\sigma}_f (L) G/E_f \quad \text{(A5)}$$

where

$$G = (L/\delta)^{1/\omega} \left[(\omega e)^{-1/\omega} / \Gamma(1+1/\omega) \right] . \tag{A6}$$

Thus for composite test lengths greater than δ and reaction conditions characterized by constant ω, composite axial fracture strain should be directly proportional to the average tensile strength of individual reacted fibers extracted from the composites. The proportionality constant G is, however, weakly dependent on fiber diameter through the δ parameter (cf. Eq. A3).

References

1. A. G. Metcalfe and M. J. Klein: Interfaces in Metal Matrix Composites, Vol. 1, Composite Materials, A. G. Metcalfe, ed., p. 125, Academic Press, New York, 1974.

2. M. J. Klein and A. G. Metcalfe: AFML TR-71-189, 1971.

3. J. A. DiCarlo and R. J. Smith: NASA TM-82806, 1982.

4. H. T. Corten: Modern Composite Materials, L. V. Broutman and R. H. Krock, eds., pp. 27-105, Addison-Wesley, Reading, Mass., 1967.

5. M. KH. Shorshorov, L. M. Ustinov, A. M. Zirlin, V. I. Olefirenko, L. V. Vinogradov: J. Mater. Science, 1979, vol. 14, pp. 1850-1861.

6. W. Kim, M. J. Koczak, and A. Lawley: in ICCM/2, Second International Conference on Composite Materials, B. Noton, R. Signorelli, K. Street, and L. Phillips, eds., pp. 487-505, The Metallurgical Society of AIME, New York, 1978.

7. J. A. DiCarlo and T. C. Wagner: NASA TM-82599, 1981.

8. J. Thebault, R. Pailler, G. Bontemps-Moley, M. Bourdeau and R. Naslain: J. Less Com. Metals, 1976, vol. 47, pp. 221-233.

9. B. W. Rosen: AIAA J., 1964, vol. 2, pp. 1985-1991.

10. M. A. Wright and B. D. Intwala: J. Mater. Science, 1973, vol. 8, pp. 957-963.

11. H. H. Grimes, R. A. Lad and J. E. Maisel: Met. Trans. A., 1977, vol. 8A, pp. 1999-2005.

12. G. C. Olsen and S. S. Tompkins: NASA Technical Paper 1063, 1977.

13. M. K. White and M. A. Wright: ONR #N00014-75-C0352 NR 031-760, 1976.

14. J. A. DiCarlo: J. Comp. Mater., 1980, vol. 14, pp. 297-314.

15. F. E. Wawner, Jr.: Boron, Vol. 2, Preparation, Properties, and Applications, G. K. Gaule, ed., pp. 283-300, Plenum Press, New York, 1965.

16. A. S. Telelman and A. J. McEvily, Jr.: Fracture of Structural Materials, John Wiley, New York, 1967.

TENSILE DEFORMATION AND FRACTURE BEHAVIOR OF UNIDIRECTIONAL DEPLETED URANIUM/TUNGSTEN COMPOSITES

A. Pattnaik, C. Kim, and R. J. Weimer

Composite Materials Branch
Naval Research Laboratory
Washington, D.C. 20375, USA

Uniaxial longitudinal and transverse tensile properties of uranium/tungsten composites were studied in detail. These composites were fabricated by unidirectionally reinforcing a U-0.75Ti depleted uranium alloy (DU) with either pure tungsten (W) or tungsten-rhenium alloy (W-3Re) fibers. The W-3Re fiber reacted with the matrix to a lesser extent than pure W fiber. Parametric studies on DU/W composites established a baseline with which to compare the heat-treated DU/W-3Re composites. The structure and strength of the diffusion-affected zone in the fiber varied with the composition of the fiber and controlled both longitudinal and transverse tensile strength properties. Effects of the following variables on the tensile behavior of these composites have been analyzed: fiber type, fiber diameter, volume fraction of fibers, and heat treatment which influences yield strength and residual stresses.

Introduction

Composite materials are now well-known for their high specific stiffnesses and strengths - these properties are being utilized effectively in the aerospace industry today. These composites are made by the incorporation of high-strength fibers like Gr and B into low-density matrix materials like Al, Mg, and a variety of polymers. However, the concept of fiber strengthening of high-density matrix materials is relatively new. Although the specific mechanical properties improve only marginally, the failure mode of the high-density matrix material can be altered in a significant manner by the reinforcing fibers. For applications that are based on fracture-mode control, this type of composite holds great promise and the present study focused on characterizing the strength properties of such high density composites.

Tungsten-reinforced uranium composites were selected for investigation because expected property improvements might benefit the performance of depleted uranium (DU) in several well-known applications (1). Moreover, tungsten fibers were commercially available in a variety of sizes and compositions that facilitated parametric studies of the composite system. In addition to high density, tungsten fibers have good thermal stability which is critical in the retention of fiber strength during high temperature fabrication. Specifically, a depleted uranium alloy (U-0.75Ti) was unidirectionally reinforced with pure tungsten (W) or tungsten-rhenium alloy (W-3Re) fibers by a process of liquid metal infiltration followed by controlled solidification (2), so that the effects of fiber type, diameter, volume fraction, and composite heat treatment on tensile behavior of the composite could be evaluated.

Other composite systems with comparable mechanical properties to those of DU/W composites are steel/W (3), superalloy/W (4), Cu/W, and Cu-alloy/W composites (5). With negligible diffusional reactions, the Cu/W system is ideal for studying mechanical properties. However, alloying elements in copper diffuse into the tungsten fiber rendering it weak through the following diffusional processes: (i) diffusion-penetration followed by partial or complete recrystallization, (ii) diffusion-penetration followed by solid solution formation, and (iii) diffusion-penetration followed by a two-phase formation. Each of these three processes reduces the strength of *in situ* fibers and, consequently, the strength of the composite. In the steel/W system brittle intermetallics form at the interface, and fail prematurely to reduce composite strength drastically (3). All of these processes may be operative in the DU/W composite system.

In the present study the behavior of unidirectional DU/W composites in both longitudinal and transverse tension is analyzed and correlated with microstructural and fractographic features. A concurrent study dealing with longitudinal and transverse compression is presented elsewhere (6). In particular, the behavior of DU/W composites is compared to that of DU/W-3Re composites which were heat-treated to improve the matrix strength. The W-3Re fiber has greater thermal stability and ductility than the W fiber, and reacts with DU matrix to a lesser extent; hence, the comparison.

Experimental Procedure

Specimen Fabrication

The unidirectional DU/W composites were fabricated by a liquid metal infiltration technique, followed by controlled solidification (2). Predetermined numbers of commercial W (designation NS-55) or W-3Re fibers were held in molds and infiltrated with the molten DU alloy to produce composites having nominal fiber volume fractions of 35, 45 and 55 percent. Moreover, 45 volume percent composites were prepared with different fiber diameters, namely, 127, 305, and 762 μm. Control specimens with no fibers were also fabricated following the same procedure used for the composites. The cast composites could be machined and ground into specimens with special tools (1). A complete list of test materials and assigned nomenclature is presented in Table I.

Table I - Test Materials

Matrix (Control)	Fiber	Composites	
		d_f = Constant	V_f = Constant
DU*	W	DU/35W(305)**	DU/45W(127)
		DU/45W(305)	DU/45W(305)
		DU/55W(305)	DU/45W(762)
	W-3Re	DU/45W-3Re(305):HT***	

* DU is U-0.75Ti depleted uranium alloy.
** Numbers in parentheses indicate fiber diameters in μm and the numbers preceding the fiber designations are volume percent of fibers.
*** HT: Heat-treated composite.

A preliminary study (6) showed that the diffusion-affected zone that led to partial recrystallization of the fibers was much smaller in the DU/W-3Re composites than in the DU/W composites. Moreover, the interfacial bond strength of the DU/W-Re composite appeared superior to that of the DU/W. Consequent to these observations, the DU/W-3Re composites were heat-treated to strengthen the matrix. The heat treatment consisted of the following steps: the composites were (i) soaked in argon at 800°C for 1 hour, (ii) oil quenched, (iii) aged in vacuum at 400°C for 4 hours, and then (iv) cooled in vacuum.

Both the longitudinal tension(LT) and transverse tension(TT) specimens were of flat dog-bone shape. In order to utilize the minimum representative amount of material, the sizes of LT and TT specimens were different. The LT specimens had the following dimensions: gage length of 38 mm, widths of 10 mm in the gage area and 15 mm in the gripping area with a transitional radius of 13 mm, gripping length of 25 mm, and thickness of 6.4 mm. In contrast, the TT specimens had the following dimensions: gage length of 18 mm, widths of 8 mm in the gage area and 15 mm in the gripping area with a transitional radius of 13 mm, gripping length of 15 mm, and thickness of 6.4 mm.

Preliminary tests indicated that failure of the TT specimen occurred in the gage-length area, whereas, failure of the LT specimen initiated at the shoulder. The gage section of the LT specimens had to be tapered by grinding with a contoured wheel to ensure that the failure occurred in the gage length. The required taper was more than twice that recommended by the ASTM for homogenous materials, i.e., 0.10 mm for a width of 13 mm.

Testing Procedure:

Room temperature (~22°C) tension tests were conducted on all materials using a closed-loop servo-hydraulic testing system operated under displacement control at a quasi-static strain rate of $\sim 1.4 \times 10^{-4}$ s^{-1}. Two strain gages were mounted on the opposite sides of the specimen to eliminate the effects of bending on the strain measurement. The specimens could be pulled to failure, using wedge-type grips with fine serrations; shoulder tabs on the specimens were unnecessary for these composites. The strain gages were protected by a flexible silicon rubber coating and the testing was carried out soon after mounting the strain gages on freshly prepared surfaces, because DU oxidizes rapidly in air. For the same reason, the fractured specimens were preserved in vacuum desiccators and examined in the scanning electron microscope (SEM) as soon as possible.

Metallography:

Metallographic specimen preparation required careful handling and disposal of the ground DU particles to prevent any possibility of heavy-metal contamination. Other than these precautions, standard metallographic specimen preparation for composite materials was followed. In particular, vertical loads were needed during grinding on silicon carbide papers or on diamond-loaded cloths to keep the fiber and matrix surfaces at the same level.

The DU matrix was etched with a solution containing 70 percent phosphoric acid, 25 percent sulphuric acid and 5 percent nitric acid. The tungsten fibers were etched with either the Murakami's agent or with a solution containing 60 percent lactic acid, 20 percent nitric acid, and 20 percent hydrochloric acid; the Murakami's agent was sensitive to alloying elements, whereas, the second one showed the extent of diffusion processes along the grain boundaries. Photomicrographs were taken soon after etching, again to overcome the DU oxidation problem.

Experimental Results and Observations

The results of the composite microstructure analysis, the longitudinal tension tests, and the transverse tension tests are presented separately for clarity and convenience. Test materials and nomenclature were presented in Table I.

Composite Microstructure:

Extensive microstructural analysis was conducted using both SEM fractography and optical metallography to evaluate the integrity of the as-fabricated composites and to correlate microstructure with the mechanical behavior. Longitudinal and transverse sections of failed composites were obtained from regions that were 2 to 5 mm away from the fracture surfaces, and subsequent metallography was combined with fracture surface analysis to elucidate the fracture process.

The microstructures of as-fabricated(AF) DU/W composites with fiber diameters of 127, 305, and 762 μm are presented in Figs. 1(a), 1(b), and 1(c), respectively. Corresponding micrographs of specimens pulled to failure in longitudinal tension are shown in Figs. 1(d), 1(e), and 1(f), respectively. The importance of presenting the deformed microstructure will become evident in the discussion section. The diffusion-affected zone in the fiber appears dark in the micrographs and partial recrystallization in this area was evident at higher magnifications. Observation at relatively low magnification indicated that the fiber distribution was random and that some fiber contacts were observed at a 55 percent volume fraction of fibers (V_f). All test specimens were X-rayed and the radiographs showed random fiber misalignments ($\theta < 5°$); some specimens exhibited casting defects (pores).

Semi-quantitative electron-probe microanalysis and etching of the fibers with the Murakami's agent indicated interesting diffusional processes, but those results had no direct impact on the mechanical behavior of the composites. In summary, the following observations were made. Both Ti and U diffuse into the W fiber primarily through grain boundary diffusion; this was confirmed with an estimate of the penetration distance calculated on the basis of volume versus grain boundary diffusion at 1200°C. Penetration distances estimated on the basis of grain boundary diffusion matched closely the experimentally observed ones shown in Fig. 1. Titanium has a tendency to form a two-phase region while diffusing through W fibers (5). As a result, the diffusion rate of Ti into W fiber is expected to be lower than that of U into W (7); U forms a solid solution with W at ~1200°C. Consequently, Ti was partitioned while U diffused into W, leaving a Ti concentration peak. The recrystallization front followed behind the U penetration, which is seen as a boundary in Fig. 1.

Microstructures of the U-0.75Ti(DU) matrix in the AF and HT conditions are shown in Figs. 2(a) and 2(b), respectively. In the AF condition, Fig. 2(a) shows a two-phase, lamellar structure of α(0.0%Ti) and δ (U_2Ti, 9.3%Ti) phases as a result of the eutectoidal transformation from the γ phase (8). In contrast, the heat-treated DU matrix exhibits a single-phase, tempered, acicular martensitic (α_a') structure. Microhardness and yield strength of the DU matrix in both the AF and HT conditions and microhardness of the W fibers, are presented in Table II.

Longitudinal Tension:

In the initial stages of this research program, commercially available W fibers were selected that would be most suitable for strengthening and heat-treating the composites. Diffusion-affected zones and fractographs of the LT specimens showed that the alloy fibers, like W-3Re and W-$1ThO_2$, reacted with the matrix to a lesser extent than the pure W fibers (6). Moreover, the interfacial bond strength of DU/W-3Re composite appeared superior to that of DU/W-$1ThO_2$. Consequently, the DU/3-Re composite was selected for heat treatment while the less costly DU/W system was used for parametric studies, namely, the variation of strength properties with V_f and fiber diameter (d_f) under longitudinal and transverse tensile loads. In addition, the DU/45W(305) and DU/45W-3Re(305):HT composites were compared to evaluate the effects of heat treatment.

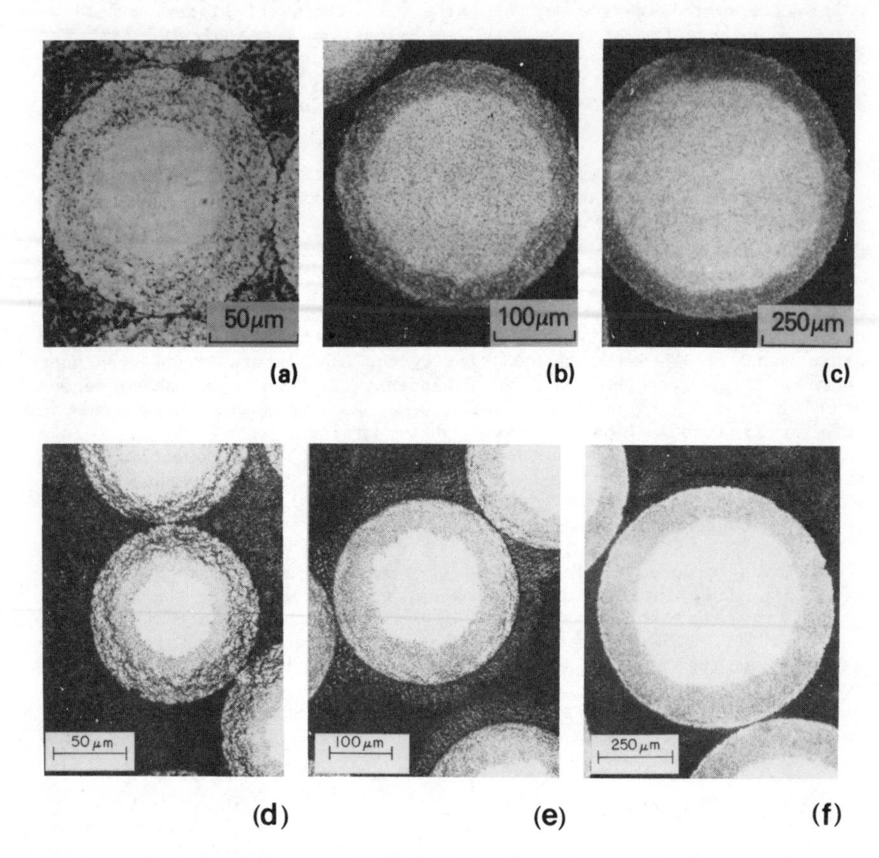

Fig. 1 - Optical micrographs of DU/45W composites as fabricated (top row) and after testing to failure in longitudinal tension.
Fiber diameters are 127, 305, and 762 μm, (left to right).

Fig. 2 - Microstructure of matrix in: (a) as-fabricated DU/45W and (b) heat-treated DU/45W-3Re composites.

Table II - Effects of Heat Treatment on the DU/W Composite

Property	Fiber	Matrix	
		As-fabricated	Heat-treated
Hardness*	645V VHN	400 VHN	625 VHN
Yield Strength, σ 0.2%	**	415 MPa(T) 627 MPa(C)	1 255 MPa(T)***

* Average values.
** Yield strength somewhat less than the 2344 MPa measured for virgin W fibers; T=tension; C=compression.
*** Estimated from Ref. 8, using hardness and heat treatment parameters.

Before presenting the composite properties, the fiber properties will be described. During fabrication of these composites, the fibers are exposed to temperatures in the range of 1100 to 1200°C for approximately 30 minutes, opening the possibility of severe thermal degradation, in addition to the diffusional effects. In an effort to separate the two effects, annealing experiments on W and W-3Re fibers were conducted in vacuum, and the effects of this annealing on the room-temperature tensile strengths are shown in Fig. 3. The strength of both fibers decreased after annealing at ≥800°C. However, the W-3Re fibers retained their strength more effectively than the W fibers, and the elongation to fracture actually increased for the W-3Re fiber after high-temperature exposure. The elongations to fracture, shown in Fig. 3, are approximate values because they were determined from ram displacement on the testing machine.

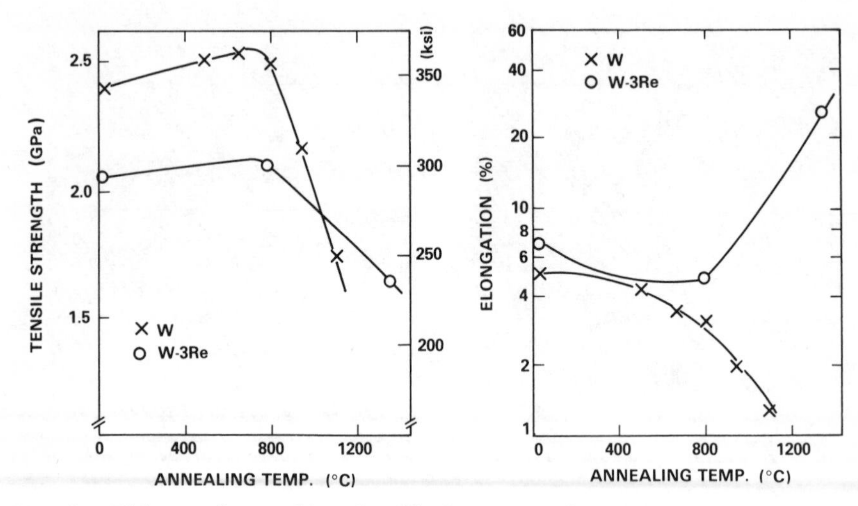

Fig. 3 - Effects of annealing for 30 minutes on the room-temperature tensile strength and elongation of W and W-3Re fibers.

The ductile, pure W fibers fail in tension with the usual cup-and-cone fracture at room temperature. However, they failed in a brittle manner exhibiting a flat surface after annealing at temperatures above 800°C. In contrast, the W-3Re fiber exhibited a ductile failure mode after similar annealing treatments, as expected from the plots in Fig. 3.

Strain to failure of the pure W fibers also varied with fiber diameter, in the absence of annealing effects. The 127 and 305 μm tungsten fibers failed in a ductile manner whereas the 762 μm fiber behaved unpredictably, sometimes failing in a ductile manner and other times failing in a brittle manner. Variation of ductility and fracture mode of W fibers with fiber diameter has been studied systematically by Leber, et. al.(10), who observed a transition from ductile to brittle fracture at a $d_f \sim 600$ μm. The transition was explained on the basis of fiber outer-layer texture which varied with fiber diameter as would be expected from a wire-drawing operation.

Stress-strain curves of DU, DU/45W(305) and DU/45W-3Re(305):HT are presented in Fig. 4. A complete list of relevant strength properties of all composites tested is given in Table 3. The DU matrix strain to failure was much higher than that of either the DU/W or DU/W-3Re composites, indicating that composite failure was initiated by fiber fracture. The experimental Young's moduli of the composites were within 5 percent of the theoretical rule-of-mixture(ROM) predictions. The composites exhibited typical three-stage tensile behavior. The stage I yield strength, σ_L^{ty} (I), and stage II yield strength, σ_L^{ty} (II), of DU/W and DU/W-3Re:HT composites at a V_f of 45 percent were of equal magnitude, Fig. 4. However, their ultimate strengths differed significantly.

Longitudinal tensile strengths of DU/W composites generally increased linearly with V_f as shown in Fig. 5; however, a strength reduction at a V_f of 55 percent was observed and attributed to the fiber contacts observed at this level of fiber loading.

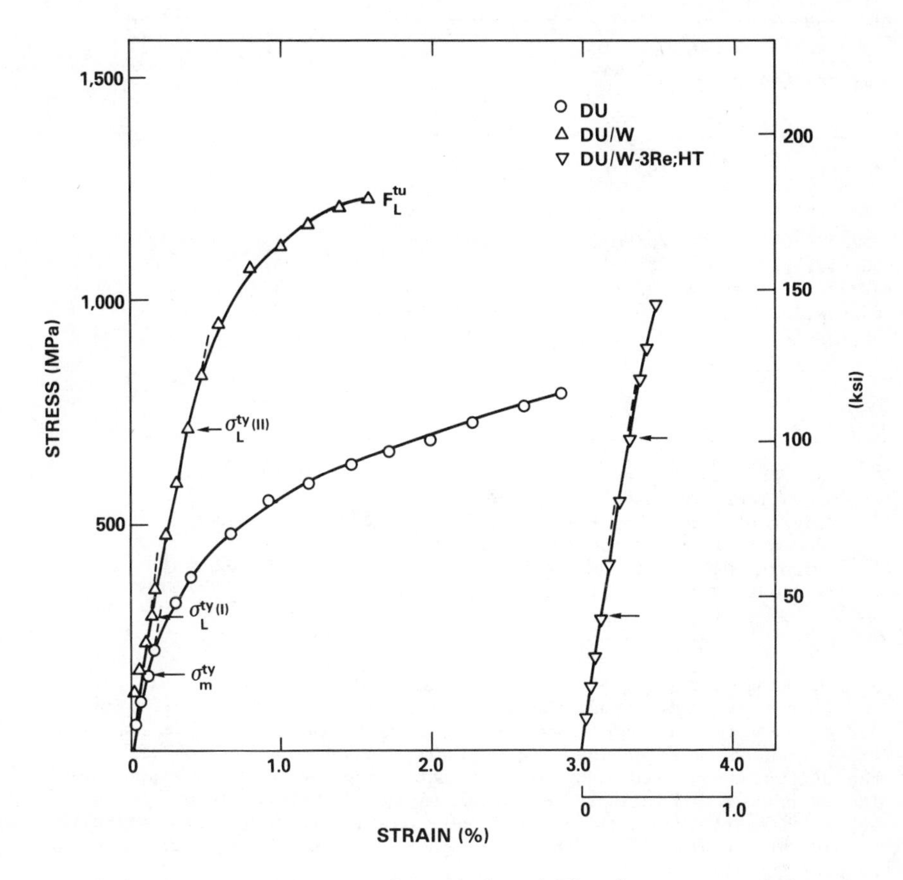

Fig. 4 - Deformation behavior of unreinforced DU and tungsten-reinforced DU in longitudinal tension. V_f = 45%; d_f = 305 μm.

The large scatter in strength of the heat-treated DU/W-3Re composite specimens could not be explained with confidence. Strength predictions based on ROM are also shown in Fig. 5. The experimental strengths of DU/W composites were respectively 10, 15, and 25 percent lower than ROM predictions for 35, 45, and 55 percent volume fraction of fibers; in comparison, the strength of the DU/45W-3Re:HT composite was 53 percent lower. The strengths of the fibers, and thus the strengths of the composites, depended on the fiber diameters as shown in Fig. 6. Virgin fiber data was used in ROM calculations due to lack of data on extracted fibers.

Table III - Longitudinal Tensile Mechanical Properties*

Material Designation	Young's Modulus** E_L^t (GPa)	Yield Strength σ_L^{ty} (I)(MPa)	Ultimate Strength F_L^{tu} (MPa)	Strain to Failure ε_L^{tu} (%)
DU(matrix)	187	160***	793	2.4
DU/35W(305)	258	293	1 127	1.0
DU/45W(127)	281	219	1 424	1.6
DU/45W(305)	292	324	1 239	1.1
DU/45W(762)	290	213	962	0.6
DU/55W(305)	311	432	1 214	0.8
DU/45W-3Re(305):HT	279	275	768	0.4

* Data presented as average values from three tests; standard deviations are shown in figures.

** Moduli of fibers: E_W = 414 GPa; E_{W-3Re} = 379 GPa, taken from Ref. 11, used for ROM estimates.

*** Proportional limit; 0.2% offset yield strength of DU matrix is 415 MPa in tension.

Looking at the failure processes from a strain-to-failure viewpoint, the curves in Fig. 7 indicate that the DU/W composite failure strains were closer to the approximate elongations to failure of the annealed wires than to those of the virgin fibers. Accounting for the effect of the diffusion-affected zone would probably improve that correspondence. The mechanisms of failure initiation in the DU/W-3Re:HT composite were quite different from those in DU/W composites, because the elongation to failure of annealed W-3Re fiber is ~12 percent whereas, the corresponding composite failure strain was only 0.4 percent.

Fractographic analysis of the failed LT specimens also supported the aforementioned observations. SEM fractographs of DU/45W(305) composites are presented in Fig. 8. The diffusion-affected zone failed in a ductile manner with chisel-shaped features, typical of ductile W, whereas, the remaining core region failed in a quasi-cleavage mode, which is characteristic of annealed W fibers (9). The quasi-cleavage fracture of the core region cannot be simply attributed to a notch effect in this composite. Some indications of a notch effect were observed in the 762 μm fibers in the DU/45W(762) composites. In contrast, the core region and the diffusion-affected zone in the 127 μm fibers failed independently of one another in the DU/45W(127) composites; in fact, a physical separation between these regions was observed. On the other hand, the core region of W-3Re fiber failed in a quasi-cleavage mode as shown in Fig. 9(a), and the failure initiation point was close to the interface between the core and the diffusion-affected zone, indicating a notch effect, as shown in Fig. 9(b).

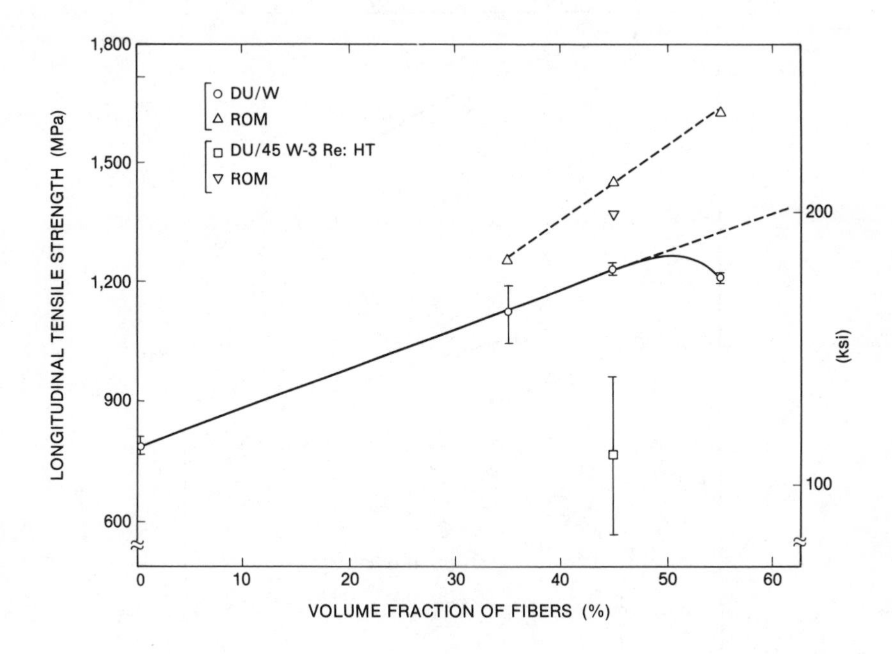

Fig. 5 - Longitudinal tensile strength of DU/45W(305), and DU/45W-3Re(305) composites as a function of volume fraction of fibers.

The dimensions of the diffusion-affected zone (also called fiber recrystallized area) seen in Fig. 1 matched closely with those observed in Figs. 8 and 9. Hence, the percentages of the fiber areas that recrystallized could be calculated from the fractographs. These percentages were, respectively, 71, 58, and 47 percent for fibers having diameters of 127, 305, and 762 μm. For the 305 μm W-3Re fibers, the recrystallized area was only 34 percent.

The effects of fiber recrystallized area on fiber strength were analyzed quantitatively by plotting the retention of fiber strength, defined as the ratio of _in situ_ fiber strength to virgin fiber strength, versus the recrystallized area, as shown in Fig. 10. The _in situ_ fiber strength was calculated from the rule-of-mixture equation for longitudinal tensile strength, which, though obviously a simplification, makes possible useful comparisons. Because the strength and the percentage recrystallized area both increase with decreasing fiber diameter, and because these factors have opposite effects on composite strength, a curve of the form shown in Fig. 10 for the DU/W composites is to be expected. However, the strength retention for W-3Re fiber was considerably lower than that of W fibers, Fig. 10. Strength, strain to failure, and fractographic analysis all indicated that the W-3Re fiber was severely embrittled by the diffusion-affected zone at the periphery.

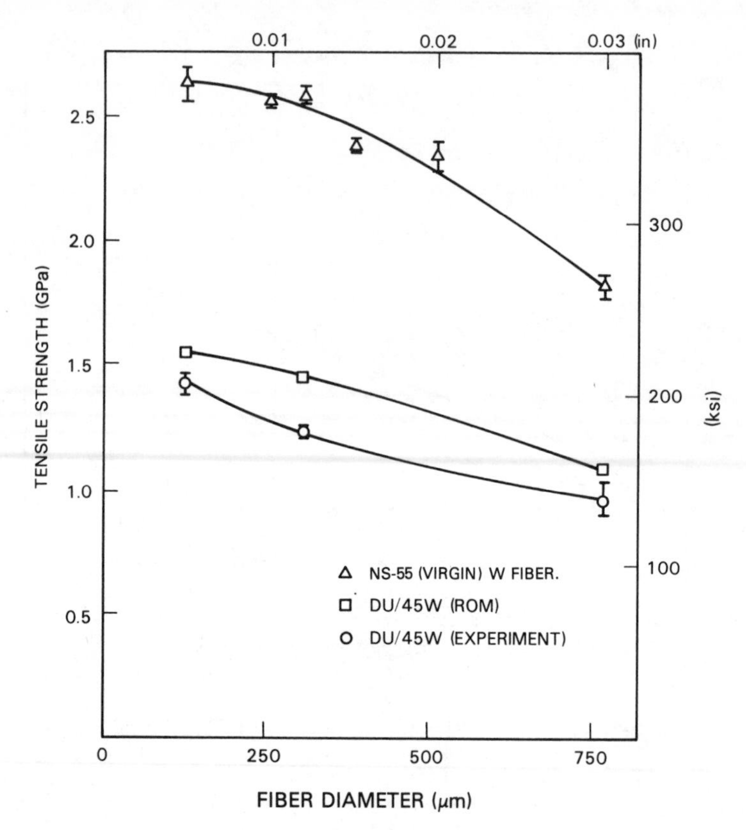

Fig. 6 - Tensile strength of DU/W composites and virgin W fibers as a function of fiber diameter.

Transverse Tension:

Mechanical behavior of unidirectional composites in transverse tension (TT) is quite different from their behavior in longitudinal tension (LT). Typical TT stress-strain curves for DU/45W and DU/45W-3Re:HT composites are compared in Fig. 11. The deformation curve is relatively straight with two slopes of nearly equal magnitude, unlike that of a brittle fiber-ductile matrix composite in which a large, parabolic elasto-plastic region is observed. This behavior of DU/W composites occurred because the strengths and hardnesses of the W fibers were comparable to those of the DU matrix (Table II). Under transverse loading the fibers act like rod inclusions in the matrix. Both the yield strength and the ultimate strength of the DU/45W-3Re(305):HT were somewhat greater than those of DU/45W(305) composites as shown in Table IV and Fig. 12, which also shows the variation of strength of DU/W composites with V_f. As expected, the transverse strength decreases rapidly with increasing V_f. However, the strength increased with decreasing fiber diameter at a V_f of 45 percent as shown in Fig. 13. A complete list of relevant strength properties is given in Table IV.

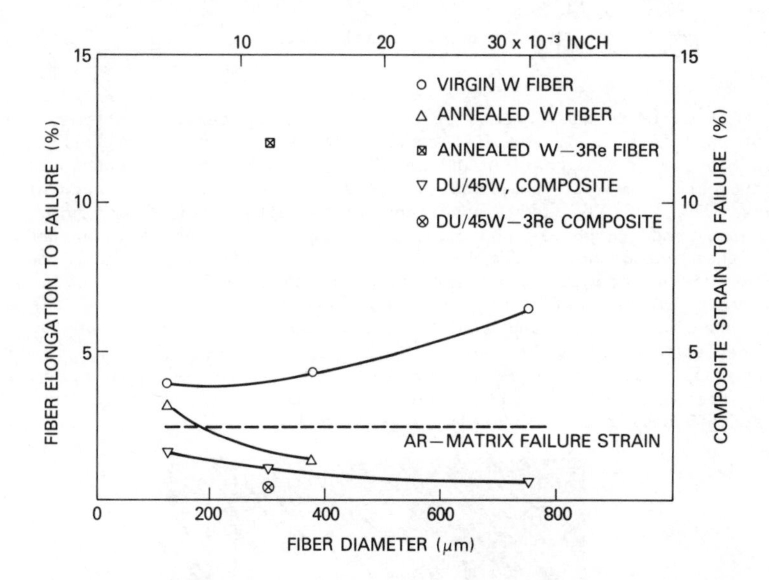

Fig. 7 - Fiber elongations and composite failure strains as a function of fiber diameter. Fibers were annealed for 30 min. at 1100°C.

Fractographic analysis can be used to explain qualitatively the dependence of transverse tensile strength on fiber diameter. Other investigators have shown that the tendency of boron fibers to split controls the transverse tensile strength of Al/B composites (12,13). Similar fiber splitting was observed in the DU/W composites. SEM fractographs (Fig. 14) showed that the tungsten fibers tended to split less frequently as the fiber diameter was decreased. In fact, the 127 μm diameter tungsten fibers showed virtually no splitting in the DU/45W composites. These observations on tungsten fiber splitting tendencies reflect the behavior of fibers with large diffusion-affected zones and not necessarily the behavior of the virgin tungsten fibers.

Discussion

The tensile behavior of DU/W composites is analyzed in detail in this section. Longitudinal tension and transverse tension are discussed separately for clarity and convenience.

Longitudinal Tension

Deformation behavior of DU/W and DU/W-3Re:HT composites is analyzed here in detail. The average proportional limits (stage I yield strengths), σ_L^{ty} (I), of DU/45W(305) and DU/45W-3Re(305):HT composites were 325 and 276 MPa, respectively (see Fig. 4); the heat treatment did not increase σ_L^{ty} (I) as originally expected. This discrepancy may be clarified by considering the well-known equation:

$$\sigma_L^{ty}(I) = E_f V_f \varepsilon_f + \sigma_m^{ty}(1-V_f) \quad (1)$$

where

$\sigma_L^{ty}(I)$ = proportional limit of the composite,

σ_m^{ty} = proportional limit of the *in situ* matrix, and

ε_f = strain in the fiber (where $\varepsilon_f = \varepsilon_m = \varepsilon_c$).

Eq.1 neglects the effects of the longitudinal residual stress, which may be calculated from the difference between the proportional limits of the composite in compression and tension (14). In the present work, the proportional limit in compression could not be determined accurately, because most of the test specimens had length-to-diameter ratios of ~1, which produced severe end constraints (6). The problem is compounded by another phenomenon. The U-0.75Ti alloy exhibits an initial Young's modulus of 241 GPa up to a microyield stress of ~95 MPa followed by a relatively large elastic region characterized by a modulus of 187 GPa (not shown in Figs. 4 and 11). Similar behavior has been observed in pure depleted uranium for which a microyield stress of 30 MPa was reported (15). For the present analysis, the initial modulus and microyield stress of U-0.75Ti alloy have been neglected in order to study the composite behavior at relatively large strains.

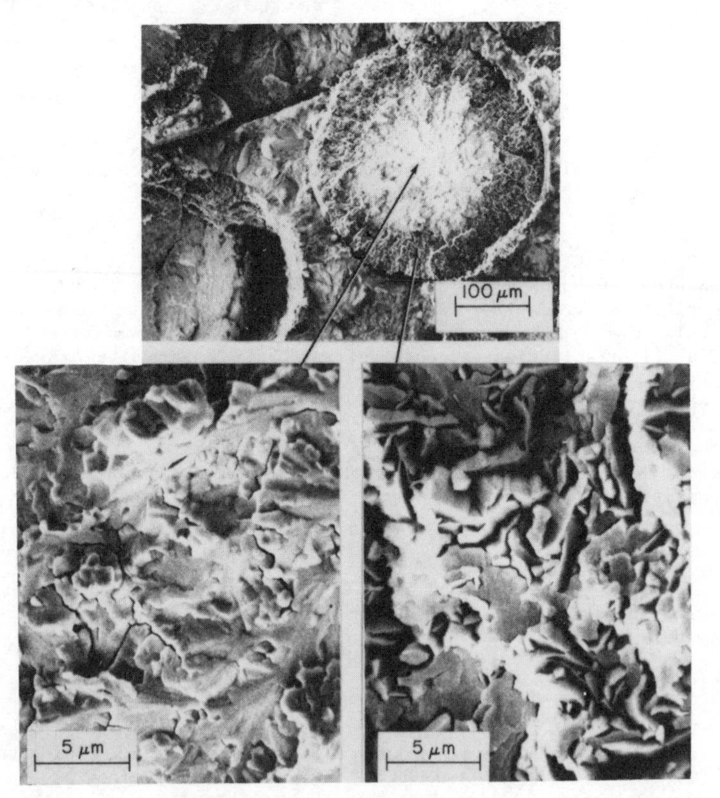

Fig. 8 - SEM fractographs of a DU/45W(305) composite that failed in longitudinal tension.

Moreover, Kim, et al. (3) reported interesting comparisons of 1095 St/W composites in the AF and HT conditions. The residual stress was less in the HT condition than in the AF condition, because of the specific volume expansion that accompanied the austenite-to-martensite transformation in 1095 steel. That volume expansion relieves some of the longitudinal residual stress that arises primarily from the differences in the thermal expansion of the fiber and matrix. However, in the DU-0.75Ti alloy, a specific volume contraction is expected in the γ to α_a' transformation and in the aging processes (16), both of which should increase the tensile residual stress in the matrix of the HT composite above that to be found in the AF composite.

The proportional limits for DU/45W(305) and DU/45W-3Re(305):HT composites were calculated (Eq. 1) to be 390 MPa and 565 MPa, respectively. Comparison of these values with experimental values demonstrates the significant role played by residual stresses in determining composite properties. Heat treatment increased the residual stress in the DU/45W-3Re(305):HT composite. The tensile proportional limit of that material was considerably lower than was predicted by Eq. 1, because the residual stress reduced the in situ proportional limit of the matrix far below that which would be measured for the unreinforced matrix. Although they reduce the composite proportional limit in tension, these residual stresses increase that limit in compression, which was the primary design application.

Dependence of the longitudinal tensile strength of DU/W composites on V_f and d_f is shown in Figs. 5 and 6, which also show that theoretical (ROM) strengths were not attained. The loss of strength was traced to thermal degradation of fiber strength during the fabrication process; the sources of degradation were: (i) thermal exposure in the range of 1100-1200°C (Fig. 3), (ii) partial recrystallization of the fiber periphery (Fig. 1) and (iii) grain boundary diffusion of matrix elements. Because the precise thermal history of the fiber during fabrication was not known, the data in Fig. 3 were useful only as a guide for interpreting behavior and not as quantitative values for reconciling the discrepancy between the experimental and ROM strengths (Fig. 6).

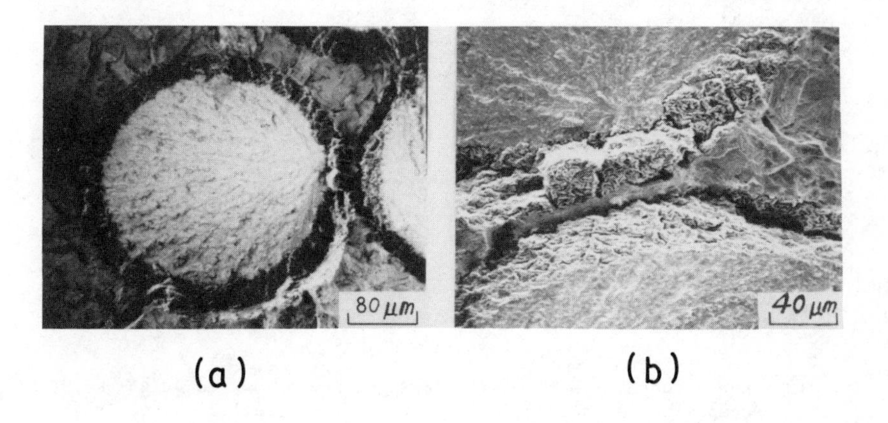

Fig. 9 - SEM fractographs of a DU/45W-3Re(305):HT composite that failed in longitudinal tension.

The complications in the interpretation of the strength and strain to failure of DU/W composites due to possible notch effects, arising from a premature failure of the partially recrystallized layer, were resolved by analyzing longitudinal sections of failed specimens. There was no evidence of premature failure of this layer, which was loosely attached to the fiber core as shown in Fig. 15(b). However, the behavior of the DU/W-3Re composite was such that only a notch effect due to premature failure of the recrystallized layer could explain the results presented in Figs. 4, 5, 7, 9 and 10.

The strength of the peripheral, diffusion-affected zone could not be ascertained. However, microhardness evaluation of 762 μm diameter fibers in the DU/W composites indicated that the average hardness of the core region was ~580 VHN and that of the recrystallized layer was 525 VHN. In summary, the partially recrystallized zone is somewhat softer than the remaining core and fails at a lower strain than the virgin W fiber, giving rise to a notch effect (4, 17).

In addition to showing the difference in the behavior of DU/W and DU/W-3Re composites, Fig. 10 also shows a minimum in the strength retention of W fibers as a function of percent recrystallized area in the fibers. Such a minimum is expected (4) in the above relationship if one assumes that diffusion penetration distance is constant for fibers of different sizes and analyzes the two opposing trends of the strength versus diameter (top curve of Fig. 6) and strength versus recrystallized area relationships.

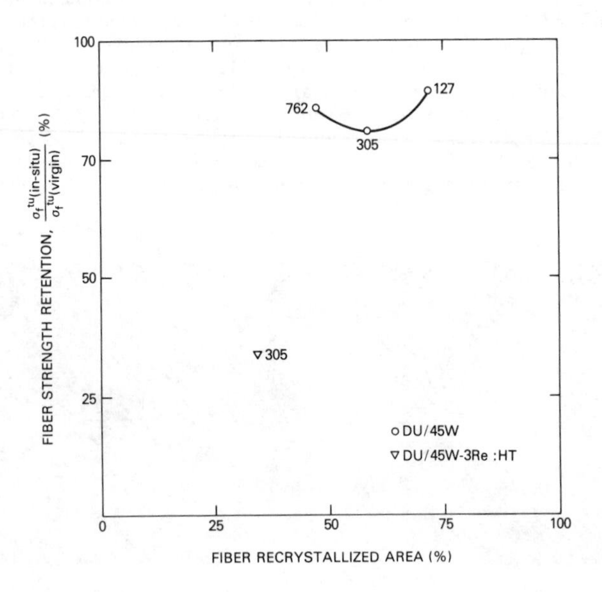

Fig. 10 - Retention of fiber tensile strength in DU/45W and DU/45-3Re:HT composites as a function of percent recrystallized area in the fiber. Numbers associated with data points refer to fiber diameters in μm.

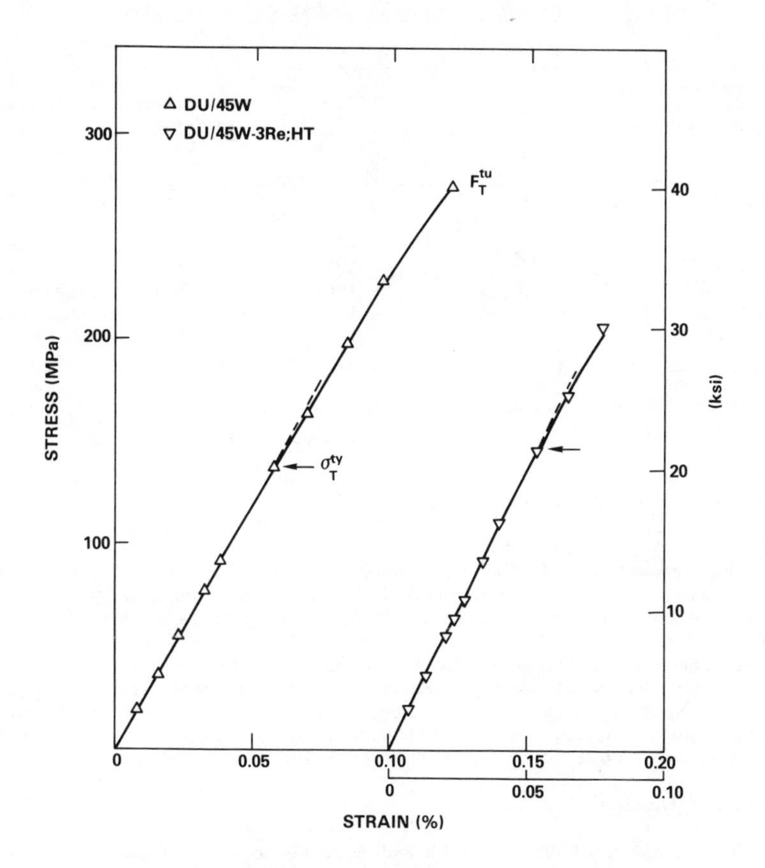

Fig. 11 - Stress-strain curves for DU/45W(305) and DU/45W-3Re(305):HT composites in transverse tension.

However, the actual average diffusion distances, i.e., penetration distances, for 127, 305, and 762 μm W fibers were determined to be 28, 59, and 107 μm, respectively, Fig. 1. This result is contrary to what is predicted analytically for volume diffusion into cylinders of different diameters, keeping all other variables constant (18). Moreover, grain boundary area per unit cross-section increases with decreasing fiber diameter due to the finer grains resulting from the wire-drawing operations. As a result, the trend predicted by volume diffusion is enhanced by grain boundary diffusion. To reconcile the discrepancy between the predicted and the observed penetration distances, it is presumed that diffusion processes actually slow down with time, because of extensive grain boundary decohesion brought about by the release of circumferential, tensile residual stress that exists for heavily-drawn wires (19). The deformed structure shown in Fig. 1(d) supports the notion of excessive grain boundary decohesion, particularly when compared to Fig. 1(a). The decohesion process is, of course, somewhat exaggerated by the etching process. In other words, a structure with weakly-bonded grains develops around the fiber periphery as the grain boundary diffusion proceeds and this structure impedes further diffusion.

Table IV. Transverse Tensile Mechanical Properties*

Material Designation	Young's Modulus E_T^t (GPa)	Yield Strength σ_T^{ty} (MPa)	Ultimate Strengths F_T^{tu} (MPa)	Strain to Failure ε_T^{tu} (%)
DU(matrix)	187	215	851	1.85
DU/35W(305)	256	110	158	0.06
DU/45W(127)	259	204	335	0.14
DU/45W(305)	251	160	184	0.07
DU/45W(762)	275	109	130	0.05
DU/55W(305)	266	137	170	0.05
DU/45W-3Re(305):HT	256	167	202	0.06

* Data presented as average values from three tests; standard deviations are shown in various figures.

The parameters of the wire drawing processes (i.e., percent reduction per pass, number of passes, intermediate annealing treatments, etc.) were not known for the 127, 305 and 762 μm diameter fibers. Moreover, the variation of residual stresses with those parameters is not characterized as well as it is for a simple, single-pass process (19, 20). On the basis of what is known, it is reasonable to assume that residual stresses decrease with increasing diameter of the W fibers of interest here. Consequently, the tendency to form a weak structure at the fiber periphery would decrease with increasing fiber diameter.

Transverse Tension

The strength of DU/W and DU/W-3Re composites in transverse tension is important in determining the overall performance of the composites and is usually dependent on the quality of the fiber/matrix bond. The comparison of the two composites at a V_f of 45 percent, Fig. 11, indicates that the transverse yield strength, σ_T^{ty}, increased marginally with heat treatment (Table IV) in a manner quite similar to that observed in longitudinal specimens, Fig. 4. The residual stress analysis for the TT orientation is not as straightforward as for the LT orientation (14). Although residual stresses affect composite transverse yield strength to some degree, stress concentration due to fibers is of greater importance and is discussed in the following.

Variation of the transverse tensile strength of DU/W composites with V_f shows that the experimental data points fall well below the theoretical estimates of this strength based on a fiber distribution either of a square array (21) or of a hexagonal array (22). These theoretical estimates are based on the assumption that the fiber/matrix bond strength makes no contribution to the composite strength, and the experimental strengths were even less than these crude, low estimates of the transverse strength probably as a result of fiber splitting. Many such split fibers were observed in composites made with 305 μm and 762 μm diameter fibers, Figs. 14(b) and 14(c).

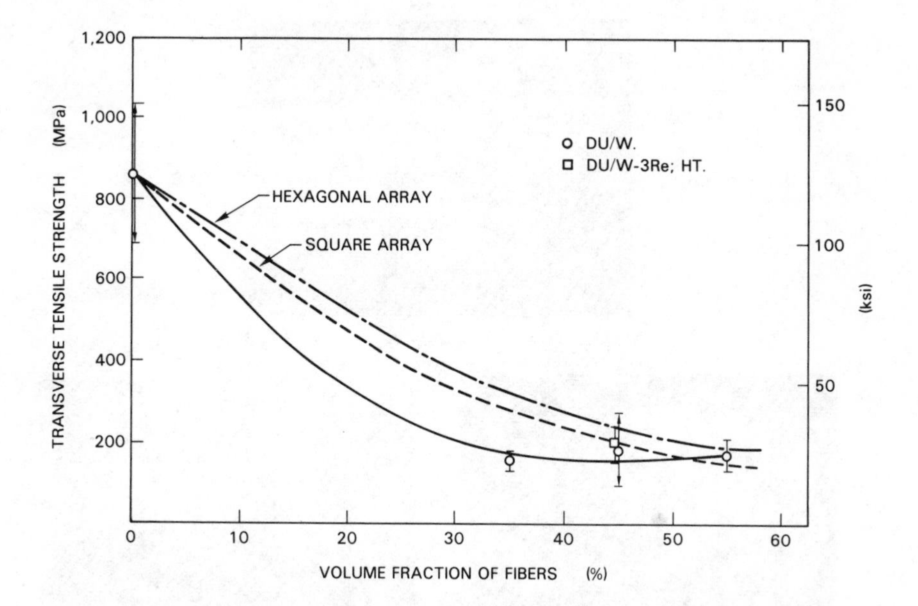

Fig. 12 - Transverse tensile strength of DU/W composites as a function of V_f. The strength of DU/45W-3Re is shown for comparison. Fiber diameter: 305 μm.

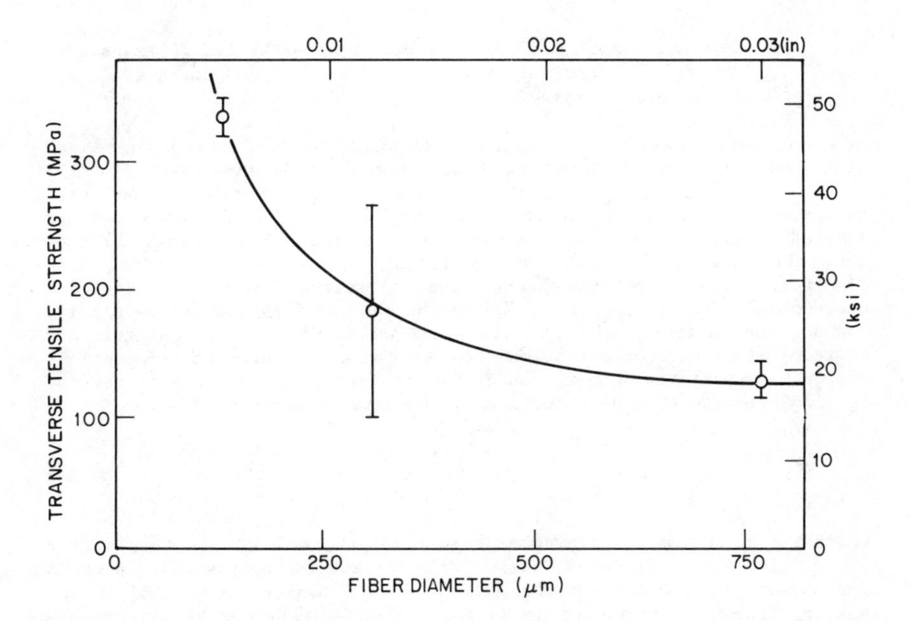

Fig. 13 - Transverse tensile strength of DU/45W composites as a function of fiber diameter.

Fig. 14 - SEM fractographs of DU/45W composites that failed in transverse tension. Fiber diameters: (a) 127 μm, (b) 305 μm, and (c) 762 μm; V_f = 45%.

However, the transverse strength of DU/W composites increased significantly with decreasing fiber diameter as shown in Fig. 13. At least two factors contribute to this effect. First, although the tendency of W fibers to split as a function of fiber diameter was not studied independently, the fracture surfaces presented in Fig. 14 show that the tendency toward fiber splitting decreases with decreasing diameter. One therefore expects the transverse strength to increase correspondingly, as has been shown in the case of Al/B composites (12,13). Second, the variation of F_T^{tu} with d_f shown in Fig. 13 can be explained in terms of stress concentration effects. Using the strength-of-materials approach to predict stress concentration factor (SCF), the transverse tensile strength of a unidirectional composite is given by (23):

$$F_T^{tu} = \frac{\sigma_m^{tl}}{(SCF)} \qquad (2)$$

where F_T^{tu} is the transverse tensile strength, and σ_m^{tl} is the "matrix limit stress" as defined by Chamis (23). The SCF for composites, in which the transverse modulus of the fiber is much higher than that of the matrix, increases gradually up to a V_f of 40 percent and then increases rapidly thereafter. A similar variation would be expected for DU/W composites in which the fiber modulus is nearly twice that of the matrix.

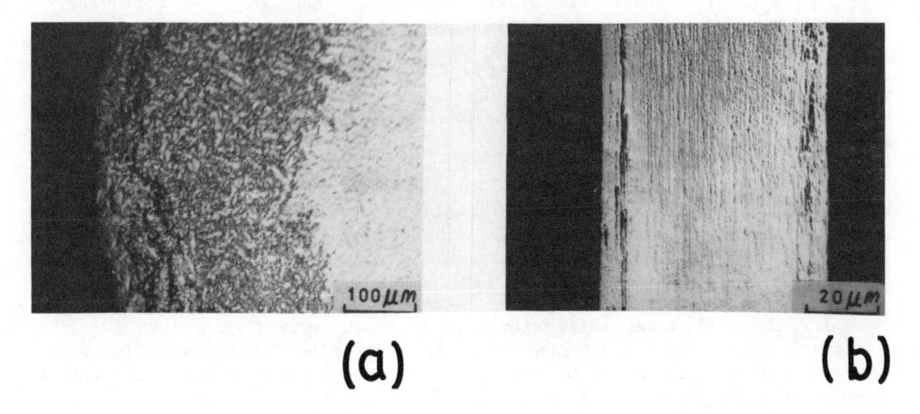

Fig. 15 - Optical micrographs of DU/45W composites that failed in longitudinal tension. Transverse (a) and longitudinal (b) sections are from a region near the fracture surface.

The variation of F_T^{tu} with V_f, shown in Fig. 12, can also be rationalized from Eq. 2. More importantly, the variation of F_T^{tu} with d_f at a V_f of 45 percent, shown in Fig. 13, can be explained by Eq. 2. It was concluded earlier that the diffusion-affected zone, particularly for fibers with a diameter of 127 μm, is a loosely-packed, soft layer on the remaining core, Fig. 1(d). Using photoelasticity, Marom and Arridge (24) have shown, that a soft interlayer at the fiber/matrix interface reduces stress concentration around the fiber. It is presumed here that the softness of the interlayer increases with decreasing fiber diameter, Figs. 1(f), 1(e), and 1(d), thereby decreasing the stress concentration. Consequently, the transverse tensile strength increases with decreasing diameter at a V_f of 45 percent, Eq. 2 and Fig. 13.

The argument expressed in the previous paragraph can also be extended to interpret the similar magnitude of σ_T^{ty} and F_T^{tu} for the DU/45W (305) and DU/45W-3Re(305):HT composites. Since the matrix yield strength increased with heat treatment, σ_m^{tl} is expected to increase along with σ_m^{ty} (they are assumed proportional). Then, from Eq. 2, one expects a higher F_T^{tu} for the heat-treated composite. But the stress concentration due to W-3Re fibers is higher than that due to W fibers at a V_f of 45 percent because the W fiber has a relatively soft interlayer, Fig. 1(e), whereas, the diffusion-affected zone for the W-3Re fiber is expected to be hard from our earlier data interpretation. Hence, the effects due to σ_m^{tl} and stress concentration counterbalance each other and the magnitudes of σ_T^{ty} and F_T^{tu} for DU/W and DU/W-3Re:HT composites are approximately the same.

Conclusions

This study has shown that the high-density, depleted uranium alloy, U-0.75Ti, can be successfully strengthened by tungsten-fiber reinforcement. The effects of fiber diameter and fiber volume fraction on composite tensile strength were studied parametrically using commerical W fibers. The effects of changing fiber alloy composition and composite heat treatment were also investigated. Quantitative interpretation of the experimental results was often difficult. However, the following conclusions could be drawn from this study.

1. Properties of the reacted W fibers in the DU/W composites are controlled by two equally-important degradation processes: (i) thermal exposure above 800°C and (ii) diffusional processes leading to a diffusion-affected zone.

2. The diffusion of U and Ti into W fibers results in partitioning of Ti, because of solid solution formation between U and W and of two-phase formation between Ti and W. These processes are influenced by the amount of cold work, the texture, and the residual stress pattern in the fibers, all of which vary with fiber diameter.

3. The peripheral, diffusion-affected zone acts like a soft layer for the W fiber and a hard layer for the W-3Re fiber; this layer influences stress concentration due to the fiber and, therefore, controls the transverse strength of the composite.

4. Heat treatment of the composite resulted in marginal improvements in tensile properties of the composite, but did significantly improve compressive properties, which are reported elsewhere (6).

5. Heat treatment also influences the residual stress pattern and, consequently, the yield strength of the composite, particularly in the longitudinal direction.

Nomenclature

$\sigma_L^{ty}(I)$	Composite longitudinal, stage I yield strength (also called proportional limit)
$\sigma_L^{ty}(II)$	Composite longitudinal, stage II yield strength
σ_T^{ty}	Composite transverse yield strength
σ_m^{ty}	Matrix proportional limit
σ_m^{tl}	Matrix limit stress, defined according to strength-of-materials approach to predict strength (21)
F	Ultimate strength
F_L^{tu}	Composite longitudinal tensile strength
F_T^{tu}	Composite transverse tensile strength
ε	Engineering strain
d_f	Fiber diameter

	V_f	Volume fraction of fibers
Subscripts		
	L	Direction parallel to fiber direction, called "longitudinal"
	T	Direction perpendicular to fiber direction, called "transverse"
	f,m	Fiber, matrix
Superscripts		
	c	Compression
	t	Tension
	y	Yield
	u	Ultimate

Acknowledgements

This research program was sponsored by the Defense Advanced Research Projects Agency under DARPA order No. 3388.

References

1. T. Olofson, G.E. Meyer, and A. L. Hoffmanner, "Processing and Application of Depleted Uranium Alloy Products", Report MCIC-76-28, Metals and Ceramic Information Center, Battelle Columbus Labs., Columbus, Ohio. Sept. 1976.

2. Battelle Columbus Laboratories, Columbus, Ohio, under contract to the Naval Research Laboratory, Contract No. N00014-78-C-0745.

3. C. Kim, A. Pattnaik, and R. W. Weimer, "Matrix/Fiber Interface Effects on Tensile Fracture of Steel/Tungsten Composites", paper to be published in Failure Modes in Composites - VI, AIME, Publ. (see this volume).

4. R. A. Signorelli, "Wire-Reinforced Superalloys", in Composite Materials: Vol. 4: Metallic Matrix Composites, pp. 229-267, K. G. Kreider, Ed., Academic Press, 1974.

5. D. W. Petrasek and J. W. Weeton, "Effects of Alloying on the Room-Temperature Tensile Properties of Tungsten-Fiber-Reinforced-Copper-Alloy Composites", Trans. AIME, vol. 230, 1964, p. 977.

6. A. Pattnaik, C. Kim, and R. J. Weimer, "Deformation Behavior of Depleted Uranium/Tungsten Fiber Composites in Compression", paper to be published, in Proceedings, ASTM Symp. on Compression Testing of Homogenous Materials and Composites, held at Williamsburg, VA., March 1982.

7. G. Roberts, "Diffusion with Chemical Reaction", Metal Science Journal, vol. 13, 1979, p. 94.

8. K. H. Eckelmeyer, "Microstructural Control in Dilute Uranium Alloys", in Microstructural Sciences, vol. 7, p. 133, I. Le LeMay, Ed., Elsevier North Holland Publ., New York, 1979.

9. C. Kim., W. L. Phillips, and R. J. Weimer, "Fracture of Tungsten Wire in Metal Matrix Composite", in Fractography and Materials Science, ASTM STP 733, p. 314, L.N. Gilbertson and R. D. Zipp, Ed., American Society for Testing of Materials, 1981.

10. S. Leber, et al., "Fracture Modes in Tungsten Wires", Journal of the Less-Common Metals, vol. 48, 1976, p. 119.

11. B. Harris and E. C. Ellison, "Creep and Tensile Properties of Heavily Drawn Tungsten Wires", Trans. ASM, vol. 59, 1966, p. 744-754.

12. K. G. Kreider and K. M. Prewo, "Failure Mechanisms in Transversely Loaded Boron-Aluminum", Journal of Phys. D: Appl. Phys., vol. 5, 1972, p. 2075.

13. L E. Dardi and K. G. Kreider, "Effect of Salt Water and High-Temperature Exposure on Boron-Aluminum Composites", pp. 269-283, in Composite Materials: Testing and Design (Third Conf.), ASTM STP 546, American Soc. for Testing of Matls., 1974.

14. E. R. Thompson, D. A. Koss, and J. C. Chestnut, "Mechanical Behavior of a Carbide Reinforced Co-Cr Eutectic Alloy", Met. Trans., vol. 1, 1970, p. 2807.

15. A. N. Holden, Physical Metallurgy of Uranium, Chapter 5, pp. 59-77, Addition-Wesley, 1958.

16. A. M. Ammons, "Precipitation Hardening in Uranium-Rich Uranium-Titanium Alloys", in Physical Metallurgy of Uranium Alloys, pp. 511-585, J. J. Burk, et al., Ed., Brook Hill Publ. Co., Chestnut Hill, Mass., 1976.

17. J. W. Pugh, "Tensile and Creep Properties of Tungsten at Elevated Temperatures, Proc. of ASTM, vol. 57, 1957, p. 906.

18. J. Crank, The Mathematics of Diffusion, 2nd Edition, Chapter 5, pp. 69-88, Clarendon Press, Oxford, 1975.

19. G. E. Dieter, Mechanical Metallurgy, 2nd Edition, p. 580, McGraw-Hill, 1976.

20. W. M. Baldwin, "Residual Stresses in Metals", Proceedings of the Amer. Soc. for Testing of Matls., vol. 49, 1949, p. 570.

21. G. A. Cooper and A. Kelly, "Role of the Interface in the Fracture of Fiber-Composite Materials", in Interfaces in Composites, pp. 90-106, ASTM STP 452, American Society for Testing of Matls., Publ., 1969.

22. G. Marom and E. F. T. White, "Improvements in Transverse Properties of Composites", Journal of Matls. Science, vol. 7, 1972, p. 1299.

23. C. C. Chamis, "Micromechanics Strength Theories", in Composite Materials: Vol. 5: Fracture and Fatigue, Chapter 3, p. 126-131, L. J. Broutman, Ed., Academic Press, 1974.

24. G. Marom and R. G. C. Arridge, "Stress Concentrations and Transverse Modes of Failure in Composites with a Soft Fiber-Matrix Interlayer", Materials Science and Engineering, vol. 23, 1976, p. 23.

METALLURGICAL AND TENSILE PROPERTY ANALYSIS OF SEVERAL SILICON CARBIDE/TITANIUM COMPOSITE SYSTEMS

William D. Brewer

NASA-Langley Research Center
Hampton, VA 23665
USA

Jalaiah Unnam
Virginia Polytechnic Institute and
State University
Blacksburg, VA 24061
USA

Summary

Several silicon-carbide fiber reinforced titanium matrix composite systems were investigated to determine composite degradation mechanisms and to develop techniques to minimize loss of mechanical properties during fabrication and in service. Emphasis was on interface control by fiber or matrix coatings. Fibers and matrix materials were sputter coated with various metals to determine the effects of the coatings on basic fiber properties, fiber-matrix interactions, and on composite properties. The effects of limited variations in fabrication temperature on composite properties were determined for composites consolidated by standard press-diffusion-bonding techniques.

Introduction

Titanium matrix composites have unique properties which make them attractive for a variety of aerospace applications for which high strength and stiffness at elevated temperatures are required. However, these composites have been marked by strengths well below those expected, at least in part because of deleterious fiber-matrix reactions during fabrication. It is well known that the fiber matrix interactions play a major role in determining composite properties. Any reactions other than those required for good bonding are usually undesirable.

In general, there are three ways to minimize the effects of interface reactions: adjust fabrication parameters (time, temperature, pressure) to reduce interactions; apply fiber or matrix coatings to prevent interactions (diffusion and reaction barriers); or modify fibers and/or matrix materials to tolerate interactions (sacrificial layers, new fibers or matrices, etc.). Successful application of either or all of these controls is expected to result in a material with improved properties.

Several silicon-carbide fiber reinforced titanium matrix composite systems were investigated to develop a more fundamental understanding of composite degradation mechanisms and to develop techniques to minimize the degradation. X-ray diffraction studies were conducted to see if the transport of elements that cause detrimental reactions could be minimized by the application of appropriate interfacial layers. Fibers and matrix materials were coated with various metals and then consolidated into composite panels by press-diffusion-bonding. The effects of the coatings on basic fiber properties, reaction-zone characteristics, and composite properties were determined.

Materials

Three types of silicon-carbide fibers were used in this study. The simplest fiber, referred to as the SiC fiber, is beta silicon carbide with a nominal diameter of 142 μm (1). The strength of this fiber is relatively low and varies over a wide range. The second type of fiber, referred to as the SiC-C fiber, is the well known silicon-carbide fiber with the carbon-rich surface layer (1). Except for the 1 μm thick outer layer, this fiber is the same as the SiC fiber. Although the strength of the SiC-C fiber also varies considerably, the SiC-C is generally much stronger than the SiC. The third fiber is designated SCS-2 and is essentially the SiC-C fiber in which a silicon enriched layer has been added to the carbon surface (2). However, the outer surface layer remains silicon deficient relative to stoichiometric silicon carbide.

Two matrix materials were used in this study: TiA55 and Ti-6Al-4V. The TiA55 is a commercially pure titanium, readily available in foil form and relatively inexpensive. The Ti-6Al-4V, referred to as Ti(6-4), is the widely used titanium structural alloy.

Metal coatings investigated were aluminum, molybdenum, and vanadium. These materials were selected because they are common alloying elements for titanium and have been shown to inhibit titanium reactions with silicon carbide under some conditions (3,4). Aluminum and Ti_3Al were used in the x-ray diffraction studies because of the well known aluminum rejection (or exchange reaction) mechanism observed in SiC/titanium systems (5,6).

Experimental Procedures

Planar Samples

It has been reported (6,7) that the formation of various titanium silicides can be detrimental to composite properties. X-ray diffraction is a convenient technique for detecting small quantities of these reaction products. Therefore, x-ray studies were conducted to see if the transport and subsequent reactions of silicon and titanium in SiC/Ti systems could be minimized by a metallic barrier. Planar samples were used instead of fibers or composites to simplify data analysis and interpretation. Reference samples were made by sputtering 2 μm of silicon carbide onto a 22 mm diameter, 3 mm thick pure titanium disk. Other samples were made by first sputtering 1 μm of aluminum on titanium disks. Silicon carbide was then sputtered onto some of these aluminum coated disks to form SiC/Al/Ti samples for x-ray analysis. The remaining aluminum coated samples were heat treated to develop a Ti_3Al surface and then coated with SiC to form SiC/Ti_3Al/Ti samples. All samples were then exposed to temperatures typical of composite fabrication (875C) and examined after various exposure times to determine the relative amounts of titanium silicide formed. (See (8) for full details of x-ray analysis procedures.)

Fiber Studies

Primarily because of the high cost and long turn-around time associated with composite fabrication, fiber studies were conducted in which conditions at the fiber-matrix interface were approximated by sputtering titanium or aluminum and titanium onto the fiber and exposing the fiber to composite consolidation conditions. The effects of these treatments on fiber properties were determined by conducting tensile tests on individual fibers before and after treatments. From 30 to 50 fibers of each type were tested.

Because the tensile properties of all the fibers tested varied over wide ranges, a projected composite strength based on customary rule-of-mixtures and average material properties did not seem appropriate. A rule-of-mixture calculation gives realistic composite properties only if individual fiber strengths do not differ significantly from the average strength. Therefore, a simple analysis was developed to calculate a projected composite strength based on actual fiber strength distributions (9). This analysis takes into account the effects of individual fiber failures, matrix plasticity and residual stresses due to differential thermal expansion of the fibers and the matrix. This analysis was used to predict composite strengths using the measured properties of the treated and untreated fibers.

Composite Panels

Composite panels were fabricated with various combinations of matrix materials, fibers and coatings to determine the effects of coatings on actual interface structure and composite properties. TiA55 foils were sputter-coated with either aluminum, molybdenum, or vanadium and then consolidated into composite panels with the SiC-C (10) or the SCS-2 fibers. Some of the aluminum coated foils were heat treated prior to consolidation to develop a Ti_3Al layer. Panels were also made with Ti(6-4) matrix and aluminum-coated SCS-2 fibers. Baseline panels were made with no fiber or matrix coatings or treatments. In addition, panels were made with no reinforcing fibers at all in order to determine the effects of the fabrication process on the matrix materials.

All panels were 10 cm x 10 cm x 4-ply thick and were consolidated by customary press-diffusion-bonding. Fabrication temperatures from 760C to 930C and pressures from 41 MPa to 138 MPa were used. Fabrication time for all panels was 30 minutes. The panels were machined into 1.3 cm x 10 cm rectangular samples for tensile testing and metallurgical analysis. All tensile samples were unidirectional and tested in the longitudinal direction. All tensile tests were performed at room temperature. Three to five samples from each panel were tested. Polished cross sections and fracture surfaces of tested composites were examined by scanning electron microscopy and electron microprobe analysis.

Results and Discussion

The effects of Al and Ti_3Al interfacial barriers on the formation of titanium silicides are shown in figure 1. For heating times up to 60 minutes both materials appeared to significantly reduce silicide formation with the Ti_3Al apparently being the more effective. These results indicate that appropriate metallic coatings may be effective in reducing fiber-matrix interaction in composite materials.

Figure 2 shows the effects of coating TiA55 foil on the reaction zone thickness in SiC-C/TiA55 composites. After fabrication the coated foil panels were exposed to 875C for 25 hours in order to develop thick reaction zones to facilitate measurements and comparisons. The Mo and V coatings resulted in reaction zone thicknesses slightly greater than those in the panel with no coating, whereas the Al coating seemed to significantly restrict reaction zone growth.

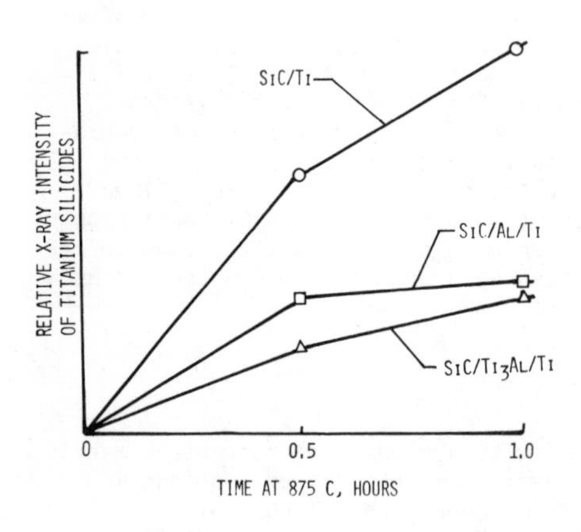

Figure 1.- Effect of metallic barriers on titanium silicide formation. (Planar samples)

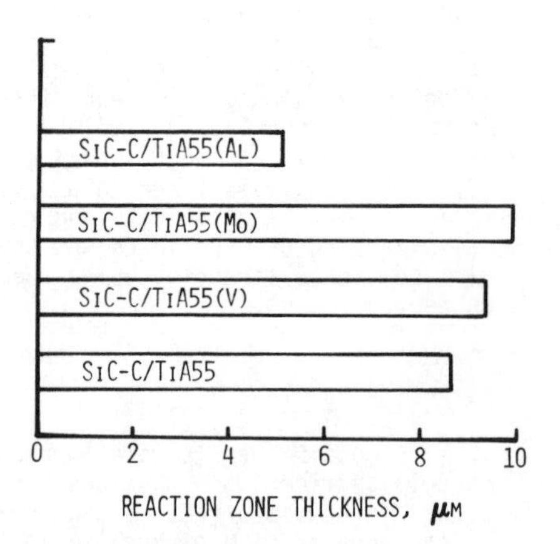

Figure 2.- Effect of metal coatings on reaction-zone thickness. (25 hrs. at 875C)

Photomicrographs of polished cross-sections of SiC-C/TiA55 composites (Figure 3) showed the reaction zone thickness in the as-fabricated Al-coated panel to be about one-third that in the uncoated panel. Unfortunately, for this particular composite system, the reduction in interactions did not result in a corresponding increase in composite strength, most likely because of the characteristics of the SiC-C fiber itself. For the uncoated panel, the high reaction rates caused the carbon rich layer to be essentially consumed, resulting in a graded, integral, interface region with relatively good fiber-matrix bonding even though the fiber may have been weakened. For this panel, photomicrographs of leached fibers (Figure 4a) showed significant fiber break-up adjacent to composite fracture surfaces indicating that, because of the relatively good bonding, the matrix was able to transfer some of the loads away from the fracture surface. On the other hand, the small amounts of reaction in the coated panel (SiC-C/TiA55(Al)) left most of the carbon rich layer undisturbed resulting in relatively poor bonding, as evidenced by essentially no fiber break-up near the composite fracture surface (Figure 4b).

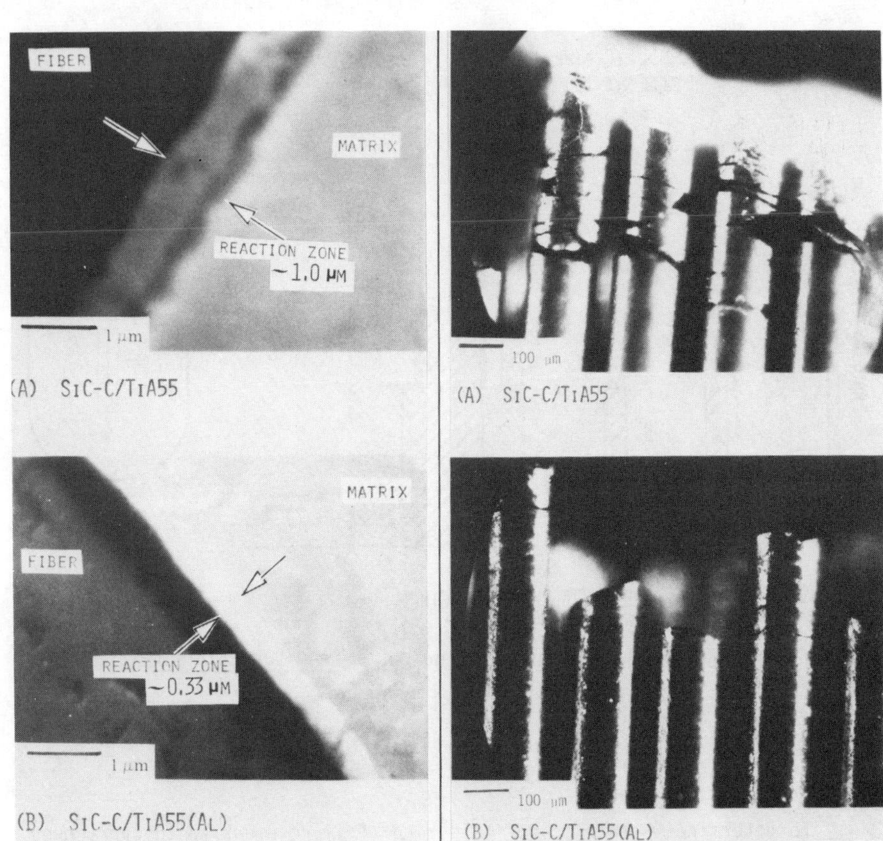

Figure 3.- Fiber matrix interface in as-fabricated SiC/TiA55 composites.

Figure 4.- Leached fibers near fracture surface of SiC/TiA55 composites.

Because of these results, the SiC-C fiber was discounted as a viable candidate for high strength titanium composites, and research emphasis was placed on the other fibers. However, aluminum as an interfacial coating did show some promise and was investigated further.

SiC fibers and SiC fibers coated with 1 μm of aluminum were tested in the as-fabricated condition to determine whether or not the aluminum could be used as a replacement for the carbon rich layer on the SiC-C fibers. The data for the SiC fibers (Figure 5) showed a wide range in tensile strength (0.5-3GPa) indicating a highly variable fiber. Coating the fiber with aluminum, however, reduced the variability considerably by eliminating the very low strength fibers. This behavior is significant because these low strength fibers play a major role in determining the composite properties. Therefore, it appeared that the aluminum served in somewhat the same capacity as the carbon layer in improving fiber properties.

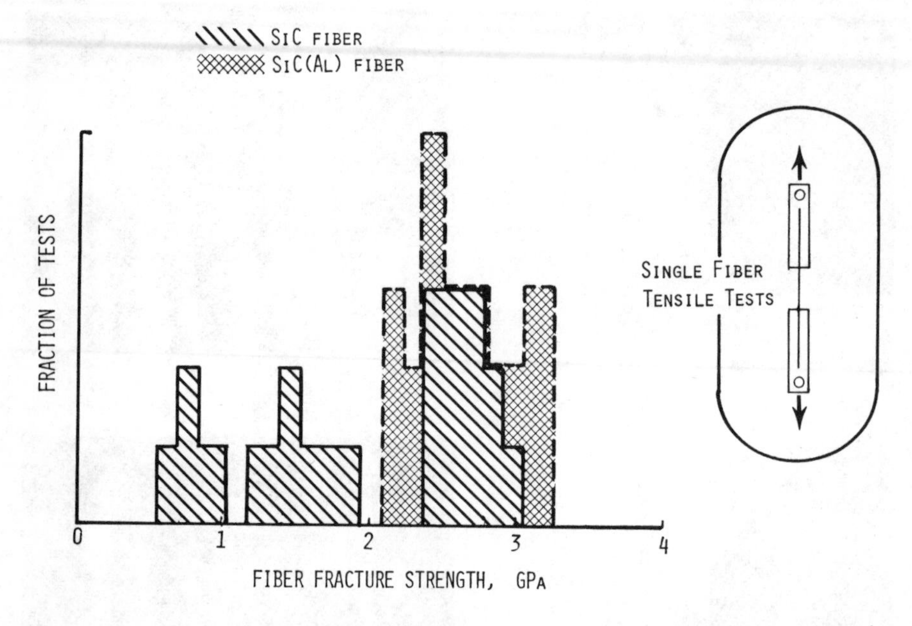

Figure 5.- Effect of aluminum coating on as-received SiC fiber strength.

To determine whether or not this kind of improvement would carry over to composite properties, SiC and SCS-2 fibers were coated with titanium or aluminum and titanium, exposed to composite fabrication conditions and then tested for tensile property changes. The analysis discussed earlier was then used to calculate projected composite strengths based on the measured strength distributions.

Figure 6 shows the fiber strength distributions. As seen previously the SiC fibers showed a wide range of fracture strengths including some very low strengths. As expected, depositing titanium directly onto the fiber and heat treating the fiber reduced the strength even further, showing the severe fiber degradation caused by reaction with the titanium. However, the strengths of heat-treated, aluminum-titanium coated fibers were about 3 times those of the fibers with no aluminum coating. Similar tests were run for the SCS-2 fibers, and although the strength distribution was also fairly wide, the fiber was generally much stronger than the SiC.

The projected composite strength (Figure 7) was significantly higher for the SiC (Al)/TiA55 (0.96 GPa) than for the SiC/TiA55 (0.41 GPa). Coating the SCS-2 fiber with aluminum and titanium and exposing it to composite consolidation conditions had a minor effect on the fiber strength distribution. The projected composite strength for the SCS-2 system was considerably greater than that for the SiC system.

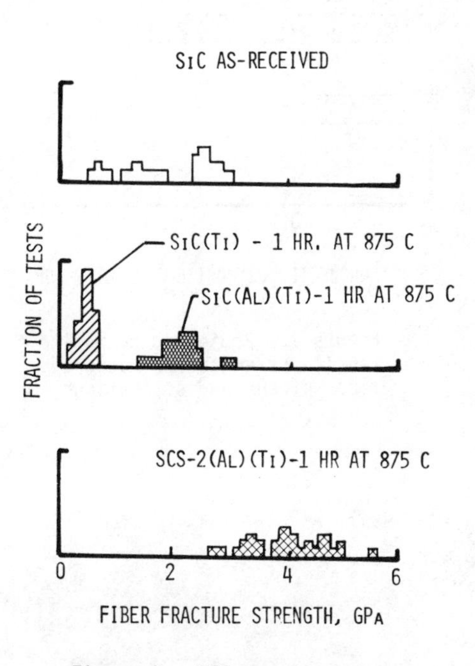

Figure 6.- Effect of coatings and heat treatment on fiber tensile strength.

Because the SCS-2 fiber was found to be superior to the SiC fiber, it was used to make composites for testing and analysis. TiA55 foils were coated with aluminum, heat treated to develop a stable Ti_3Al surface layer, and then consolidated into composite panels with SCS-2 fibers. Polished and etched cross-sections of typical panels are shown in Figure 8. Analysis showed that the aluminum (or Ti_3Al) was uniformly distributed around the fibers and confined to the fiber-matrix interface region and to where the coated foils joined together between fibers. Therefore it appeared that, if a coating is required for interface control, applying the coating to the matrix is an effective technique.

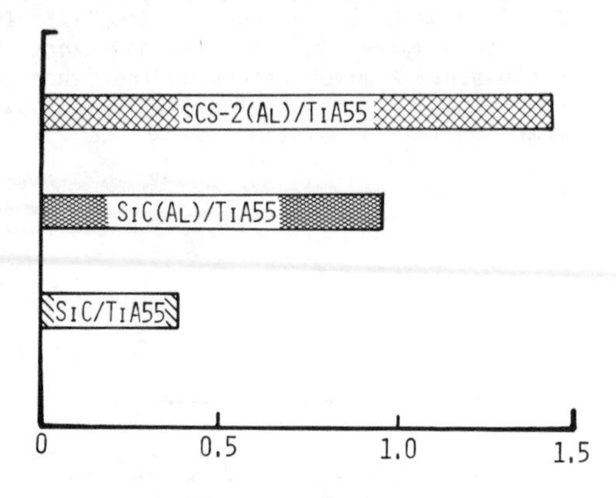

Figure 7.- Projected composite tensile strength for measured fiber strength distributions.

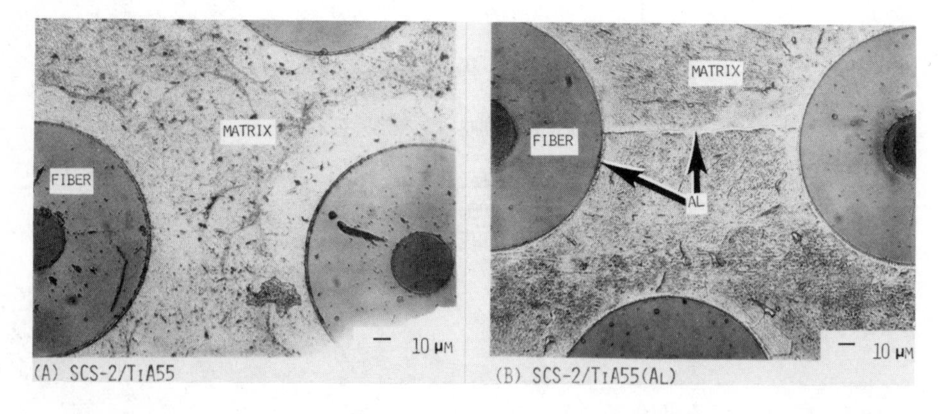

Figure 8 - Polished cross-sections of SCS-2/TiA55 composites.

Electron microprobe traces of the interface regions in coated and uncoated panels (Figure 9) showed somewhat different interface structures. The figure shows elemental distributions in the area of the interfaces. In the uncoated panel, the titanium, silicon, and carbon interdiffused to form the generally detrimental silicide interface whereas in the aluminum coated panel, the carbon and silicon were pretty much contained within the original confines of the fiber by the Ti_3Al barrier thus significantly reducing the potential for undesirable reactions.

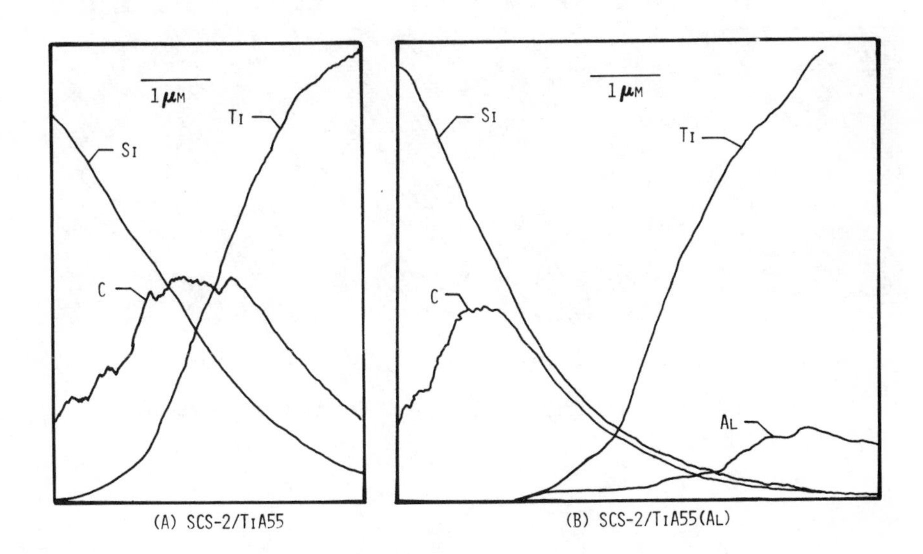

Figure 9.- Elemental distributions across SCS-2/TiA55 interfaces

Polished cross-sections and fracture surfaces of coated and uncoated panels are shown in Figure 10. In contrast to the SiC-C fiber, the SCS-2 fiber-matrix bonding seemed to be better for the coated system than for the uncoated system. Polished cross-sections of the uncoated system (Figure 10a) showed more fiber-matrix disbonds and the fracture surfaces showed somewhat more fiber pull-out.

Typical stress-strain data for the SCS-2/TiA55 system (Figure 11) showed that the coated panels had higher strengths as well as higher strains to failure. Although the initial modulus was about the same for each material, as the material strained, the modulus of the coated panel remained constant out to relatively high stresses and strains, whereas the modulus of the uncoated panel decreased throughout the range.

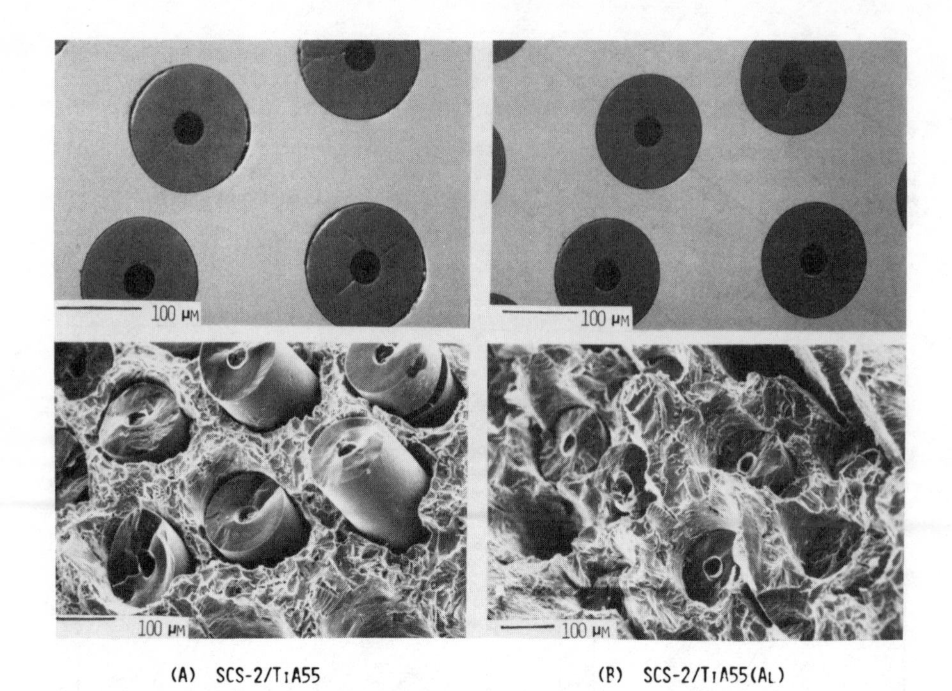

Figure 10.- Polished cross-sections and fracture surfaces of SCS-2/TiA55 composites.

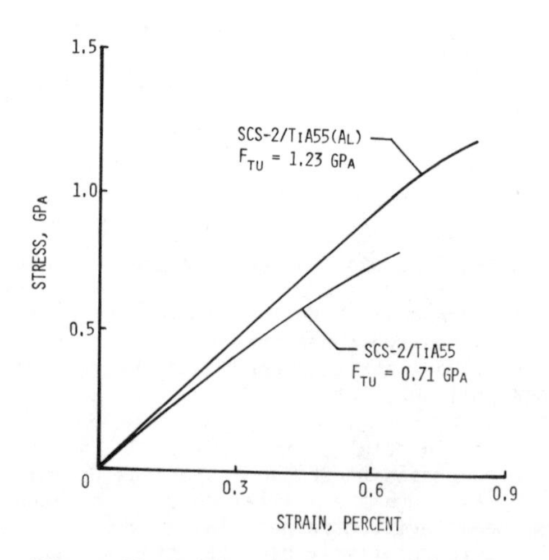

Figure 11.- Effect of matrix coating on tensile behavior of SCS-2/TiA55 composites.

A summary of the SCS-2/TiA55 tensile strengths is given in Figure 12. As indicated by the stress-strain curves, coating the A55 foil with aluminum significantly improved the composite strength (from 0.8 GPa to 1.2 GPa) for a given fabrication temperature, in this case 816C. Reducing the fabrication temperature from 816C to 760C appeared to have about the same effect on the composite strength. The 760C temperature is well below the 870-930C customarily used to fabricate titanium composites, but this low temperature appeared to produce well consolidated panels with good properties. However, fabrication of highly formed structural components or parts with complex shapes may require the higher temperatures in which case the interfacial barrier would be important.

For the Ti(6-4) matrix studies, instead of coating the matrix, SCS-2 fibers were coated with aluminum and then consolidated into composites with the uncoated alloy foil. Tensile strengths for the SCS-2/Ti(6-4) are shown in Figure 13. As for the TiA55 coated panels, the coated fibers tended to yield higher strength composites for a given fabrication temperature. Again, reducing the fabrication temperature from 927C to 816C also produced a significant improvement in composite-strength.

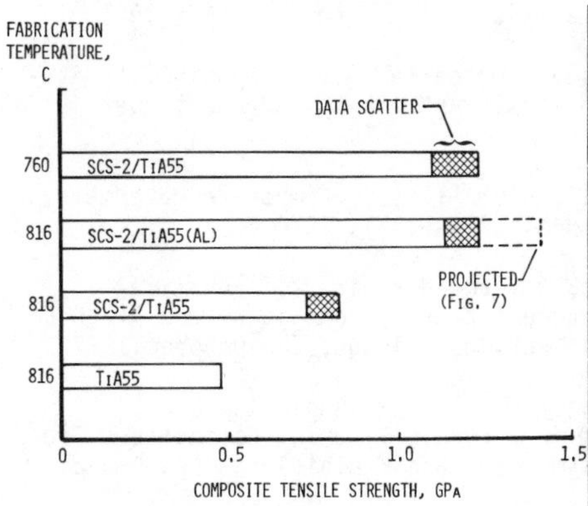

Figure 12.- Measured tensile strength of SCS-2/TiA55 composites.

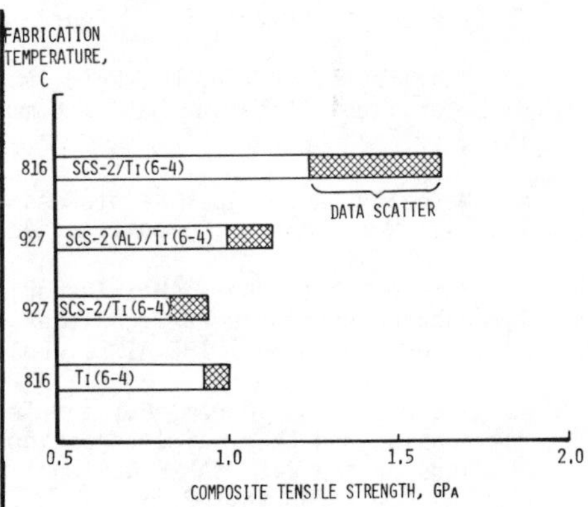

Figure 13.- Measured tensile strength of SCS-2/Ti(6-4) composites.

Concluding Remarks

Several silicon-carbide/titanium composite systems were investigated to determine composite degradation mechanisms and to develop techniques to mimimize loss of mechanical properties during fabrication and in service. X-ray diffraction studies on planar samples showed that certain interfacial reactions in silicon carbide/titanium systems can be controlled by introducing appropriate intermetallic layers into the SiC/Ti system. Insights gained from the x-ray work led to fiber studies which showed that the as-fabricated fiber strength was improved and the fiber degradation during fabrication was reduced by coating the fiber with a 1 μm thick layer of aluminum. These improvements in fiber properties were seen to carry over into composite properties. A new approach of applying an interfacial barrier by coating the matrix foils instead of the reinforcing fiber was found to be effective in improving composite strength. For a given fabrication temperature, coating the matrix or the fibers with aluminum improved the composite strength in SCS-2/TiA55 and SCS-2/Ti(6-4) systems. Reducing the fabrication temperature also resulted in significant improvements in composite strength. Good, well consolidated composites were fabricated at temperatures well below those customarily used for SiC/Ti composite fabrication.

References

1. Debolt, H.; and Krukonis, V.: Improvements of Manufacturing Methods for the Production of Low Cost Silicon Carbide Filament. AFML-TR-73-140, April 1973.

2. Cornie, J. A.: A Review of SiC Filament Composite Production and Fabrication Technology. Proceedings, 4th Metal Matrix Composite Technology Conference. Institute for Defense Analysis. Washington, DC, May 1981.

3. Schmitz, G.; Klein, M.; Reid, M.; and Metcalfe, A.: Compatibility Studies for Viable Titanium Matrix Composites. AFML-TR-70-237, September 1970.

4. Metcalfe, A.: Fiber Reinforced Titanium Alloys. Composite Materials, Vol. 4: Metal Matrix Composites, Academic Press, NY, 1974.

5. Pailler, R.; Lahaye, M.; Thebault, J.; and Naslain, R.: Chemical Interaction Phenomena at High Temperature Between Boron Fibers and Titanium Metal (or TA6V alloy) TMS-AIME. Fall Meeting, Chicago, IL, October 1977.

6. Brewer, W. D.; Unnam, J.; and Tenney, D. R.: Mechanical Property Degradation and Chemical Interactions in a Borsic/Titanium Composite. Proceedings of the 24th SAMPE National Symposium and Exhibition. San Francisco, CA, May 1979, Vol. 2.

7. Rao, V. B.; Houska, C. R.; Unnam, J.; Brewer, W. D.; and Tenney, D. R.: Interfacial Reactions in Borsic/Ti-3Al-2.5V Composites. New Developments and Applications in Composites. TMS-AIME Fall Meeting, October 1978.

8. Rao, V. B.: Control of Interface Reactions in SiC/Ti Composites and an X-ray Diffraction Study of Interdiffusion Between Aluminum Thin Films and Titanium Substrates. PhD Thesis, VPI&SU, Blacksburg, VA, October 1980.

9. Landis, H.: Fiber Matrix Interactions in Ti/SiC Composites. Masters Thesis, George Washington University, July 1981.

10. House, L.: A Study of Interfacial Reactions in Silicon Carbide Reinforced Titanium. Masters Thesis, George Washington University, May 1979.

THERMAL DEGRADATION OF THE TENSILE PROPERTIES OF UNDIRECTIONALLY

REINFORCED FP-Al_2O_3/EZ 33 MAGNESIUM COMPOSITES

R. T. Bhatt
Propulsion Laboratory
AVRADCOM Research and Technology Laboratories
Lewis Research Center
Cleveland, Ohio 44135

and

H. H. Grimes
National Aeronautics and Space Administration
Lewis Research Center
Cleveland, Ohio 44135

Abstract

The effects of isothermal and cyclic exposure on the room temperature axial and transverse tensile strength and dynamic flexural modulus of 35 and 55 volume percent FP-Al_2O_3/EZ 33 magnesium composites have been studied. The composite specimens were continuously heated in a sand bath maintained at 350° C for up to 150 hours or thermally cycled between 50° and 250° C or 50° and 350° C for up to 3000 cycles. Each thermal cycle lasted for a total of six minutes with a hold time of two minutes at the maximum temperature. Results indicate no significant loss in the room temperature axial tensile strength and dynamic flexural modulus of composites thermally cycled between 50° and 250° C or of composites isothermally heated at 350° C for up to 150 hours from the strength and modulus data for the untreated, as-fabricated composites. In contrast, thermal cycling between 50° and 350° C caused considerable loss in both room temperature strength and modulus. Fractographic analysis and measurement of composite transverse strength and matrix hardness of thermally cycled and isothermally heated composites indicated matrix softening and fiber/matrix debonding due to void growth at the interface and matrix cracking as the likely causes of the strength and modulus loss behavior.

Introduction

Composites of magnesium alloys reinforced with FP-Alumina are being considered for aerospace applications because of their high specific strength and modulus, and because of their relative ease of fabrication. Some of these applications require repeated high temperature exposure of the composite material for extended periods of time. In a recent study, Bhatt, et al.,[1] measured the room temperature axial and transverse strengths and moduli of three different, unidirectional FP-Al_2O_3/magnesium composites prepared by Dupont by a liquid metal infiltration technique. The matrix materials used were EZ 33 (2 to 4 percent rare earths: 2 to 3 percent Zn: 0.5 to 1 percent Zr: bal Mg), QH 21A (2 to 3 percent Ag, 0.6 to 1.6 percent Th: 0.6 to 1.5 percent rare earths; 0.4 to 1 percent Zr: bal Mg) and pure magnesium. The effect of short term (120 hr) isothermal exposure at 350° and 425° C on the above composite properties was also determined. At these temperatures all these composites showed both axial and transverse strength degradation. The major cause of this degradation was matrix softening.

The isothermal exposure temperatures used in this earlier study were chosen to insure strength degradation and therefore represent a more severe temperature condition than would be experienced by the composite in normal use. On the other hand, the cyclic environments the composites will experience in normal use may represent a severe degradation condition even at the lower temperatures. This is because under cyclic heating conditions the plastic deformation which may occur in the matrix due to thermally induced stresses will be at least in part irreversible leading to cumulative effects such as matrix cracking or void growth.

The objective of this investigation, therefore, was to characterize more fully the useful range of application of a selected FP-Al_2O_3/magnesium composite, as determined by tensile strength and modulus measurements after cyclic thermal exposure.

Of the three magnesium matrix alloys previously studied, EZ 33 was choosen for this investigation because it bonded well to the FP-Alumina fiber resulting in a composite which displayed adequate axial and transverse strengths combined with an excellent modulus. While the properties determined in that study for the FP-Al_2O_3/QH 21A composites were generally similar, this material was found to be poorly bonded so that the full advantage of the stronger matrix alloy was not realized. It was felt that the poorly bonded composite would not perform well in cyclic testing.

In this study, 35 and 55 volume percent fiber unidirectionally aligned composites were exposed in air to temperatures of either 250° or 350° C for up to 3000 cycles. Isothermal exposure at 350° C was used as a baseline to evaluate the additional effect of cycling. Metallographic and fractographic analysis and matrix hardness measurements were also conducted on untreated and thermally cycled composite material in an attempt toward understanding of the failure mechanisms involved.

Experimental

The FP-Al_2O_3/EZ 33 Mg composites used in this study were fabricated by the Dupont Pioneering Research Laboratory using molten metal infiltration techniques. The nominal FP-Al_2O_3 fiber content used was either 35 or 55 volume percent. The fibers were aligned unidirectionally in a EZ 33 magnesium alloy matrix having the composition previously given. Plates of 0.25 cm thickness were cast and cut into specimens 12.7 cm long and 1.25 cm wide. The fiber orientation was either parallel to the specimen length (0°) for axial testing or perpendicular to the specimen length (90°) for transverse testing. Thermal cycling was done by alternately dipping a frame supporting six specimens into a hot (250° or 350° C) fluidized sand bath and then into a cold bath that equilibrated near 50° C. Each complete thermal cycle lasted for six minutes. Typical time-temperature profiles of the composite specimen cycled to 250° or 350° C are shown in Fig. 1. The time at temperature during each cycle was approximately two minutes.

Similar composite specimens were also isothermally heated at 350° C in a sand bath for periods up to 150 hours, which corresponds to a time about 50 percent longer than the time at temperature for 3000 cycle experiments.

After cycling to a predetermined number of cycles or isothermal heating to set time periods, the specimens were removed from the bath and aluminum doublers were adhesively bonded to the specimen ends.

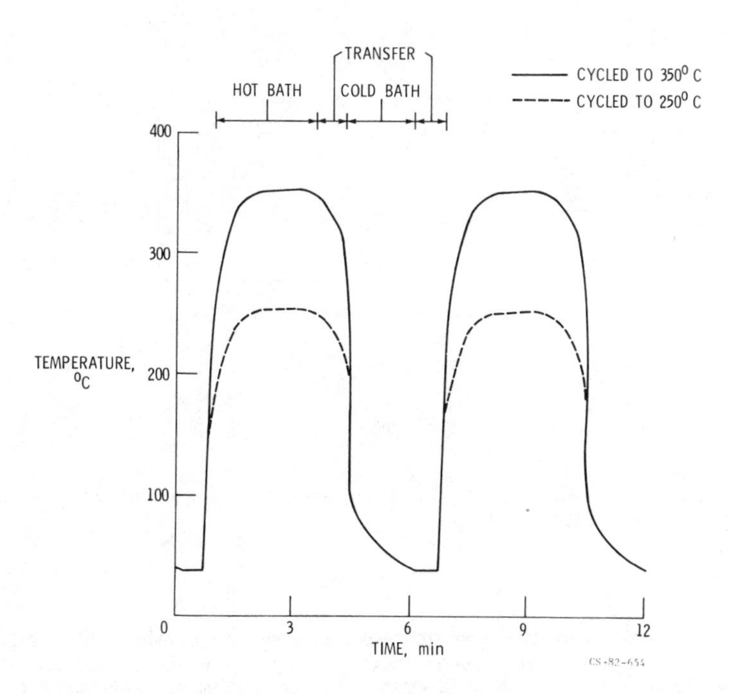

Figure 1 - Typical thermal cycle temperature-time profiles.

Tensile testing was done in an Instron testing machine equipped with wedge type grips. The specimens were pulled to failure at a constant cross head speed of 0.126 cm/min. Matrix hardness was measured in a Vickers hardness testing machine using a diamond pyramid indentor and a 25 gm load.

A flexural modulus test, similar to that of McDanels et. al[2], was used for measuring the dynamic modulus of the composite specimen. In this test, the composite specimen is supported on a pair of steel wires. The distance between the wires was set to 0.7758L, where L is the specimen length. This corresponds to the two nodal positions for the flexural deformation of a free bar of the given dimensions in flexural vibration. The specimen is driven by a piezo-electric transducer positioned at one end of the specimen. A similar transducer placed at the other end of the specimen was used as a detector. The block diagram of the transducer drive and detection system used is shown in Fig. 2.

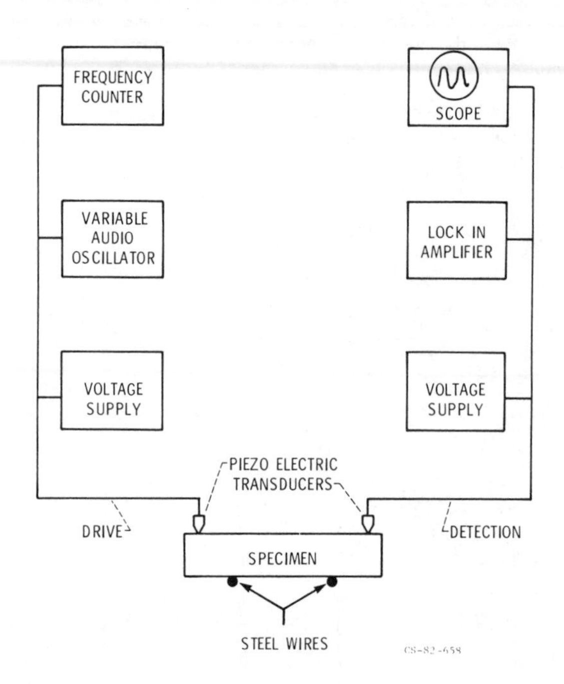

Figure 2 - Block diagram of the piezo electric transducer drive and detection system.

For determination of the dynamic flexural modulus, E, the variable drive transducer was manually tuned to that frequency which produced resonance (the maximum flexural displacement) in the specimen. For the free-free flexural modes of rectangular bar specimens with low damping, the resonant frequency, ω, is related to E by the equation:

$$E = \frac{12 \times (2\pi \times \omega)^2}{b^2}\left(\frac{M}{W}\right)\left(\frac{l}{h}\right)^3 \quad (1)$$

where b is a constant and M, h, W and l are the specimen mass, thickness, width and length respective. For $l/h > 100$, $b = 22.37$. For $l/h < 100$, $b < 22.37$. The exact value of b can be determined from specimen dimensions.(3) Thus, by the measurement of the resonant frequency ω, and the dimensions and weight of the specimen, the dynamic flexural modulus of the specimen can be calculated from Eq. (1).

Results and Discussions

The room temperature flexural moduli of the 35 and 55 fiber volume percent FP-Al_2O_3/EZ 33 Mg composites cycled to 250° or 350° C to a maximum of 3000 cycles are shown in Fig. 3. The flexural moduli of untreated composites are also shown in Fig. 3 for base line comparison. The data points represent the range and average value for at least three determinations. The modulus values for 35 and 55 fiber volume composites thermally cycled to 250° C show no significant change from the modulus values of as-fabricated unheated composites. However both composites, thermally cycled to 350° C show a loss in flexural modulus proportional to the number of thermal cycles. After 3000 cycles to 350° C, both the 35 and 55 fiber volume composites degraded to near 80 percent of the modulus values of unheated composite specimens. Also, shown in Fig. 3 is the modulus values for similar composite specimens isothermally treated at 350° C for 150 hours, a time approximately 50 percent longer than the cumulative time at temperature for the 3000 cycle test. These specimens did not show any loss in flexural modulus even after 150 hours of exposure. In the cyclic tests, while the specimen length and weight remained nearly the same, the width and thickness of the specimen increased continuously with cycling. In isothermally heated composites, however, no dimensional changes were mea- sured even after 150 hours of exposure at 350° C. The losses in the dynamic modulus of the cycled composites were also found to be proportional to the width and thickness changes.

The room temperature axial tensile strengths of the 35 and 55 volume percent composites cycled to 250° or 350° C to a maximum of 3000 cycles are shown in Figs. 4 and 5. Again, the data points indicate the range and the average value for typically three tests. The axial strength data for the 35 and 55 fiber volume percent composites without thermal treatment are also shown in Figs. 4 and 5 respectively for a base line comparison. The strength data, as seen in Fig. 4, for the 35 fiber volume percent composites cycled to 250° C show no appreciable degradation from the untreated composite strength of 0.33 GN/m^2 even after as many as 3000 cycles. However, cycling this composite to 350° C resulted in an initial drop in strength from the base line value of 0.33 GN/m^2 to 0.315 GN/m^2 after only 20 cycles, then no further drop in strength until after 2000 cycles and finally a second loss of strength to 0.285 GN/m^2.

For the 55 fiber volume percent composite data shown in Fig. 5, a somewhat different axial strength degradation behavior was observed. Composite cycled 3000 times to 250° C showed a modest loss of strength from the base line value of 0.50 GN/m^2, to 0.47 GN/m^2. Similar composites cycled to 350° C, on the other hand, showed a rapid loss of strength from the base line value of 0.50 GN/m^2 within the first 20 thermal cycles. Additional cycling resulted in a gradual decrease in strength to 0.33 GN/m^2 after 3000 cycles.

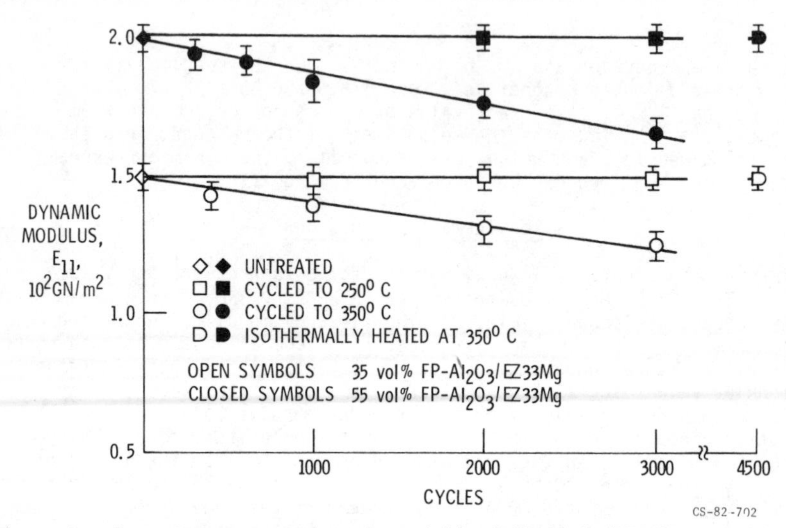

Figure 3 - Room temperature dynamic moduli of FP-Al_2O_3/EZ33Mg composites after cycling to 250°C or 350°C for indicated number of cycles. Dynamic moduli of composites isothermally heated at 350°C for 150 hours as shown. (Dynamic moduli of untreated composites are shown at zero cycles.)

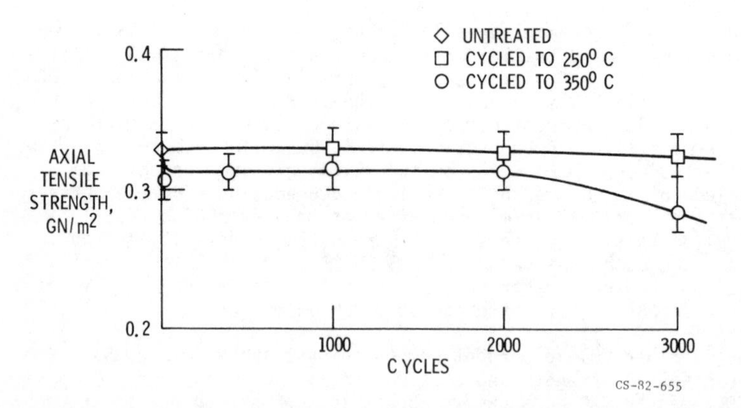

Figure 4 - Room temperature axial tensile strengths of 35 volume percent FP-Al_2O_3/EZ33Mg composites after cycling to 250°C or 350°C for indicated number of cycles. Strengths of untreated composites are shown at zero cycles.

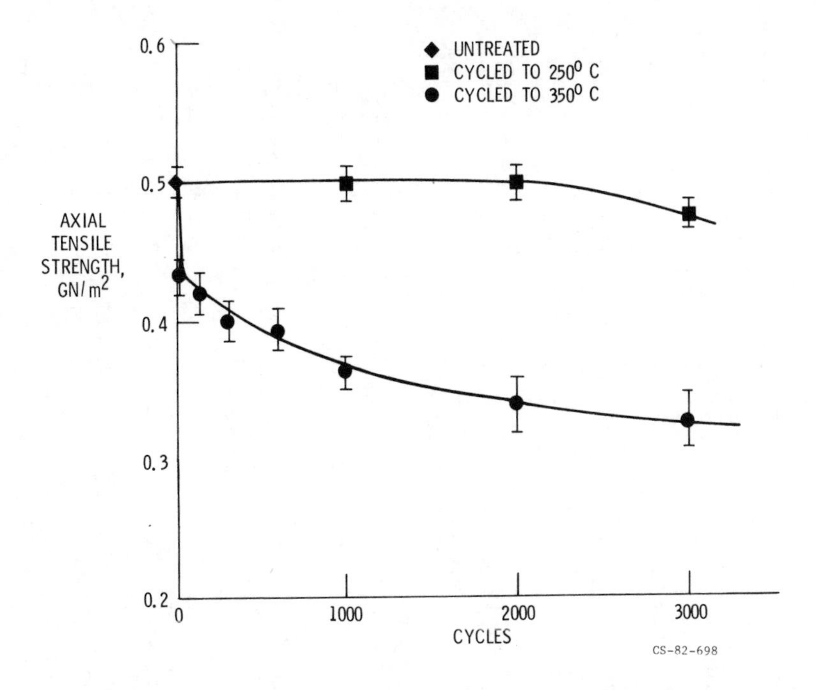

Figure 5 - Room temperature axial tensile strengths of 55 volume percent FP-Al_2O_3/EZ33Mg composites after cycling to 250°C or 350°C for indicated number of cycles. (Strengths of untreated composites are shown at zero cycles.)

The base line axial strength data and strength data measured after 1000, 2000 and 3000 cycles from Figs. 4 and 5 are replotted in Fig. 6 against maximum cycle temperature to better illustrate the temperature dependence of the strength degradation. Clearly from Fig. 6 no significant strength loss occurred for either the 35 or 55 fiber volume percent composites when cycled to 250° C, even for as many as 3000 cycles. At 350° C, however, a rapid strength loss occurs. For the 55 fiber volume percent composite this loss appears to be more a function of cycle temperature than number of cycles. Also shown in Fig. 6 is the range and average value of all the strength data of similar composite specimens which have been isothermally heated at 350° C for 150 hours a time 50 percent longer than the cumulative time-at-temperature for specimens cycled 3000 times. The 35 fiber volume percent composite specimens isothermally heated at 350° C show strength values similar to the strength values of untreated composites, whereas the 55 fiber volume percent composite specimens show a 10 percent loss of strength to a value of 0.45 GN/m². In all cases, however, the strength values of isothermally heated specimens were equal to or higher than the strength values for thermally cycled composite specimen.

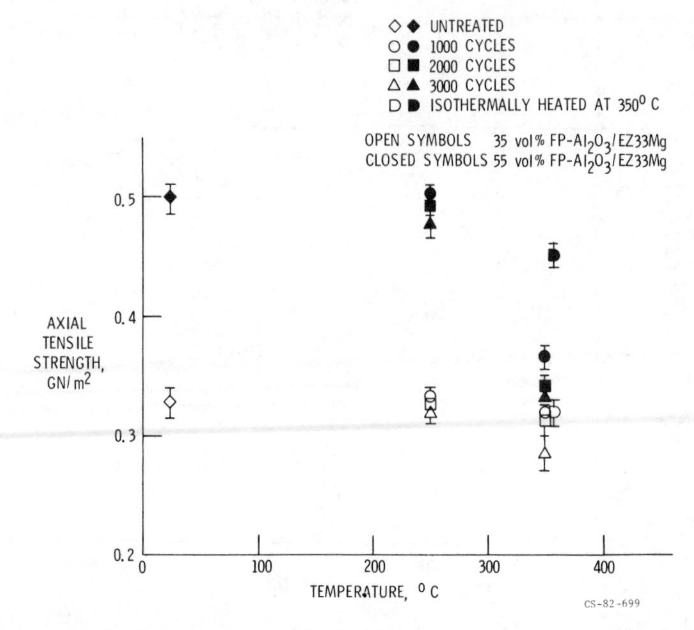

Figure 6 - Room temperature axial tensile strengths of FP-Al_2O_3/EZ33Mg composites cycled 1000, 2000, and 3000 times to indicated temperatures. (Strengths of similar composites isothermally heated at 350°C for 150 hours are also shown.)

The greater degradation observed for cyclically heated composites is seen as evidence of a mechanism involving more than a simple thermally activated process. If a single thermally activated process were operating we would expect similar behavior for specimens which were cyclically heated or isothermally heated for an equivalent time at temperature. A likely candidate for this additional mechanism is one which involves the generation of large matrix stresses due to the differences between the thermal expansions of the fiber and matrix during heating. The maximum thermally induced matrix stress in fiber composites can be obtained from the equation derived by Piggott[4]

$$\sigma_m \simeq \frac{3\,(\alpha_m - \alpha_f)\,(\Delta T)\,V_f E_m E_f}{E_c + E_f\,(1 + \gamma_m V_m) - \gamma_f E_m} \tag{2}$$

where α is the thermal expansion coefficient, V is the fiber volume fraction, E is the elastic modulus, γ is poisson's ratio, and ΔT is the cyclic temperature range. The subscripts m, f, and c refer to the matrix, fiber and composite, respectively. From this study we use; $\Delta T = 200^\circ$ or 300° C and measured values of E_c. We also use $\alpha_m = 25.4$ μ/cm/cm/°C, $\alpha_f = 5.7$ μ/cm/cm/°C, $E_f = 379$ GN/m^2 and $E_m = 44.8$ GN/m^2 [5], and $\gamma_f = 0.2$, $\gamma_m = 0.33$ [6,7]. The maximum matrix thermal stresses, σ_m, calculated for 35 and 55 fiber volume percent composites are shown in Table I.

Table I. Calculated Matrix Thermal Stresses in FP-Al_2O_3/EZ33Mg Composite

$(T_2 - T_1) = \Delta T$, °C	σ_m, GN/m^2	
	35 Vol percent	55 Vol percent
200	0.119	0.174
300	.179	.262

These stresses exceed even the room temperature yield stress value of 0.11 GN/m^2 for the EZ 33 magnesium alloy.[7] In the thermally cycled composite, therefore plastic deformation will occur during each cycle. The cumulative effect of such repeated deformation has been observed to produce voids in the matrix in highly constrained regions where the matrix deformation cannot be reversed.[8,9] Evidence of similar void growth and matrix cracking in this composite is seen in Fig. 7 which shows microphotographs of the 55 volume percent FP-Al_2O_3/EZ 33 Mg composite before thermal exposure and after cyclic heating to 350° C. The voids here appear to extend to the fiber/matrix interfaces. The presence of voids or cracks at the interface will result in a loss of fiber/matrix bonding.

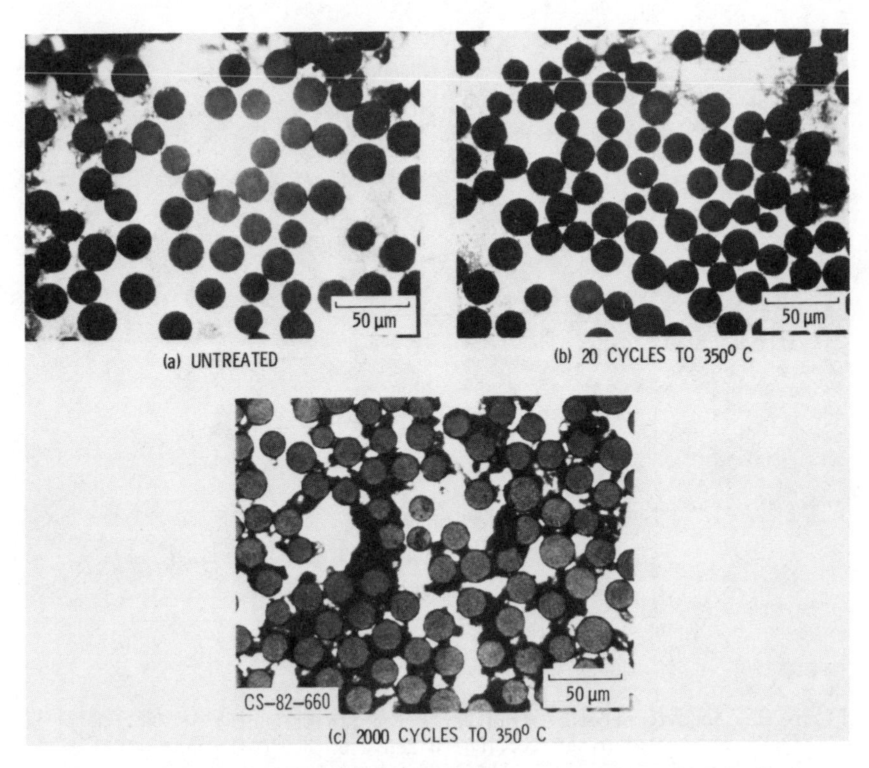

Figure 7 - Photomicrograph of untreated and thermally cycled composites after indicated numbers of cycles.

Whether the loss of axial strength results from fiber/matrix debonding or from a weakening of the fiber or the matrix cannot be determined from these micrographs. However, some insight may be obtained from the results of transverse strength of the composite and from matrix hardness measurements. Figure 8 shows the transverse strength data for the 35 and 55 fiber volume percent composites thermally cycled to 250° or 350° C. It is obvious from this figure that the transverse strength decreases significantly within the first 20 cycles at both temperatures. Further thermal cycling to a maximum of 3000 cycles resulted in a more gradual loss of strength to values as low as 0.055 GN/m^2 for 55 fiber volume FP-Al_2O_3/EZ 33 Mg composite. The magnitude of the initial loss of strength, as seen in Fig. 8, increases markedly with maximum cycle temperature and with increased fiber volume fraction for the higher temperature tests. Also shown in Fig. 8 are the transverse strength data for similar composites isothermally heated at 350° C from 1 hour to 150 hours. As with the data for cycled composites, a rapid initial loss in strength was observed at 350° C for both 35 and 55 fiber volume composites. However, for these composites no further degradation was measured for exposure up to 150 hours. Thus, again it appears that an additional degradation mechanism is occurring during cycling.

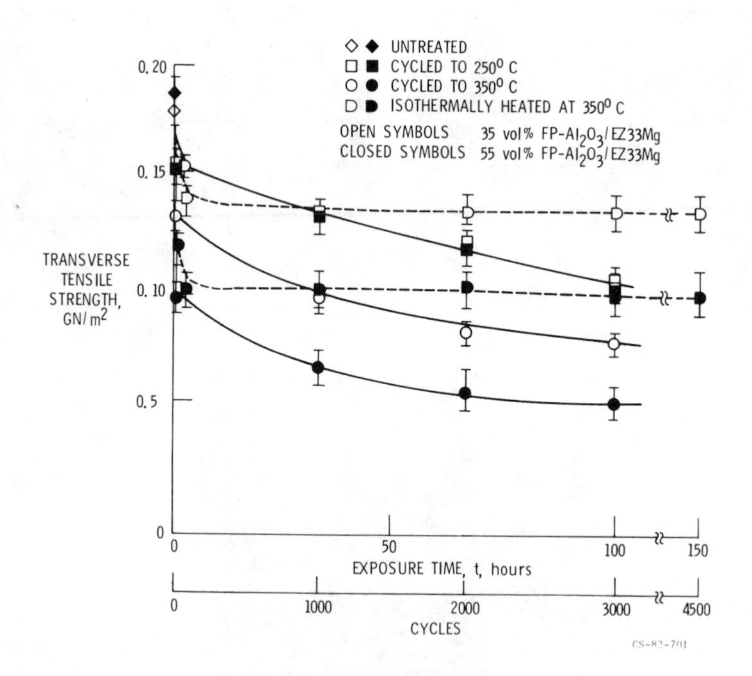

Figure 8 - Room temperature transverse tensile strengths of FP-Al_2O_3/EZ33Mg composites after cycling to 250°C or 350°C for up to 150 hours. are shown. (Strengths of untreated composites are shown at zero cycles.)

The rapid strength loss in the first 20 cycles or within one hour of isothermal exposure at 350° C, appears to correlate better, however, with hardness measured on the void free regions of matrix before and after heat treatment. The matrix hardness data for the 55 volume percent FP-Al_2O_3/EZ 33 Mg composites cycled to 350° C or isothermally heated at 350° C up to 150 hours are plotted in Fig. 9. Each data point is an average of at least ten measurements. The hardnesses of both thermally cycled and isothermally heated composites decreased to nearly one half of the value for unheated as fabricated material within 20 cycles or within one hour of heating. Further cycling to 3000 cycles or heating to 150 hours resulted in no further hardness decrease. This matrix softening correlates exactly with the transverse strength loss for the isothermally heated composites and is assumed to be responsible for this loss.

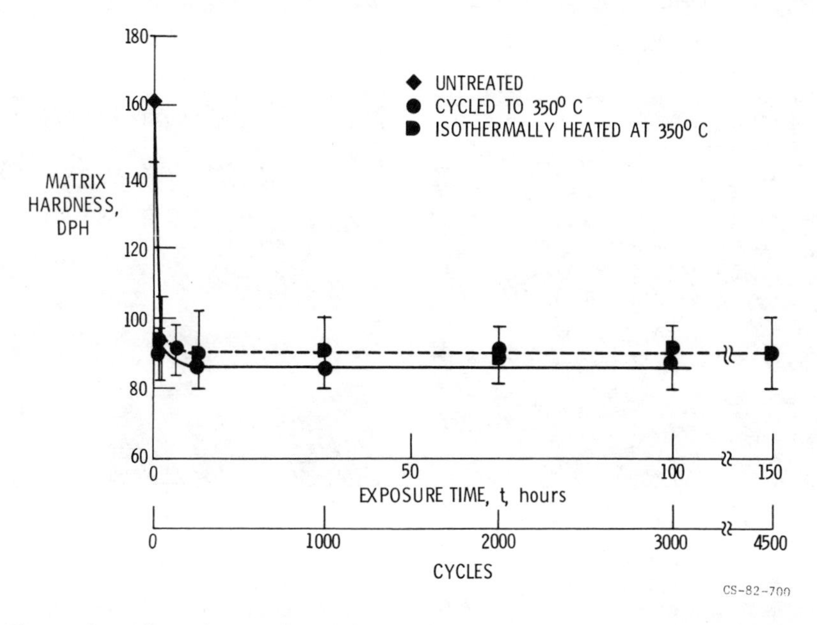

Figure 9 - Diamond pyramid hardness of the matrix in 55 volume percent FP-Al_2O_3/EZ33Mg composites after cycling to 350°C for indicated number of cycles. Matrix hardness values of similar composites isothermally heated at 350°C for up to 150 hours are also shown. (Matrix hardness values of untreated composites are shown at zero cycles.)

Matrix softening, however cannot explain the additional gradual transverse strength loss between 20 and 3000 cycles for thermally cycled specimens. An additional degradation mechanism is indicated for the cycled composites. Further insight into this additional mechanism results from examination of the transverse fracture surfaces of the cycled composites. Typical fracture surfaces of the untreated and thermally cycled composites for 20, 1000 and 2000 cycles are shown in Fig. 10. As previously observed

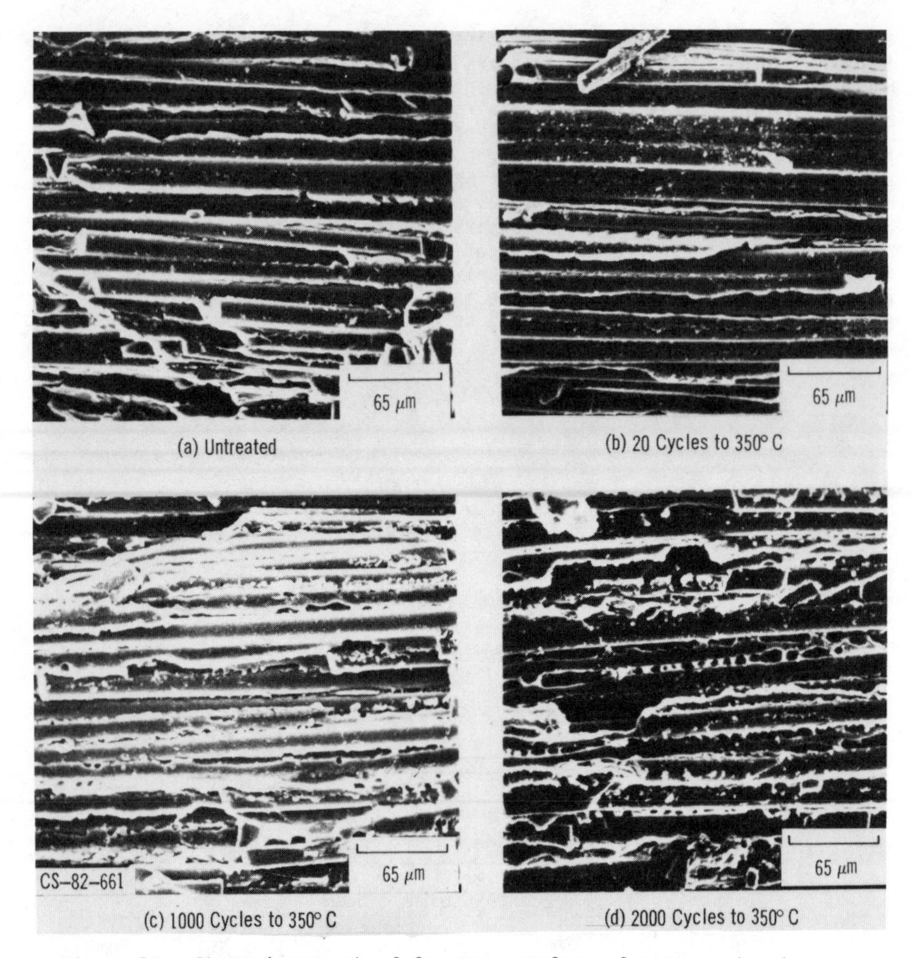

Figure 10 - Photomicrograph of fracture surface of untreated and thermally cycled FP-Al_2O_3/EZ33Mg composites, stressed in transverse direction showing effect of cycling to 350°C for indicated number of cycles.

in Fig. 7 for unfractured specimens, there is an increasing tendency for void growth at the fiber/matrix interface with cycling. Since isothermally heated composites showed no voids even after 150 hours, we associate the gradual strength decrease with the number of cycles as shown in Fig. 8 with the growth of voids.

Matrix softening and void growth can also explain the axial strength and modulus loss behavior. The loss of fiber/matrix bonding caused by void growth at the interface will have greater effect on the axial strength and modulus properties than that due to the matrix softening. Indeed these effects were observed in the axial strength and modulus data of cycled or isothermally heated composites.

Summary

The effects of cyclic and isothermal heat treatments on the axial and transverse tensile strengths and axial moduli of 35 and 55 volume percent FP-Al_2O_3/EZ 33 Mg composites have been evaluated to understand the cause of thermally induced strength degradation and to help determine the limiting use conditions for these composites. Specific findings are as follows:

1. Thermal cycling of FP Al_2O_3/EZ 33 Mg composites to 250° C for 2000 cycles did not cause any appreciable room temperature strength or modulus loss compared with base line data for untreated composites. In contrast composites thermally cycled to 350° C showed considerable loss in both strength and modulus, with the 55 volume percent composites degrading more than the 35 volume percent composites.

1. Measurement of the transverse strength and matrix hardness, and fractographic analysis of thermally cycled composites indicated interface void formation, matrix cracking and matrix softening as prime contributors to observed strength and modulus losses. These results and the observed fiber volume dependence are consistant with a degradation mechanism based on thermal induced stresses in the matrix.

3. No appreciable loss in the axial dynamic modulus of these composites was observed after isothermal exposure. Small strength losses observed in isothermally heated composites were attributed to matrix softening.

4. The high temperature mechanical properties were not measured for the composite in this study. This would be required to properly design using FP-Al_2O_3/EZ 33 composite. However, aside from the usual matrix softening at higher temperatures, this study indicates that we would not expect additional axial strength or modulus degradation resulting from either isothermal or cyclic exposure below 2000 cycles or 250° C.

References

1. R. T. Bhatt and H. H. Grimes: "Modulus, Strength and Thermal Exposure Studies of FP-Al_2O_3/Aluminum and FP-Al_2O_3/Magnesium Composites," paper presented at the 5th Annual Conference on Composite Materials, American Ceramic Society, Cocoa Beach, Fl, Jan., 1981.

2. D. L. McDanels and R. A. Signorelli: NASA TN D-8204, 1976.

3. J. DiCarlo and J. E. Maisel: Conference on Composite Materials; Testing and Design, 5th, pp. 201-227, ASTM STP 674, 1979.

4. M. R. Piggott: Load Bearing Fiber Composites, p. 208, Pergamon Press., Oxford, 1980.

5. A. K. Dhingra and W. H. Krueger: New Engineering Material - Magnesium Castings Reinforced with Dupont Continuous Alumina Fiber FP," paper presented at the International Magnesium Association 36th Annual World Conference on Magnesium, Oslo, Norway, June 25-26, 1979.

6. J. F. Bacon, K. M. Prewo and R. D. Veltri: ICCM/2, Proceedings of the Second International Conference on Composite Materials, B. Noton, R. Signorelli, K. Strect, and L. Phillips, eds., pp. 753-769, Metallurgical Society of AIME, Warrendale, PA, 1978.

7. Metals Handbook, 8th Edition, Vol. I, American Society for Metals, Metals Park, OH, 1961.

8. P. Shahinian: SAMPE Quart., Vol. 2, 1970, pp. 28-35.

9. W. H. Kim, M. J. Koczak and A. Lawley: New Developments and Applications in Composites, D. Kuhlmann - Wilsdorf and W. C. Harrigan, eds., pp. 40-53, Metallurgical Society of AIME, Warrendale, PA, 1979.

INFLUENCE OF INTERFACE DEGRADATION AND ENVIRONMENT ON THE THERMAL AND FRACTURE FATIGUE PROPERTIES OF TITANIUM-MATRIX/CONTINUOUS SiC-FIBER COMPOSITES

Y.H. Park and H.L. Marcus

Mechanical Engineering/
Materials Science and Engineering
The University of Texas
Austin, Texas 78712

Summary

Interface degradation from thermal cycling in the environments such as vacuum, air and sulfur was investigated using SEM fractography and AES. SEM fractography shows that thermal fatigued specimens produce more damage to the fiber interface than isothermally treated specimens. Liquid nitrogen treatment before thermal treatment gave a consistent result due to the reproducibility of the initial condition of the samples. Based on an elastic solution, the interface stress was estimated and at room temperature, the stress state was tensile with a value of approximately 100 ksi after thermal cycling between 550°C and RT. From the sulfur environment experiments, the interface diffusion coefficient was found to be about 10^{-8} cm^2/sec and the sulfur concentration vs. distance is expressed as $\ell n C_S = kx$.

Introduction

Filamentary metal-matrix composites have generated a considerable amount of interest in the materials field because of their potential applications in dynamic structures. A metal-matrix composite system (MMC) carries certain potential advantages over other non-metal containing composite systems. High strength, modulus, toughness, reproducibility of properties, surface durability, and low notch sensitivity when compared to non-metal matrix composites are just a few of these potential advantages. The MMC systems are characterized by heterogeneity, anisotropy, strengthening by load transfer and interfaces. In order for the full potential of these systems to be used, a better understanding of the above-mentioned effects must be acquired. While analytical tools like stress analysis and linear elastic fracture mechanics are useful in understanding the mechanical properties of matrix and reinforcement together, they cannot be used directly to find out the effect of interface properties on the mechanical response of the system.

Studies have been done to obtain tensile, toughness, and fatigue properties of various MMC systems. Early investigations in fatigue were limited to cyclic stress strain behavior (1). Limited studies have been done on fatigue behavior in environments as a function of temperature. Composite materials present a variety of complex problems in thermomechanical behavior due to differences in properties of matrix and reinforcement. Recently published work (2-7) has treated the case of thermal cycling of composites consisting of elastic reinforcement and a plastic matrix in order to analyze the distribution of stress and strain, obtain expansion properties and investigate possible damage mechanisms. Thermal cycling will obviously be important in many elevated temperature applications. It is, therefore, necessary to know more about the thermal fatigue properties of MMCs and to relate them to the interface chemistry. Better understanding of this relationship can assist in improved composite design and advancement in processing of composites.

The purpose of this investigation is to study the interaction between environment and the metal matrix-fiber interface in MMCs and how this interaction influences the overall performance of the materials under thermal cycling.

A group of the most promising of the current MMCs are SiC, B_4C and Borsic continuous fibers in the Ti-6Al-4V titanium alloy (8-9). The problems associated with the interface are some of the more significant in the potential application of these MMC systems.

Experimental

Materials

The material used in this study is the SiC fiber reinforced titanium $\alpha+\beta$ alloy (Ti-6Al-4V) matrix composite system. Some of the SiC was on a graphite substrate fiber. Fiber volume fraction was nominally 40%. The diameter of the fibers was about 150 μ. Four ply panels of 0.033 inches (0.08 cm) were used in the mechanical test. Tensile test and Auger electron microscopy (AES) specimens were cut from the panels by electrodischarge machine (EDM) and polished 1.25 inch (3.2 cm) long by 0.25 inch (0.63 cm) or 0.125 inch (0.32 cm) wide. Samples were tested in both the longitudinal and the transverse mode.

Experiments

Isothermal and Thermal Fatigue Tests. Specimens were either isothermally aged at 550°C or thermally cycled between RT and 550°C. The maximum temperature of 550°C was chosen because this approximately represents the upper potential temperature for application of titanium matrix composites. Heating was done in the electrically heated alumina fluidized bath and cooling by removal of the sample from the furnace combined with forced air cooling. A complete cycle was 12 minutes with five minutes for heating or cooling and the remainder for transport. (See Figure 1.)

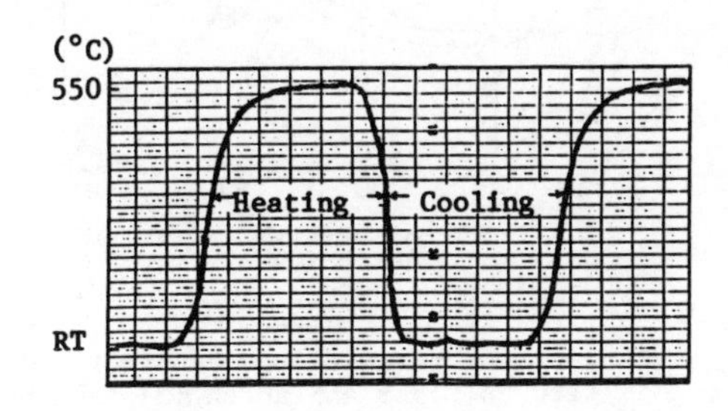

Figure 1 - Typical temperature change during thermal cycle.
(5 min for heat or cool)

Experiments were also made on samples sealed in a pyrex glass tube in the presence of sulfur vapor and in vacuum. These samples were also thermal fatigued as described above. The possibility of sulfur degradation of the titanium metal matrix composite was investigated to simulate an aggressive operational environment containing sulfur. The oxygen containing environment was established by cycling in air.

Tensile Test. All mechanical tensile tests were performed on an Instron at a cross-head speed of 0.5 mm/min. Titanium alloy tabs were glued to both ends of the specimen to prevent premature fracture of the specimen. The samples were smooth and unnotched. A minimum of two specimens were tested for each condition and the results are the average.

AES Analysis. The MMC composites were fractured in-situ in a vacuum of 10^{-10} torr (1.3×10^{-8} Pa). Auger mapping of the fracture surface was made to see the elemental distributions. Several points on the interface were selected for AES and then the surface was Argon-ion sputtered to determine the depth profile of the various elements at the interface.

Results and Discussion

In an air environment, the longitudinal fracture strength was lower for the thermal cycling specimen than for the as-received or isothermally aged specimen (10). (See Figure 2.) Metallography shows a thicker interface reaction layer for thermal cycling a specimen for a ten day exposure in air resulting in the formation of an alpha case in the titanium matrix around the fibers. An effort was made to minimize the composite degradation due to long thermal treatments in air. The fracture morphologies of the matrix from all conditions were similar.

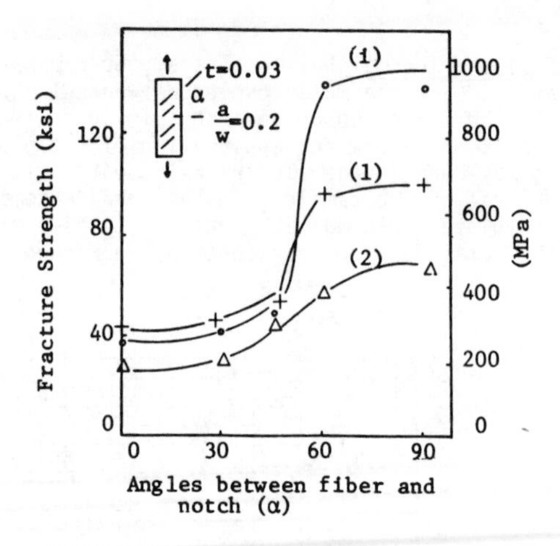

Figure 2 - Fracture strength vs. angles between fiber and notch (α) for thermal fatigued at 550°C and 750°C. (i) As-received; (1) TF 550, 10d; (2) TF 750, 10d.

The interface fracture mode showed differences in different environments and thermal conditions (Fig. 3a, b, c). For some cases, the fibers were broken and the interface was branch-cracked. The interface was investigated by AES to see how the interface chemistry is modified and it was found that the oxygen was present over the interface (Fig. 4).

To establish if there was a variation in oxygen distribution due to cyclic or isothermally treated fracture surfaces, Auger mapping techniques were employed. It was shown that oxygen in the specimen thermal fatigued in vacuum was distributed throughout the fracture interface on the matrix side. In the vacuum isothermally treated case a low oxygen content was observed over the fracture surface. Similar Auger maps were obtained for samples thermally cycled in the air. Much higher concentrations of oxygen

Figure 3(a) - As-received Ti-6Al-4V/SCS (transverse) SEM fractography, unnotched.

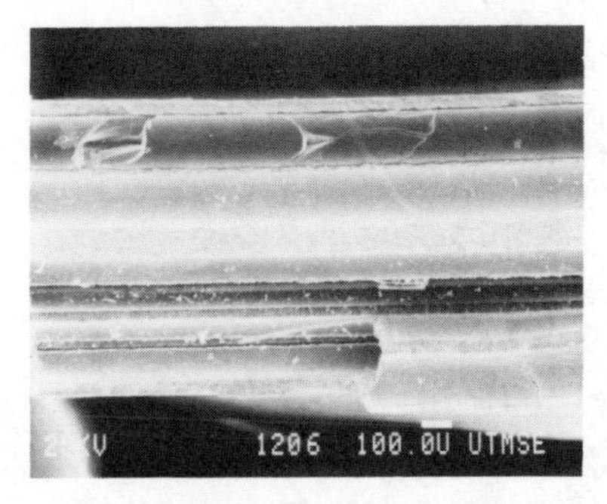

Figure 3(b) - Thermal cycled in sulfur environment Ti-6Al-4V/SCS (transverse) for one day, unnotched.

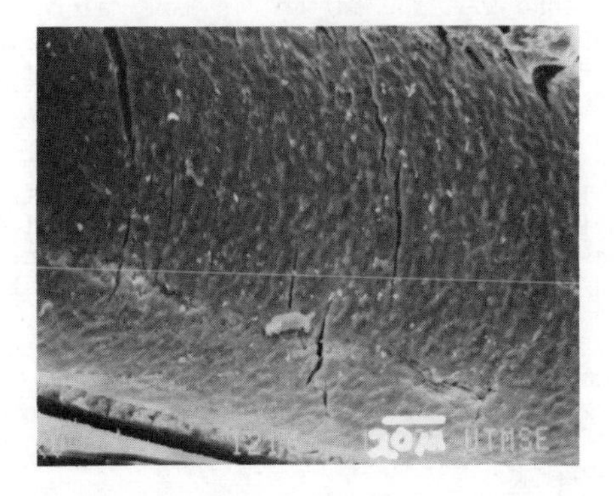

Figure 3(c) - Thermal cycled in sulfur environment Ti-6Al-4V/SCS (longitudinal); unnotched.

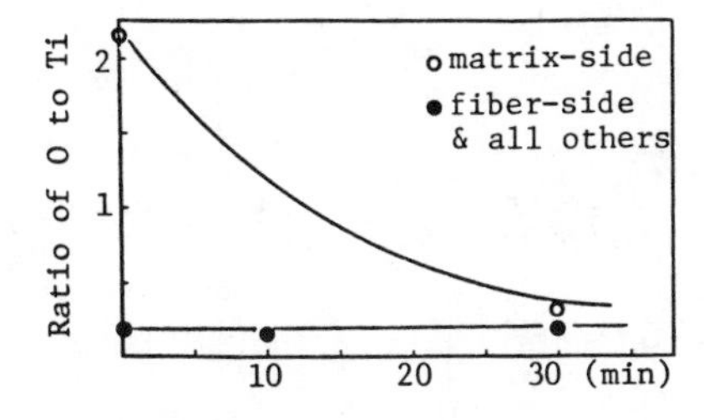

Figure 4 - AES sputtering results on matrix-side and fiber-side interfaces for thermal fatigued samples (550°C-RT, 10 days). All others include isothermal heat treatment.

were found in the broken fibers and in some matrix areas. Oxygen was present on the interface portion of the fracture surface for the isothermal case but not on the fracture surface through the matrix. It was at lower levels than for the thermal fatigued samples. This increase in fiber and matrix damage indicates that thermal cycling is much more deleterious than isothermal exposure.

It is, therefore, concluded that the interface degradation is mainly due to the diffusion of oxygen down the interface. Damage enhanced diffusion by thermal cycling is also supported from the fact that the oxide thickness of thermal cycling is larger than that of isothermal treatment.

The room temperature fracture surface of the transverse direction sample thermal fatigued in a sulfur environment showed the crack propagated along the interfaces. Branch cracks into the matrix side of the interface were observed in the presence of sulfur. When the extent of the branch cracking is measured relative to the free end and compared to the sulfur analysis, it was found that the sulfur rich area along the interface extended as far as the branch cracks. Those branch cracks were not found for isothermally exposed specimens or for the thermal fatigued specimens beyond the sulfur diffusion distance.

To determine the origin of the enhanced degradation of the interface from thermal cycling, the thermally induced stress state due to the difference of thermal expansion coefficient between the Ti-6Al-4V matrix and the SiC fibers was analyzed as follows. A plane strain elasticity analysis (11) that had been derived earlier (12) to explain the matrix deformation due to thermal stresses is used as a basis for the model. The model, shown in Fig. 5, treats the fiber of radius a and its surrounding shell of matrix of outer radius b as an independent entity. The subscripts m and f denote the matrix and the fiber respectively. ν is the poisson's ratio, α the coefficient of expansion, E the Young's modulus and ΔT the temperature cycling range.

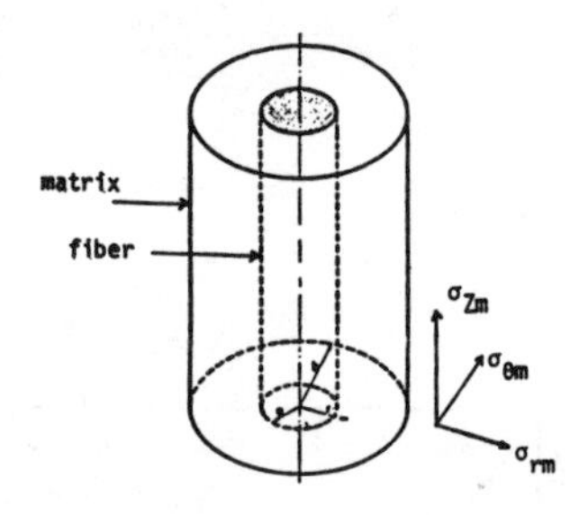

Figure 5 - The model of a unit composite of a fiber and a matrix shell.

Poritsky's elastic solution gives the principal thermal stresses in the matrix as:

$$\sigma_r = \Gamma \left(1-\frac{b^2}{r^2}\right) \tag{1}$$

$$\sigma_\theta = \Gamma\,(1+\frac{b^2}{r^2}) \qquad (2)$$

$$\sigma_Z = K \qquad (3)$$

where

$$\Gamma = -[\frac{E_m(\alpha_m-\alpha_f)\Delta T(\frac{a}{b})^2}{1+(\frac{a}{b})^2(1-2\nu)\{(\frac{b}{a})^2-1\}\frac{E_m}{E_f}}]$$

and

$$K = \frac{\Gamma}{(\frac{a}{b})^2}\,[2\nu(\frac{a}{b})^2+\frac{1+(\frac{a}{b})^2(1-2\nu)+(\frac{a}{b})^2(1-2\nu)\{(\frac{b}{a})^2-1\}\frac{E_m}{E_f}}{1+\{(\frac{b}{a})^2-1\}\frac{E_m}{E_f}}]$$

The room temperature stresses due to the thermal cycling of a Ti-6Al-4V/SiC composite system between RT and 550°C are calculated to be σ_r=-56.3 ksi, σ_θ=101.4 ksi, and σ_Z=102 ksi. During thermal cycling it is the cooling cycle that produces the major thermal fatigue damage. The reason is that the cooling cycle induces longitudinal and tangential tensile stress in the matrix. During the heating cycle, the stress states are reversed and are small due to the elastic behavior (Fig. 6).

From the above argument, it is reasonable to propose that tensile stress states are induced for every cycle in the matrix side of the interface whose magnitudes are estimated from Poritsky's elastic analysis.

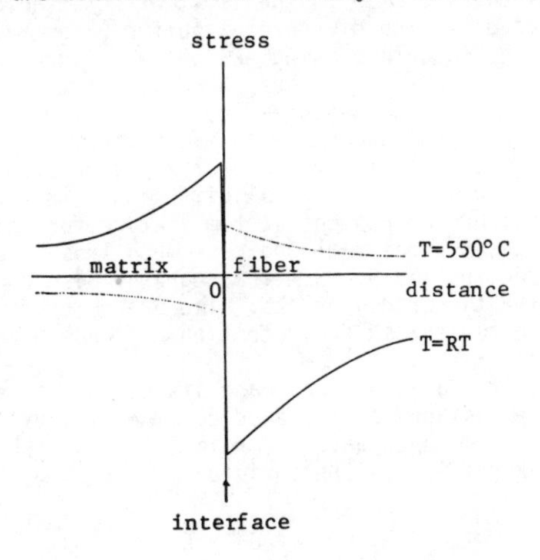

Figure 6 - Schematic representation of stress variation with temperature near the interface.

Cyclic tensile stress, now, combines with environment to speed up the formation of oxide or sulfide and degrade the interface.

To ensure a reproducible stress state during the thermal cycling, the specimen was quenched to LN_2 temperature for five minutes. From the calculation based on thermal expansion difference with Poritsky's solution and the composite fabrication temperature of 800°C with no stress at the interface at 800°C, the room temperature interface stress was estimated and shown in Fig. 7 (14).

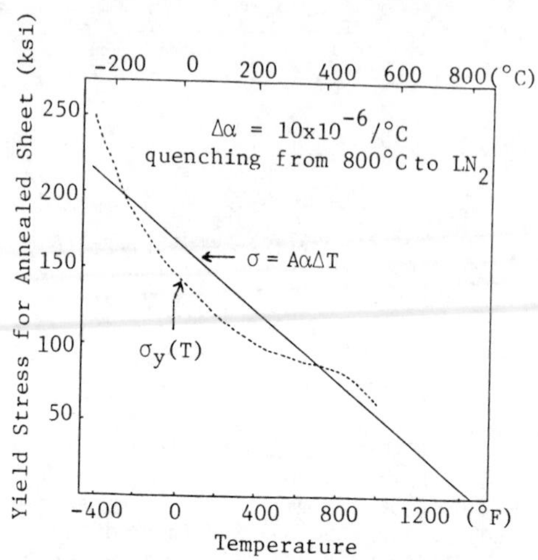

Figure 7 - Effect of temperature on the tensile yield strength of annealed Ti-6Al -4V sheet (dotted) (14).

The diffusion of sulfur can be formulated as follows. The distance from the end of the specimen to the limit of detectable sulfur down the interface is the diffusion distance of sulfur (Fig. 8). The sulfur was detected along the interface using AES. The measured diffusion distance was approximately 400μ after 24 hrs thermal treatment. A diffusion coefficient was estimated from the one-dimensional diffusion equation with an infinite source at the boundary (13). This gives a characteristic diffusion distance, $x=\sqrt{Dt}$. Diffusivity was then calculated to be on the order of 10^{-8} cm^2/sec implying interfacial diffusion. In the isothermal experiment with the sulfur environment at temperature for four times the temperature period in the thermal fatigued case much less S is diffused down the boundary as shown in Fig. 9. SEM fractography also confirms the sulfur effect by showing the branch cracks. The branch cracks extend an equivalent distance to the AES sulfur detection distance as discussed earlier.

When the sulfur analysis was made its concentration decreases monotonically as the distance from the edge. The plot of logarithm of sulfur concentration vs. distance gave a straight fitting, shown in Fig. 10 and the following expression is appropriate.

$$\ln C_s = kx$$

This is the expression for diffusion in a semi-infinite medium, x>0, when the boundary is kept at a constant concentration and the initial concentration throughout the medium is zero.

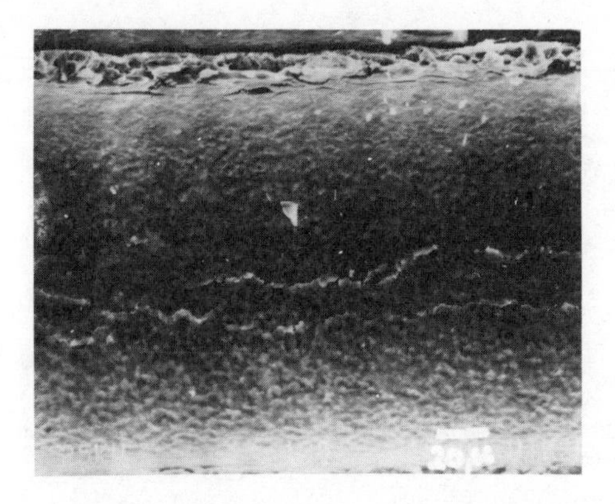

Figure 8 - Thermal cycled in sulfur environment Ti-6Al-4V/SCS (transverse) for one day; magnified the cracks from thermal fatigue, unnotched.

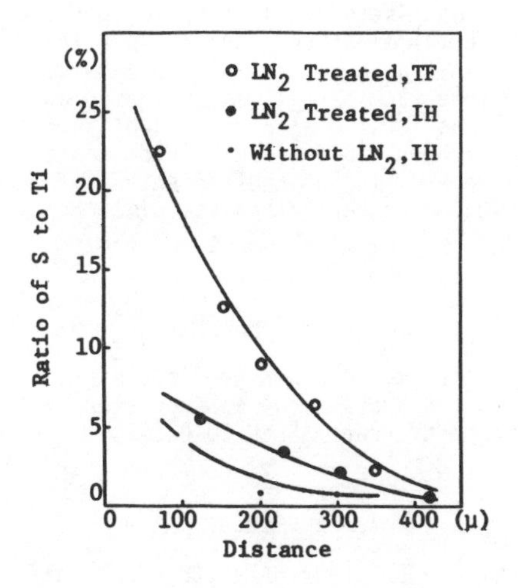

Figure 9 - AES analysis of sulfur along the matrix-side interface.

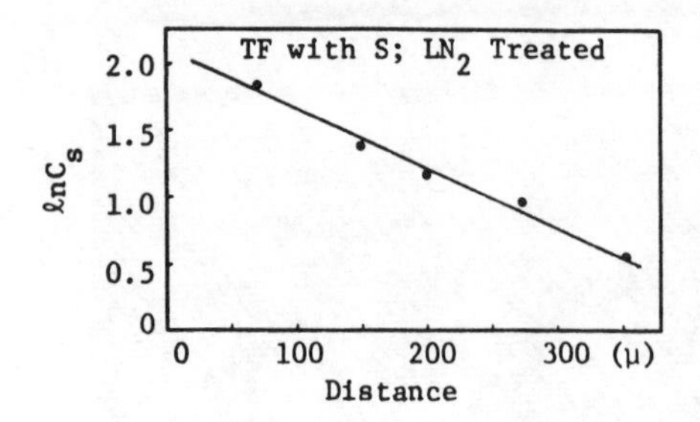

Figure 10 - Logarithmic fitting of sulfur concentration along the matrix side interface.

More rigorous analysis is now in progress. In the literature, no quantitative analysis of the thermal cycling and its effect on the interface damage in the MMC system is available.

Conclusions

Interface degradation from thermal cycling in the environments such as vacuum, air and sulfur was investigated using SEM fractography and AES. SEM fractography shows that thermal fatigued specimens produce more damage to the fiber interface than isothermally treated specimens. Liquid nitrogen treatment before thermal treatment gave a consistent result due to the reproducibility of the initial condition of the samples. Based on an elastic solution, the interface stress was estimated and at room temperature, the stress state was tensile with a value of approximately 100 ksi after thermal cycling between 550°C and RT. From the sulfur environment experiments, the interface diffusion coefficient was found to be about 10^{-8} cm^2/sec and the sulfur concentration vs. distance is expressed as $\ell nC_s = kx$.

Acknowledgments

The authors would like to acknowledge discussions with Deepak Mahulikar and Michael Schmerling. This research was supported by the Air Force Office of Scientific Research under grant AFOSR 80-0052.

References

1. J.R. Hancock, Composite Materials (Fracture and Fatigue), vol. 5, pp. 371-412; Academic Press, 1974.

2. G. Garmong, Metallurgical Trans. 5A (1974) pp. 2183-2190.

3. G. Garmong, Metallurgical Trans. 5A (1974) pp. 2191-2197.

4. G. Garmong, Metallurgical Trans. 5A (1974) pp. 2199-2205.

5. G.C. Olson and S.S. Tomkins, Failure Modes in Composites IV, TMS-AIME, pp. 1-21, 1979.

6. K.K. Chawla, Journal of Materials Science 11 (1976) pp. 1567-1569.

7. J. Billingham and S.P. Cooper, Metal Science 15 (1981) pp. 311-316.

8. A.G. Metcalfe, Composite Materials (Metallic Matrix Composites), vol. 4, pp. 269-327; Academic Press, 1974.

9. I. Ahmad, D. Hill and W. Heffernan, Proceedings of International Conference on Composite Materials, TMS-AIME, p. 85, 1976.

10. Y. Park, D. Mahulikar and H.L. Marcus,"Mixed Mode Crack Propagation," pp. 385-395, in Mixed Mode Crack Propagation, G.C. Sih and P.S. Theocaris, eds.; Sijthoff & Noordhoff, Rockville, MD, 1981.

11. H. Poritsky, Physics 5 (1934) p. 406.

12. K.K. Chawla, Proceedings of Fourth Bolton Landing Conference, p. 435, 1975.

13. J. Crank, The Mathematics of Diffusion, Oxford Press, 1979.

14. Titanium Handbook, 1972, MCIC.

INTERFACE STRUCTURE OF HEAT-TREATED

ALUMINUM-GRAPHITE FIBER COMPOSITES

James Lo, Duane Finello, Michael Schmerling and H.L. Marcus

Mechanical Engineering Department
Materials Science and Engineering
The University of Texas
Austin, Texas 78712

Summary

The interface of aluminum-graphite metal-matrix composites was characterized with TEM. The heat treatment of the composite results in carbide formation and degradation of graphite fibers. The fracture path shifts from within the oxide layer to either the fiber interface or within the fiber itself with longer aging time. Aluminum carbide was observed at aging temperatures above 550°C with the only aluminum carbide phase observed being Al_4C_3. Very coarse grains of Al_4C_3 imply preferred orientations of carbide formation due to the anisotropicity of graphite. Al_4O_4C is the only aluminum oxycarbide phase observed at the interface of aluminum graphite fiber composites. It forms a fine grain distribution at all aging temperatures.

Introduction

Aluminum and aluminum alloy metal matrices in graphite fiber reinforced composites are promising systems for structural applications. A major problem is the poor transverse tensile strength in contrast to the high longitudinal tensile strength. Recent studies (1,2) indicate that there is a close relationship between interface composition and morphology and mechanical behavior. In this study, the transverse tensile strength of the aluminum-graphite fiber composite was improved slightly when the aging temperature was raised above 500°C. Meanwhile, longitudinal tensile strength decreased noticeably as reported elsewhere (3).

The specimens in this study include G4371 and G4411,* both of which have Al-6061 as metal matrix and VSB-32 pitch fibers. They were either encapsulated in various vacuum conditions 10^{-3} - 10^{-8} torr (10^{-1} - 10^{-6} Pa) and heated to 400°C - 640°C or heated to the same temperatures without encapsulation, then naturally aged. The heat-treated specimens show the results of changes in the interface regions. Compounds at the interface that are crystalline were analyzed by using the selected area diffraction (SAD) technique in the transmission electron microscope (TEM). Crystallographic information about the interface reaction zone was obtained by examining the diffraction patterns from sections of the interface.

Experiments

The best way to get separate fibers from aluminum-graphite fiber composites without loss of major interface compounds is by using an etching solution which is prepared by dissolving 3 grams of NaOH or KOH in 100 ml of high purity methanol (containing 0.03% water) with frequent stirring. High purity methanol is used instead of water as solvent for preparing the alkali etching solution to avoid dissolution of aluminum carbide in water (4). As a test, pure aluminum carbide powders were immersed for five hours in the etching solution without indication of dissolution. The powders were washed by high purity methanol and dried completely at about 60°C. Only the spectrum of aluminum carbide appeared when the powder was examined by x-ray diffraction. At the present time, this solution is the only one that has proved to be effective for preserving the Al_4C_3 and Al_4O_4C in the aluminum-graphite fiber composites.

A JEOL 150 kV TEM was used in the SAD studies. Interface regions still attached to separate fibers were examined with applied voltages of 100 kV and 150 kV. The camera constants are 17.8 ±0.3 mm Å and 13.7 ±0.2 mm Å, respectively.

Results and Discussion

γ-Al_2O_3 or $MgAl_2O_4$ spinel and TiB_2 were observed in the interface regions of non-heat-treated specimens (5) as well as in the aged specimens. Smaller amounts of TiB_2 phase were found when the specimens were aged above 550°C. This was partly caused by displacement of the Ti-B layer away from the fiber into the matrix (2) and partly by dissolution of TiB_2 into the metal matrix during carbide formation. The specimens aged above 550°C showed aluminum carbide (Al_4C_3) and aluminum oxycarbide (Al_4O_4C) existing

---

*Supplied by the Aerospace Corporation.

in the interface. Their diffraction patterns are shown in Fig. 1 and 2.

Figure 1 - Selected area diffraction pattern of Al_4C_3 from the interface of G4371 heat-treated above 550°C.

Figure 2 - Selected area diffraction pattern of Al_4O_4C from the interface of G4371 heat-treated above 550°C.

Carbide formation seems to occur preferentially on specific orientations of the basal plane of the graphite with respect to the interface. Some orientations make nucleation of the carbide difficult and restrict their growth such that coarse grains form. The spotty nature of the Al_4C_3 diffraction pattern is due to the large grains. There seems to be no restriction on the Al_4O_4C nucleation and growth since it forms fine random oriented grains for specimens aged as high as 640°C as indicated by the continuous diffraction pattern in Fig. 2.

The higher the aging temperature, the more severe the interface reaction, and the greater the grain growth, see Figure 3. The interface region serves as diffusion path for carbon resulting in carbide formation with fiber surface pitting evident. This pitting phenomena can cause premature longitudinal failure in the fibers due to the local stress concentration. In the aged specimen, fibers are often pulled out of the fracture surface during the tensile testing. No interface compounds were found attached to graphite fibers pulled out directly from the specimens aged at temperatures above 550°C. This lack of interface material indicates that the fracture path was within the interface or in the degraded graphite fibers.

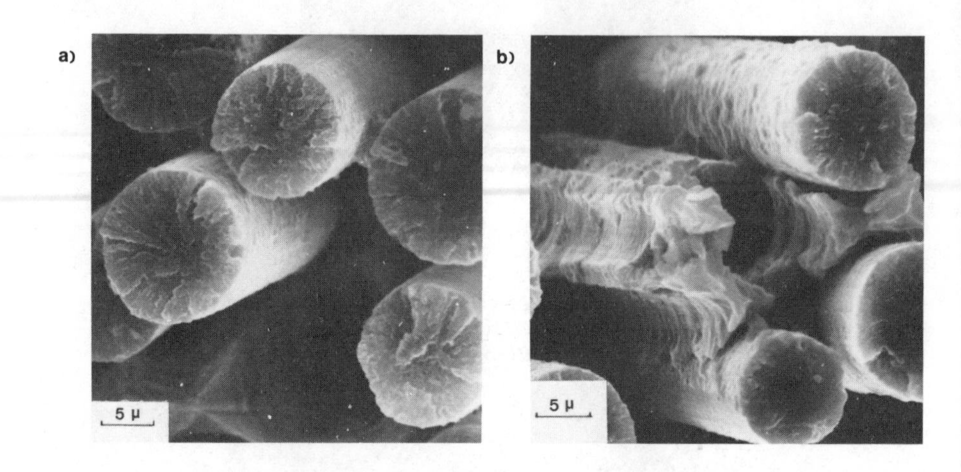

Figure 3 - Scanning electron micrographs of the interface regions of G4371 heat treated at increasing temperatures. (a) Non-heat treated. (b) 550°C at ~$5x10^{-7}$ torr for 1 day. (c) 640°C at ~$5x10^{-8}$ torr for 1 day.

Conclusions

1. Heat treatment allows carbide formation and degradation of graphite fibers to take place simultaneously.
2. The fracture path shifts from in the oxide layer to either the fiber interface or within the fiber itself with increased heat treatment.
3. Formation of aluminum carbide at aging temperatures above 550°C was observed with the only aluminum carbide phase observed being Al_4C_3. Very coarse grains of Al_4C_3 imply preferred orientations of carbide formation due to the anisotropicity of graphite.
4. Al_4O_4C is the only aluminum oxycarbide phase observed at the interface of aluminum graphite fiber composites. It forms a fine grain distribution at all aging temperatures.

Acknowledgements

The authors would like to thank Swe-Den Tsai for his discussion. This research was supported by the Office of Naval Research, contract N00014-78-C-0094.

References

1. G.L. Steckel, R.H. Flowers, and M.F. Amateau, "Transverse Strength Properties of Graphite-Aluminum Composites," Final Report for Period 1 Oct 1977 - 30 Sept. 1978 for Naval Surface Weapons Center, TOR-0078 (3726-03) -4, Sept. 30, 1978.

2. H.L. Marcus, D.L. Dull and M.F. Amateau, "Scanning Auger Analysis of Fracture Surfaces in Graphite-Aluminum Composites," in Failure Modes in Composites IV, J.A. Cornie and F.W. Crossman, eds.; Conference Proceedings, The Metallurgical Society of AIME (Fall 1977).

3. P.W. Jackson, Metals Engineering Quarterly, ASM 9, No. 3, pp. 22-30, 1969.

4. T.Y. Kosolapora, Carbides, translated by N.B. Vanghan, Plenum Press, New York, 1971.

5. Swe-Den Tsai, PhD Dissertation, The University of Texas at Austin, 1980.

SEM FRACTOGRAPHIC ANALYSIS OF METAL-MATRIX COMPOSITES EXPOSED TO ELEVATED TEMPERATURE, SHORT DURATION ENVIRONMENTS

David L. Hunn[1]

Material Science Department
University of Texas at Arlington
Arlington, Texas 76010

Elevated temperature, short time exposure tensile behavior of metal matrix composites is of high interest in advanced missile designs. Tensile tests have been performed on titanium clad boron/aluminum (B_4CB/AL) and silicon carbide particulate reinforced aluminum (SiC_p/AL) at various temperatures in air after a 2 minute soak to characterize their behavior. The temperatures that were examined include 700° F, 900° F and 1100° F for the B_4CB/AL composite and 500° F, 700° F and 900° F for the SiC_p/AL composite. This paper reflects the SEM work that was performed on each of the failure surfaces of the various tensile specimens. Failure modes were readily identified for the continuous fiber reinforced composite (B_4CB/AL) but were not as obvious for the discontinuously reinforced composite (SiC_p/AL).

[1]Currently structures engineer; Vought Corp., Advanced Metallics Group, P.O. Box 225907, Dallas, Texas 75265

Metal Matrix Composites (MMC) are an emerging material system tailored to meet the demands of a high-technology society. These materials consists of various reinforcements; fibers, whiskers, or particulates; emmersed in a metallic matrix. They offer attractive design alternatives to metals in conventional design. With MMC materials, a designer can take advantage of the anisotropic nature of the composite for efficient design and fabrication of structures. The materials supplier can tailor-make a material to meet a specific set of engineering strength and/or stiffness requirements. The stiffness, strength and thermal stability of common engineering alloys can be increased while the overall density of the material is decreased. These new material systems can also conserve the dwindling supply of strategic materials.

MMC materials exhibit advantages over resin-based composite materials as well as over conventional metals. The most obvious advantage is the resistance of the metal matrix to severe environments and the material's inherent high temperature capability. High thermal conductivity, low thermal expansion in the fiber direction, greater shear, bearing and transverse strengths, and increased toughness are also advantages in comparison to a polymeric matrix. Disadvantages include slightly higher density, matrix-filament reactions, manufacturing limitations with the continuously reinforced systems, and high cost/limited data base problems.

Because of the high temperature capabilities and the excellent specific strength/stiffness of MMC's, these materials lend themselves to advanced missile structure designs. There exists, however, a paucity of design data for eligible material systems in the high-temperature/short exposure time environment. The purpose of this study is to test selected metal matrix composites in that environment and examine the fracture surfaces with the scanning electron microscope in an attempt to identify failure modes.

The two materials systems that are examined consist of boron reinforced aluminum (B_4CB/AL) and silicon-carbide particulate reinforced aluminum (SiC_p/AL). The B_4CB/AL is a continuous fiber reinforced aluminum whereas the SiC_p/AL is a discontinuously reinforced aluminum. Tensile tests were performed at various temperatures after a 2-minute soak.

B_4 CB/AL

This material consists of boron fibers unidirectionally placed in an aluminum (6061) matrix. The boron fibers are produced by chemical vapor deposition of boron on a hot tungsten wire substrate. To avoid adverse reactions between the boron filament and aluminum during high temperature fabrication of the MMC monolayers, the surface of the boron filament is coated with a diffusion barrier of boron carbide (B_4C). Excellent fiber properties are thus maintained in the composite.

A plate of B_4CB/AL was secured from DWA Composite Specialties of Chatsworth, California. The plate was four plies thick. Tensile specimens were machined from the plate such that all fibers ran parallel to the loading axis. Tests were performed at room temperature, 500°F, 900°F, and 1100°F. Quartz lamps were used to bring the specimen rapidly up to temperature, a 2-minute soak was performed, and then the specimens were tested to failure. This environment was chosen to represent typical conditions encountered during short duration, supersonic flights.

Figure 1 illustrates a typical failure at 1100°F. Massive fiber pull-out can be seen along the fracture surface. This is due to matrix softening at high temperature and the resulting degradation of the fiber/matrix interface. This phenomenon was also observed at the 500°F and 900°F temperatures but was not as severe.

Figures 2 and 3 are SEM shots of the fracture surface of the room temperature, 500°F, 900°F, 1100°F tensile tests. At room temperature, dimpling of the aluminum matrix is seen. This morphology is characteristic of a tensile failure in a ductile metallic material. Dimpling is still evident at 500°F, but starts to disappear at 900°F and is not apparent at all at 1100°F. At these higher temperatures, the aluminum is extremely ductile and essentially carries no load and transfers no shear to the fibers. In the 1100°F case, aluminum "pile-up" is seen between the fibers. This occurs when the aluminum elongates, and then collapses on itself once the fibers fail.

As the test temperature increases from room temperature to 1100°F, the fracture surface of the boron fiber changes dramatically. The fracture surface of a typical boron filament tested at room temperature is shown in Figure 4. The failure is catastrophic, with many thin, wedge shape flakes of the broken fiber visible. This failure may be the result of the rapidly advancing shock wave which preceeds the crack during fracture. The fracture was initiated at the edge of the fiber and rapidly propagated through the fiber as the specimen failed. A radial fracture pattern can be seen starting from the edge of the fiber working inwards and was apparent on most of the room temperature specimens. The dimpling of the aluminum matrix due to micro-coalescence can also be seen. This is an indication that the matrix is functioning and does indeed carry/transfer load at the lower temperatures.

The fracture surface of a typical boron fiber tested at 1100°F is shown in Figure 5. This fracture surface is a very clean break nearly perpendicular to the fiber axis. The fracture probably initiated at the center of the fiber and propagated outward. No dimpling of the aluminum matrix is apparent in any of the 1100°F test specimens. The matrix carried/transferred very little load as the specimen failed.

This change in the fracture surface is due to the changing stress state that is experienced by the test specimens as the temperature changes. At room temperature, the micro stress state on the fiber is complex even during a pure tensile tests. Due to the differences in strength, stiffness and Poisson's ratio of the fiber and matrix, a uniaxial state of stress does not occur on the microscale. Shear is constantly being transferred through the matrix and a portion of the load is being carried by the matrix. As the temperature increases, the mechanical properties of the matrix degrade significantly more than that of the boron fiber. As a result of the degradation of the matrix properties, the fibers experience a stress state more closely to that of uniaxial than they do at room temperature. This leads to the very clean, nearly perpendicular fracture surface witnessed on the elevated tensile specimens.

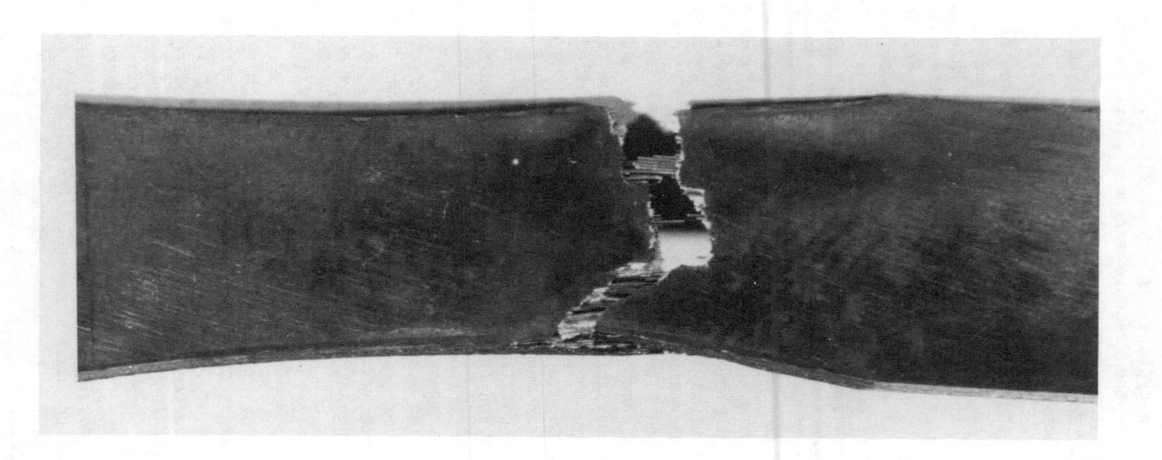

Figure 1: Massive fiber pull-out exhibited by the B_4CB/AL tensile specimen tested at 1100° F.

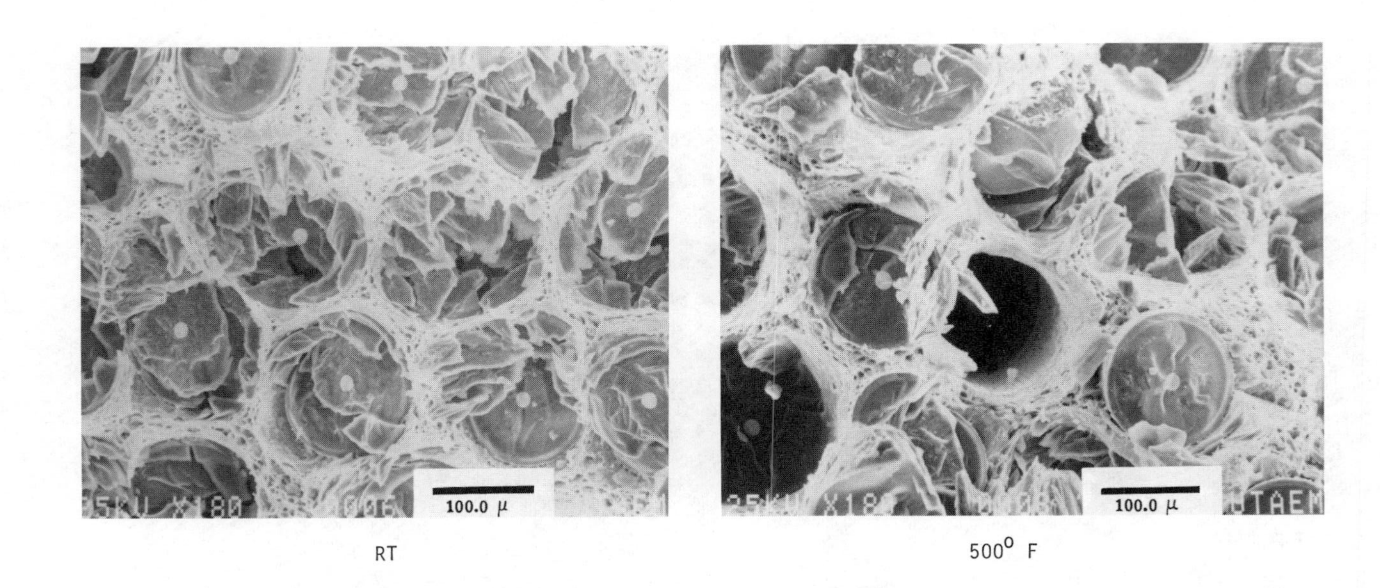

Figure 2: Fracture surfaces of B_4CB/AL tensile specimens tested at room temperature and at 500° F.

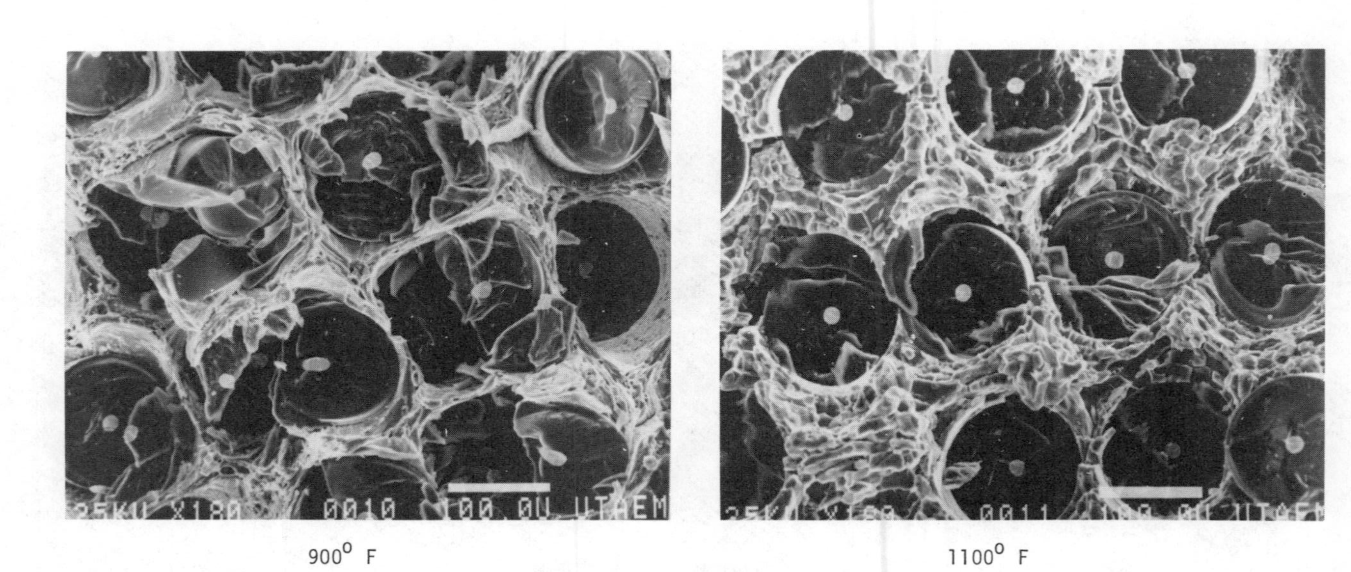

Figure 3: Fracture surfaces of B_4CB/AL tensile specimens tested at 900° F and 1100° F.

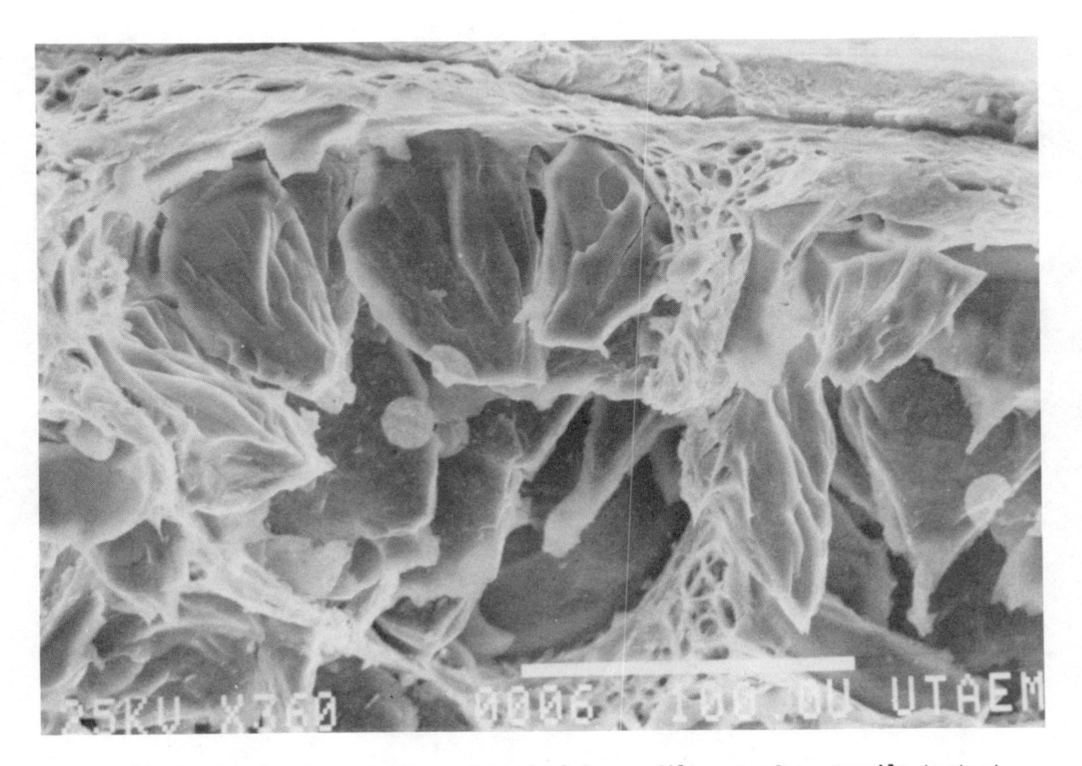

Figure 4: Fracture surface of typical boron filament after tensile test at room temperature.

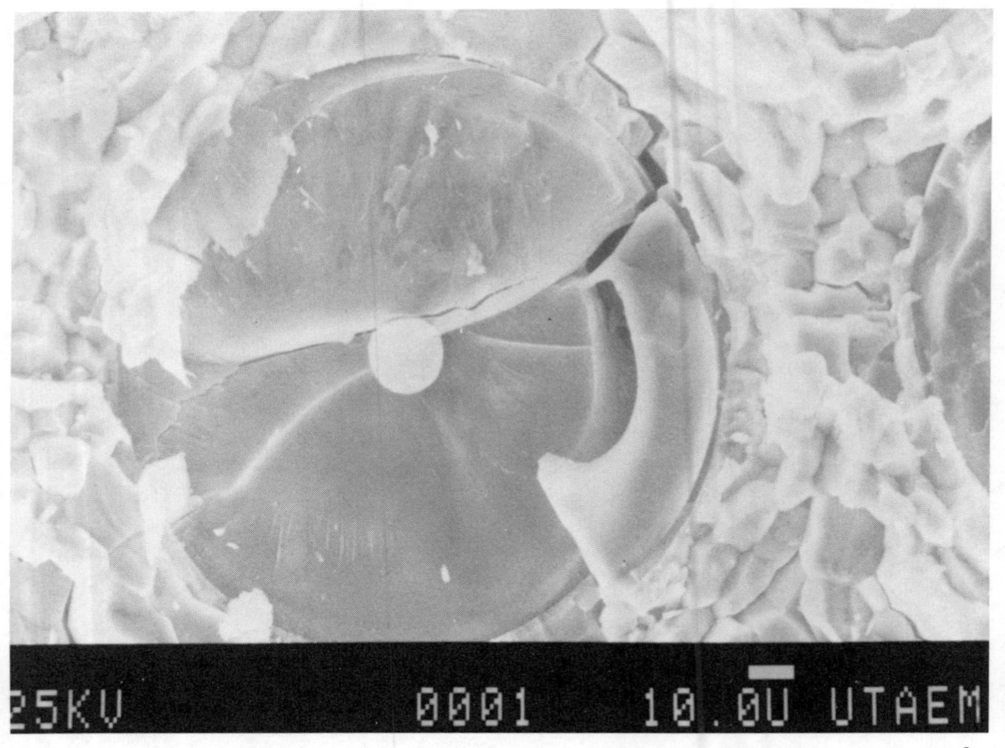

Figure 5: Fracture surface of typical boron filament after tensile test at 1100° F.

SiCpAl

This metal matrix material consists of silicon-carbide particulates in an aluminum (6061) matrix. The silicon-carbide is in a fine, platelet shaped form and is mechanically mixed with aluminum powder and then hot compacted to achieve consolidation. The resulting composite is a quasi-isotropic material with superior strength and stiffness properties over the base aluminum matrix. Being a discontinuously reinforced material, many of the fabrication difficulties and costs associated with continuously reinforced materials are circumvented.

A plate of SiC_p/Al was secured from DWA Composites Specialties of Chatsworth, California. The plate was flat forged to a thickness of 0.063 inches. Dogbone tensile specimens were machined from the plate and tensile tested at 500°F, 700°F and 900°F. The specimens were brought up to temperature rapidly in an oven, held for two minutes at temperature, and tested to failure.

Figures 6 and 7 are SEM shots of the fracture surfaces at 500°F, 700°F and 900°F. None of the dimpling associated with ductile metal tensile failures is evident. In fact, the silicon carbide particulates cannot be readily distinguished from the bulk matrix. At the 900°F temperature, drawing and smearing of the aluminum can be seen. This occurs when the aluminum is very ductile at the high temperature and massive elongation occurs upon application of the load.

Fractographic analysis of the fracture surface was performed to determine where the fracture initiated. In all cases at each temperature, the crack started at a siliconcarbide rich region of the specimen. Figure 8 illustrates this. Figure 8(b) is a view of the fracture surface away from the initiation point. A fairly uniform distribution of silicon carbide is evident. Figure 8(a) shows the region near the initiation point. A much higher concentration of the platelets can be seen.

In conclusion, SEM fractography readily identified different failure modes as test temperature changes in continuously reinforced metal matrix composites. In discontinuously reinforced materials, the role of the matrix is different and identification of different failure modes is not straightforward. Indications are that the fracture initiates at silicon-carbide rich regions and propagates along metal rich regions - reinforcing the problem in these materials of adequate matrix-reinforcement bonding and homogeneous distribution of the reinforcements. Further work should include a more in-depth look at the discontinuously reinforced material systems with an attempt to identify the failure modes and mechanisms involved.

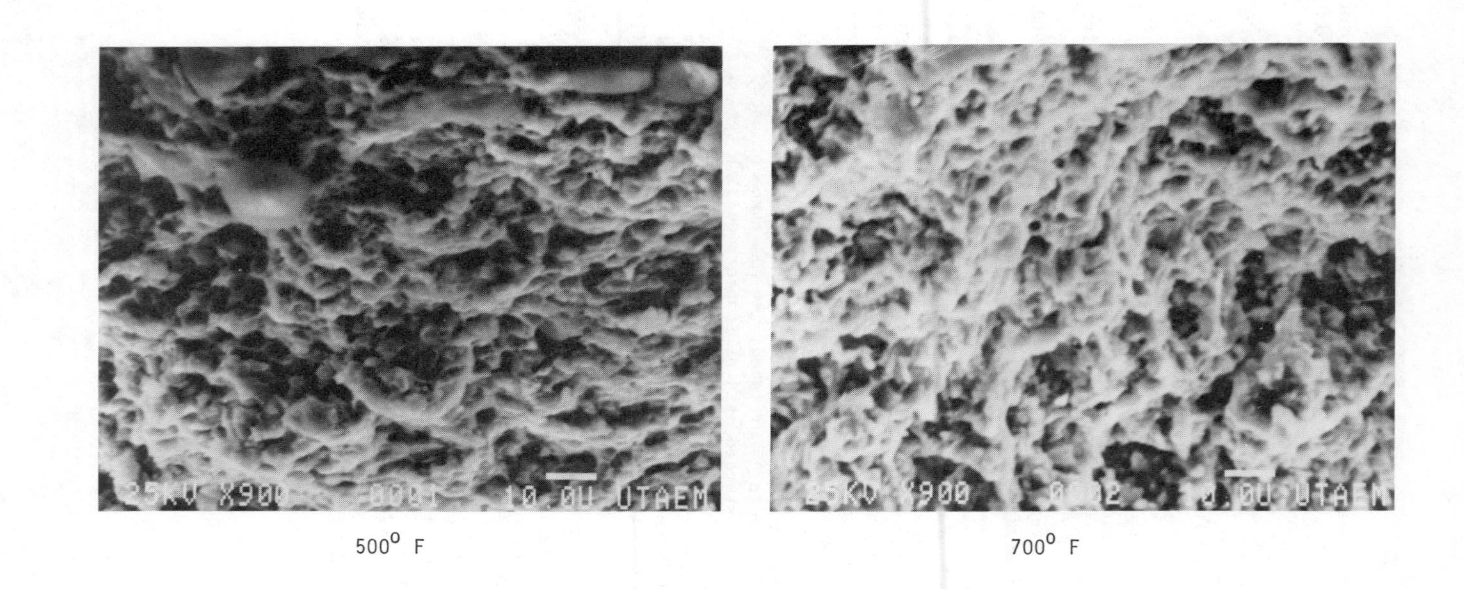

500° F

700° F

Figure 6: Fracture surfaces of SiC_p/AL tensile specimens tested at 500° F and 700° F.

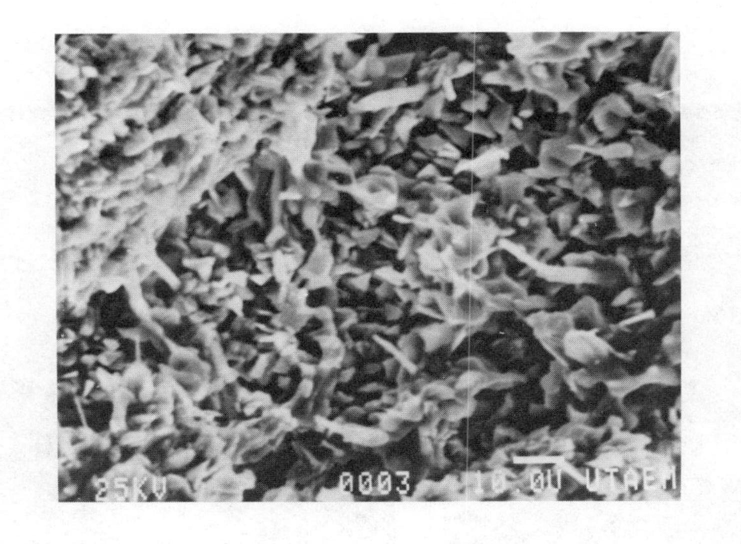

Figure 7: Fracture surface of SiC_p/AL tensile specimen tested at 900° F.

(a) (b)

Figure 8: Fracture surface of SiC_p/AL tensile specimen located (a) near the initiation site and (b) far from the initiation site.

MATRIX/FIBER INTERFACE EFFECTS

ON TENSILE FRACTURE OF STEEL/TUNGSTEN COMPOSITES

C. Kim, A. Pattnaik, and R. J. Weimer

Naval Research Laboratory
4555 Overlook Avenue, S.W.
Washington, D. C. 20375
USA

Unidirectionally reinforced composites of a mild steel (1010St) with commercial purity tungsten fibers (NS-55) and a heat-treatable, high-carbon steel (1095St) with tungsten-1% thoria fibers ($W\text{-}1ThO_2$) were consolidated by hot-press diffusion bonding to a fiber volume fraction of 45%. Both composites exhibited severe degradation, particularly in tensile strength, due to the formation of brittle interface compounds during the fabrication process. X-ray diffraction analyses of tungsten fibers extracted from the composites revealed that the interface in the 1010St/NS-55 system was primarily ditungsten carbide, whereas, the interface in the $1095St/W\text{-}1ThO_2$ system was tungsten carbide. The low fracture strain of the interface carbides provided abundant crack initiation sites along the fiber surfaces when the fibers were loaded in tension, thereby accounting for the premature failure of the composites. Structure/property relationships and matrix/fiber interface reaction kinetics were studied in detail.

Introduction

The importance of the matrix/fiber interface in composites was recognized from the earliest stages of composite material development. Due to the essential role of the interface in transferring load from one constituent of the composite to the other, the interface reaction products are often critical factors in composite performance. Metal matrix composites are particularly vulnerable to interface reactions because of the high reactivity between constituents. Except for a few "model" composites, most practical systems using matrices of the common structural metals (e.g., Al, Ti, Ni, or Fe) have resisted control of the interfaces during the necessarily high consolidation temperatures.

Specifically, in high density composite material applications, tungsten fibers are frequently incorporated into other dense metal matrices. Among others, the steel matrix has yielded exceptionally high dynamic compressive strength (1) when unidirectionally reinforced with thoriated tungsten fibers and subsequently heat treated. However, the longitudinal tensile strength was substantially degraded, because the steel/tungsten combination was conducive to embrittling of the tungsten fibers (2,3). Since tungsten is a notch sensitive material, any source of surface defects increases the brittle behavior of this material above that due to intrinsic structural changes such as polygonization and recrystallization caused by exposure to elevated temperature (4). In composites, due to inevitable reactions between constituents in the consolidation process, the extrinsic sources of degradation are introduced in the form of interface compound formation and/or solid solution (diffusion reaction) with or without recrystallization enhancement (5). In practical fabrication, the consolidation processes are optimized to minimize the source of degradation reactions while ensuring rigid bonding between elements. Identification of degradation sources in steel/tungsten (St/W) systems is not a simple task in spite of the severity of the mechanical property deterioration (3,4).

The purpose of the present study was to identify the detrimental interface products in the St/W system and to analyze composite tensile strengths through experimental investigations based on x-ray diffraction, SEM fractography, and optical metallography. Additional experiments involving elevated temperature annealing of the composites as well as individual fibers were used to clarify thermal, diffusional, and chemical effects in the matrix/fiber interactions. Finally, a new failure mechanism is proposed for the St/W composites, based on these observations.

Experiment

Composite materials and consolidation conditions for solid state diffusion bonding in vacuum are presented in Table I. All composites were unidirectionally reinforced with 45 volume percent tungsten fibers. Fiber diameters in μm are given in parentheses in composite nomenclature. The Al/W composite was chosen as a reference system, because it did not show any serious strength degradation due to its fabrication processing. The low carbon 1010 Steel/NS-55W (commercially pure tungsten fiber) and the plain carbon 1095 Steel/W-1ThO_2 systems were selected on the basis of metallurgical simplicity of the constituents for the former system and heat treatability for the latter system. Heat treating the 1095 St/W-1ThO_2 system posed a special problem, because one of the salt baths (molten potassium nitrate) dissolved the tungsten fibers. To prevent this action, the composites were encapsulated with thin steel casings during heat treating.

A variety of room temperature tension and compression tests were performed on these composites and their reinforcing fibers. Specifically, longitudinal tension tests were conducted on all four composite materials using a closed-loop servo-hydraulic testing machine operated under displacement control at strain rates of the order of 10^{-4} sec^{-1}. Test specimen configurations and test methods have been described in detail elsewhere (4). In addition, longitudinal compression strengths were determined for both as-fabricated (AF) and quenched-and-tempered (QT) St/45W-1ThO_2 (305) composites.

Table I. Materials Selection and Consolidation Condition

Materials		Designations	Consolidation Conditions		
Matrix	Fiber		Temp. (°C)	Pressure (MPa)	Time (Min.)
6061 Aluminum	NS-55W (109)*	Al/W	482	69	30
1010 Steel	NS-55W (109)	1010St/W	760	69	30
1095 Steel	W-1 ThO_2 (305)	1095St/W	970	69	60
1095 Steel (QT)**	W-1 ThO_2 (305)	1095St/W(QT)	970	69	60

* Numbers in parentheses are fiber diameters in μm.

** QT treatment: Austenitize at 794C for 30 min., oil quench, temper at 374C for 2 hrs., oil quench.

The tensile strength and ductility of the tungsten fibers were evaluated for virgin fibers to determine the effects of annealing and for extracted fibers to evaluate the effects of the fabrication processes. Tension tests for fibers were conducted at room temperature (22C) at a strain rate of $2x10^{-3}$ sec^{-1}. The paper-tab gripping method described in ASTM Test Method D3379-75 was used for the 109 μm fibers. For the 381 μm fibers, aluminum tabs (0.8 mm thick) were used. Specimen gage length was 2.5 cm.

Extensive microanalytical studies, including optical metallography, scanning electron microscopy (SEM), and x-ray diffraction analysis, were conducted to identify the interface compounds and to clarify the composite structure/property relationships.

Results

Longitudinal Mechanical Properties of Composites

The longitudinal ultimate tensile strengths of all four composite materials (Table I) are presented in Fig. 1, which also includes rule-of-mixture (ROM) calculations of composite strength. These ROM calculations were based on virgin fiber strength and matrix flow stress at the failure strain of each composite (Table II). The observed tensile strength of the aluminum/tungsten composite (Al/W) was in reasonable agreement with the ROM value. However, the tensile strengths of the St/W composites were considerably less than the expected ROM values. The improvement observed in the 1095 St/W (QT) system was due to the quench-and-temper heat treatment which more than doubled the matrix strength. The microstructure of the QT matrix was virtually all martensite, whereas, the as-fabricated matrix consisted of mostly spheroidized carbides in alpha ferrite. Isothermally transformed microstructures in this composite are discussed in detail elsewhere (1).

Table II. Data for Initial Rule-of-Mixture (ROM) Calculations

Material Designation	Stress (MPa)
NS-55 Tungsten Fiber	2882*
W-1ThO$_2$ Tungsten Fiber	2261
6061 Aluminum Matrix	276**
1010 Steel Matrix	283
1095 Steel Matrix	310
1095 Steel Matrix (Quenched and Tempered)	655

* Fiber stress was taken to be the ultimate tensile strength.
** Matrix stress was taken to be the matrix flow stress at the failure strain of the composite.

Longitudinal compression tests were conducted on St/45W-1ThO$_2$ (305) composites in both the as-fabricated and the quenched-and-tempered conditions. The cylindrical test specimens utilized diametrically opposed strain gages to reduce any effects of specimen bending on the uniaxial stress-strain relationships. The results of two tests were averaged for each material condition and are presented in Table III along with the tension test results. Compression test results were similar for AF and QT materials, but the tensile properties had improved dramatically after heat treatment. Specifically, the proportional limit stress and ultimate strength of the longitudinal QT samples were respectively 3 and 2.3 times greater than the values measured for the AF composites.

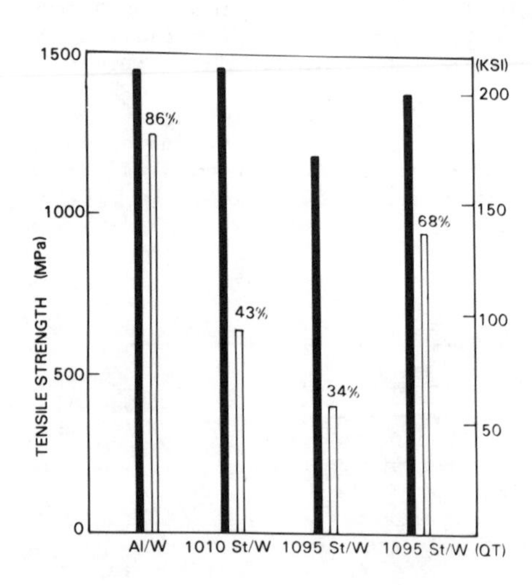

Fig. 1 Comparison of Experimental Tensile Strengths (White) with Theoretical Rule-of-Mixtures (ROM) Values (Black).

Table III. Longitudinal Mechanical Properties of 1095 St/45W-1ThO$_2$ (305) Composites

Property	Tension AF	Tension QT	Compression AF	Compression QT
Young's Modulus (GPa)	297	293	262	224
Proportional Limit Stress (MPa)	103	310	559	690
Proportional Limit Strain (%)	0.04	0.10	0.20	0.29
Ultimate Strength (MPa)	400	938	1496	1462
Fracture Strain (%)	0.20	0.34	9.70	9.00

Mechanical Properties of Tungsten Fibers

To simulate thermal effects of fabrication processes, tungsten fibers (381 μm) were annealed in 10^{-5} Torr vacuum pressure at appropriate elevated temperatures for 30 minutes. The resultant room temperature tensile strength and percent elongation are given in Fig. 2. Annealing temperatures up to 800C produced slight increases in strength. Above 800C, the strength decreased, and the fracture mode correspondingly changed from ductile to brittle. This trend was also observed in smaller diameter tungsten fibers (124 μm) (4). The thoriated tungsten fibers showed a similar trend, except that they were less susceptible to thermal

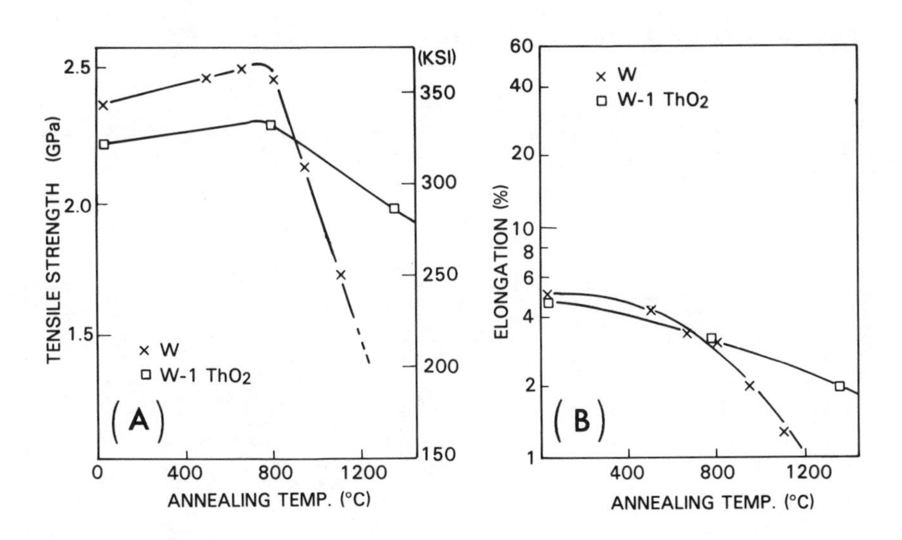

Fig. 2 Effect of Annealing Temperature (30 minute exposure) on Room-Temperature Strength (A) and Ductility (B) of Virgin Tungsten Fibers.

degradation than pure tungsten fibers. Loss of ductility (percent elongation) was also more severe for the pure tungsten fiber than for the thoriated tungsten fiber after annealing at high temperatures.

In actual composite fabrication, chemical reactions accompanying the thermal annealing effects are generally severe due to the high consolidation temperatures. To evaluate the effects of these consolidation processes on the properties of reinforcing fibers, the matrix material was dissolved away in concentrated hydrochloric acid solution at room temperature. The ultimate tensile strength and percent elongation of fibers extracted from composites depended on immersion time in acid in the manner shown in Fig. 3. Those fibers extracted from the Al/W system retained their original well-defined strength values, regardless of the length of immersion time. The percent elongation initially decreased slightly and then recovered to an ultimate steady value. The change in ductility was smaller than the test variance and was not thought to be significant. On the other hand, the fibers extracted from 1010 St/W showed a somewhat linear time dependence in the recovery of strength and ductility as function of immersion time until plateau values were reached. This distinctive behavior presented a strong indication that the effects due to surface defects or notches in this system were becoming less severe as the interface layer dissolved. The fibers extracted from 1095 St/W composite showed low, constant values of strength and percent elongation. The virgin fiber strength of W-1ThO_2 (305 μm) was 2260 MPa, whereas, the extracted fiber strength was less than 1000 MPa. Percent elongation in extracted fibers was less than a quarter of its original value. Fibers extracted from 1095 St/W (QT) composites by immersion in hot hydrochloric acid (90C) for three hours possessed identical reduced strength, thereby demonstrating that the heat treatment did not further degrade the fiber properties. However, these drastic reductions in strength and ductility were indicative of severe reactions between fiber and matrix in this system.

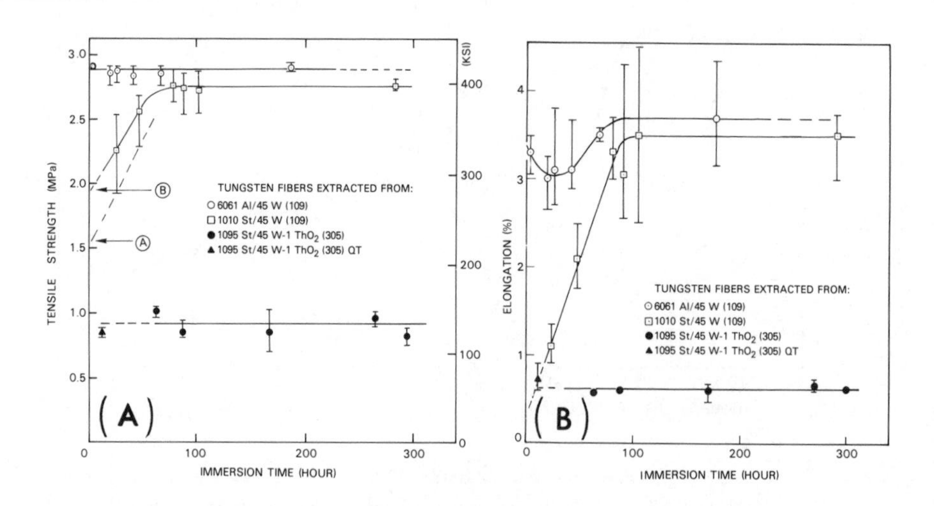

Fig. 3 Effect of Time of Immersion in HCl Acid on Tensile Strength (A) and Percent Elongation (B) of Extracted Tungsten Fibers.

Microstructural Analysis

Typical SEM fractographs of Al/W and 1010 St/W composites which were fabricated with identical tungsten fibers are compared in Figs. 4A and 4B. Failure modes of the tungsten fibers were distinctly different. In the Al/W composite, the tungsten fibers failed by ductile fracture, as did the aluminum matrix, whereas, in the St/W composite, the tungsten fiber fracture mode was brittle. The transgranular brittle fracture of the fiber in the latter system was indicative of composite strength degradation.

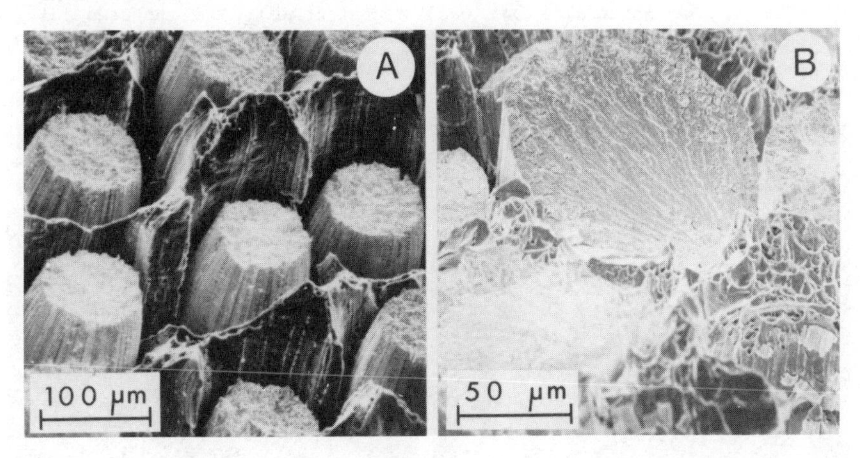

Fig. 4 SEM Fractographs of Al/W (A) and 1010 St/W (B) Composites Fractured in Longitudinal Tension.

The dependence of mechanical properties of tungsten fibers extracted from 1010 St/W composite on time of immersion in acid bears an interesting relationship to fracture morphology in failed fibers. Specifically, the SEM fractographs of fibers immersed in HCl acid for 48 hours and 103 hours, for example, exhibited quite different fracture modes, as shown in Figs. 5A and 5B. Longer times of immersion in acid apparently caused ductile fiber failure similar to the in situ tungsten fiber failure in Al/W system. Fiber necking had occurred and the area reduction was more than 50 percent for the longer immersion time. The fibrous grains in the tungsten fiber showed chisel-type ductile failure representing 100 percent area reduction (locally). On the other hand, shorter immersion times resulted in brittle fiber failure without any appreciable area reduction, although some ductile failures of individual grains were observed.

The dependence of strength and ductility recovery on time of immersion in acid suggests some sort of dissolution of an interface reaction layer. To identify the interface material, x-ray analyses were conducted with either an x-ray diffractometer or a Debye-Scherrer camera. X-ray diffractometer measurements on extracted fibers that were immersed for 24 hours in HCl disclosed ditungsten carbide (W_2C) and M_6C type eta carbides (i.e., Fe_3W_3C or Fe_4W_2C). Debye camera measurements on fibers immersed for 8 hours in acid also showed these carbide compounds after long (24 hours) exposure to Cu K_α x-rays through a Ni filter. However, fibers immersed for 290 hours indicated no trace of W_2C or $(FeW)_6C$ compounds.

The SEM pictures of fibers immersed for a short time (1 hour) revealed the interface structure shown in Fig. 6. Cracks were evident in the interface layer. The original fiber drawing markings were also clear, indicating that the interface layer was very thin (0.1-0.2 μm).

The interface layer in 1095 St/W composite was relatively easy to extract. In fact, the layer was so thick (5-10 μm) that it was extracted in large pieces by first dissolving the steel matrix in HCl acid and then dissolving the tungsten fibers in a mixed solution of HNO_3 and HF acids. A typical extracted interface layer is shown in Fig. 7, where the surface of the interface that was in contact with the tungsten fiber (Fig. 7A) may be seen to be topographically different from the surface that was in contact with the matrix (Fig. 7B). X-ray analysis of this

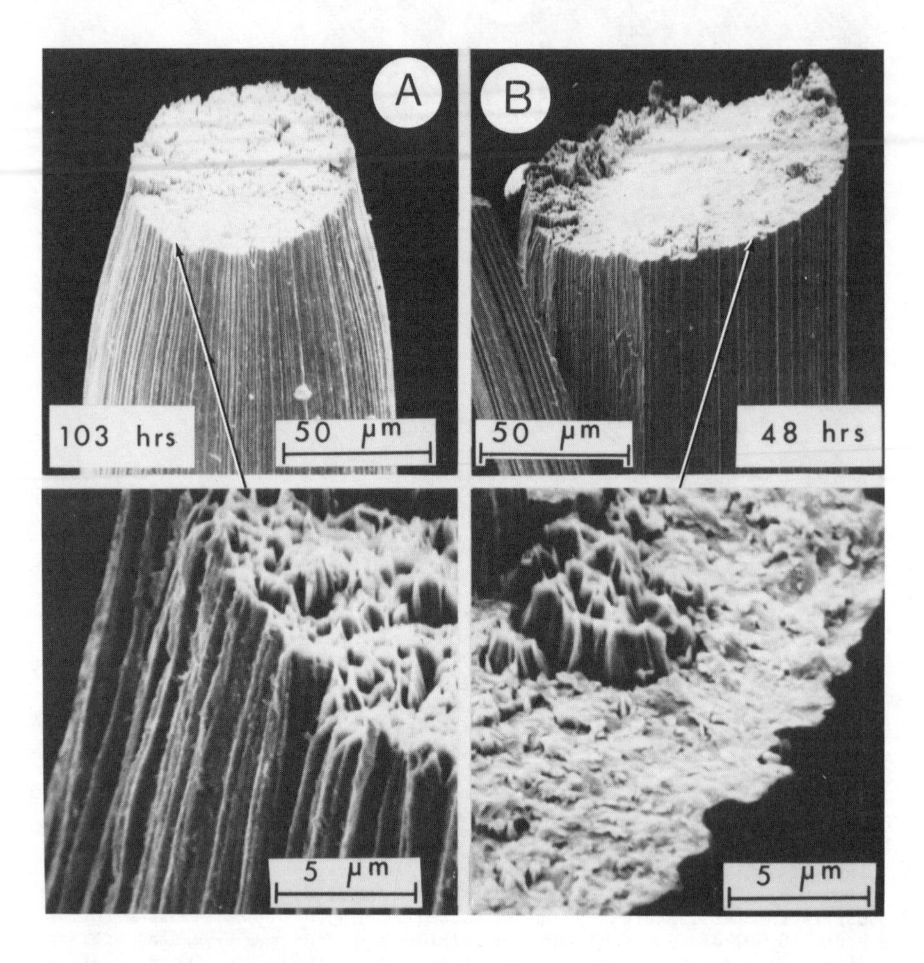

Fig. 5 SEM Fractographs Showing the Effect of Immersion Time in HCl Acid on Fracture Mode of Fibers Extracted from 1010 St/45W Composite. (A) 103 hrs. (B) 48 hrs.

interface material revealed that the primary compound was WC and that trace amounts of $(FeW)_6C$ were present. X-ray diffractometer measurements on tungsten fibers extracted by dissolving only the matrix material in HCl acid indicated the presence of W_2C and W, in addition to WC and $(FeW)_6C$. Therefore, the HNO_3 + HF acid solution dissolved the W as well as the W_2C.

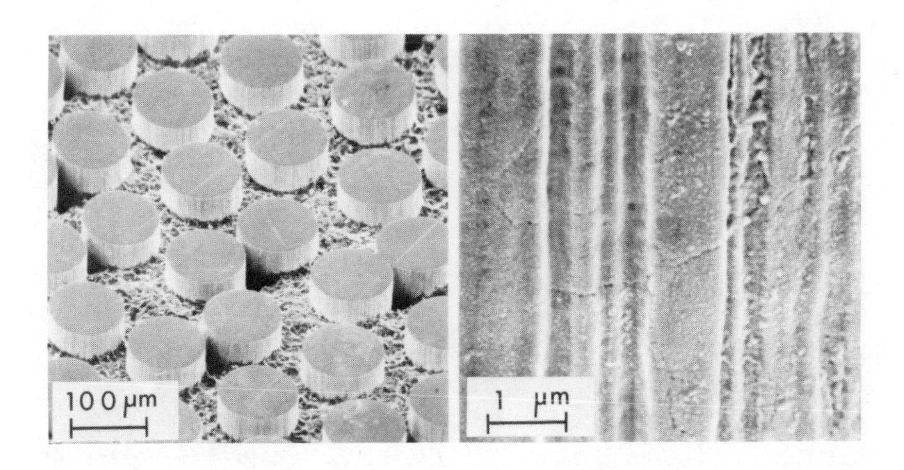

Fig. 6 SEM Photomicrographs Showing Surface Structure of Matrix/Fiber Interface in 1010 St/45W Composites Immersed for One Hour in HCl Acid.

Interface Layer Development at Elevated Temperatures

Metallographic observation of the interface layer in the 1010 St/W composite did not satisfactorily reveal the interface structure even with tapered section metallographs. Additional exposure of the composite to higher temperatures for longer times enhanced the interface structure and reaction kinetics as evident in Fig. 8. The interface of the 1010 St/W system developed progressively as the annealing temperature was increased. After annealing the composite at 930C for 1 hour, the interface layer was clearly developed to a thickness of approximately 1 μm and was identified to be $(FeW)_6C$. When the annealing temperature was raised to 1130C, the interface layer grew primarily into the fibers. In this case, the interface structure consisted of two layers, namely, an inner layer of $(FeW)_6C$ and an outer layer of Fe_7W_6. The void formations around the fibers were considered to be Kirkendall porosity. Annealing the composite at 1330C caused the interface layer to become a single compound of Fe_7W_6. The voids may have been enlarged somewhat by the metallographic polishing process.

The interface development in the 1095 St/W system was different from that in the 1010 St/W system, as might be expected in view of the greater carbon content of 1095 steel. Because the fabrication temperature of the 1095 St/W composite was 970C, additional annealing at 930C for 1 hour did not change the primary structure of the interface layer, which consisted of WC outside and W_2C inside (Fig. 9). After annealing at 1130C, however, the interface structure developed to a mixture of $(FeW)_6C$, WC, and W_2C. When annealed at 1330C, the interface layer totally transformed to a $(FeW)_6C$ ternary compound. This eta carbide also formed along the grain boundaries of the steel matrix, thereby indicating a high rate of grain boundary diffusion of tungsten in steel.

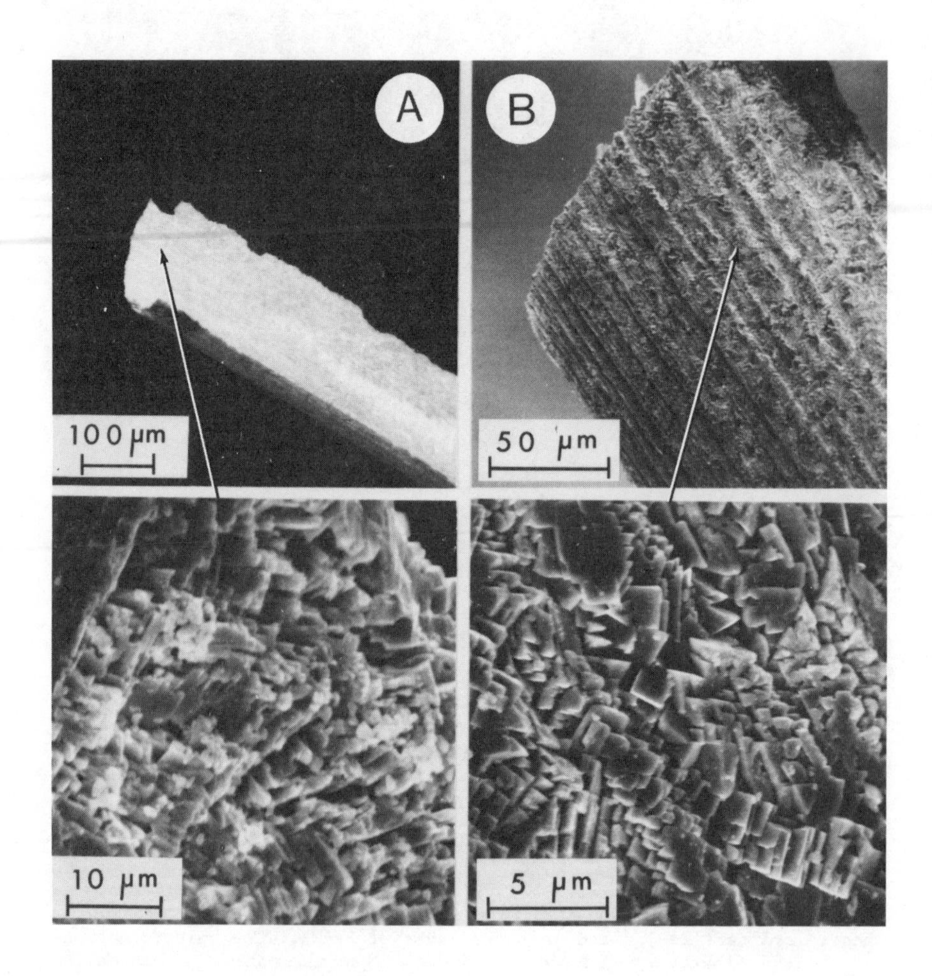

Fig. 7 SEM photomicrographs of fiber/matrix interface material extracted from 1095 St/45W-1 ThO_2 composite by immersion, first in HCl acid and then in HNO_3 + HF acid solutions. A) inside of interface, B) outside of interface.

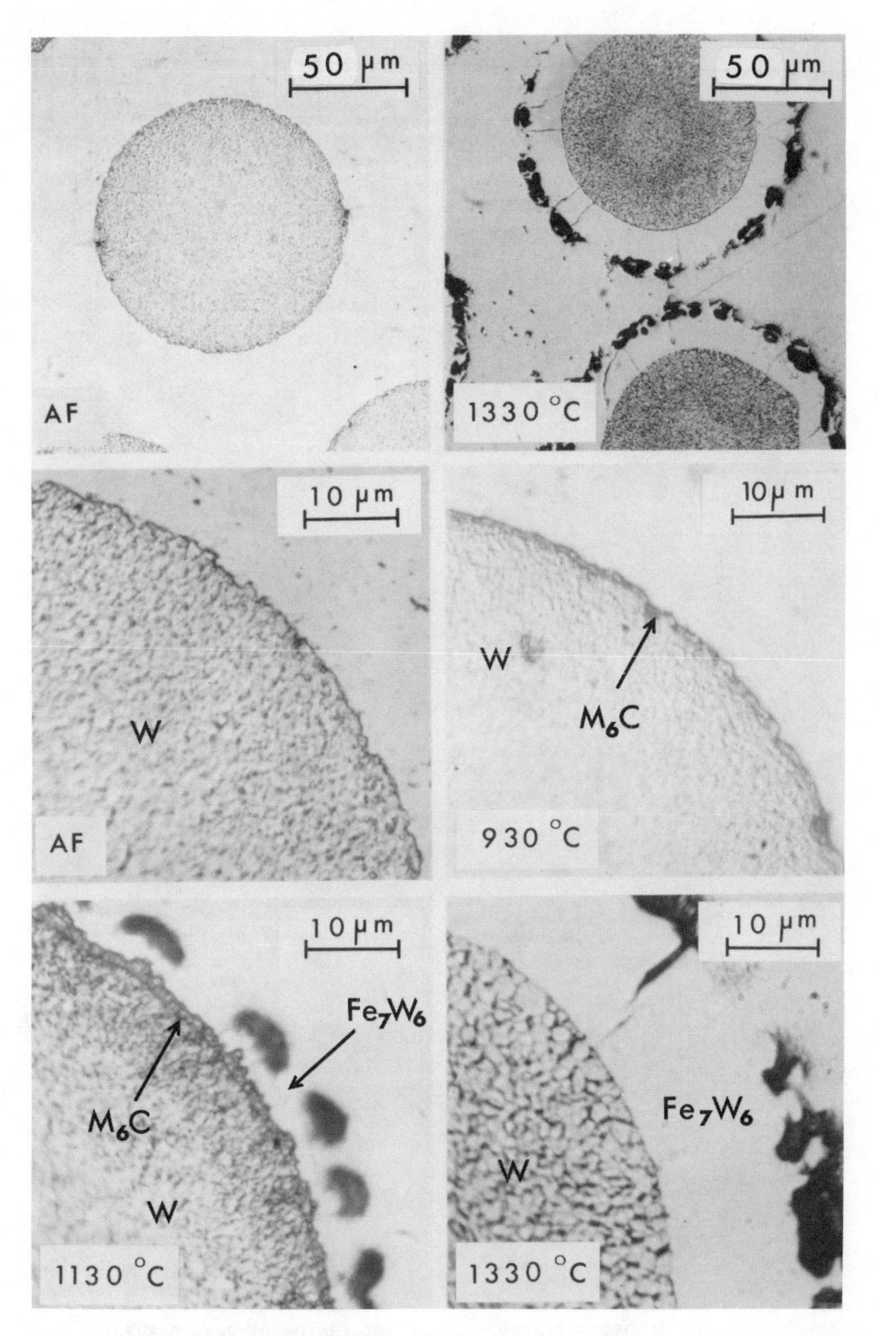

Fig. 8 Optical Photomicrographs of Interfaces Developed in 1010 St/45W Composites after Annealing for One Hour at the Indicated Temperatures.

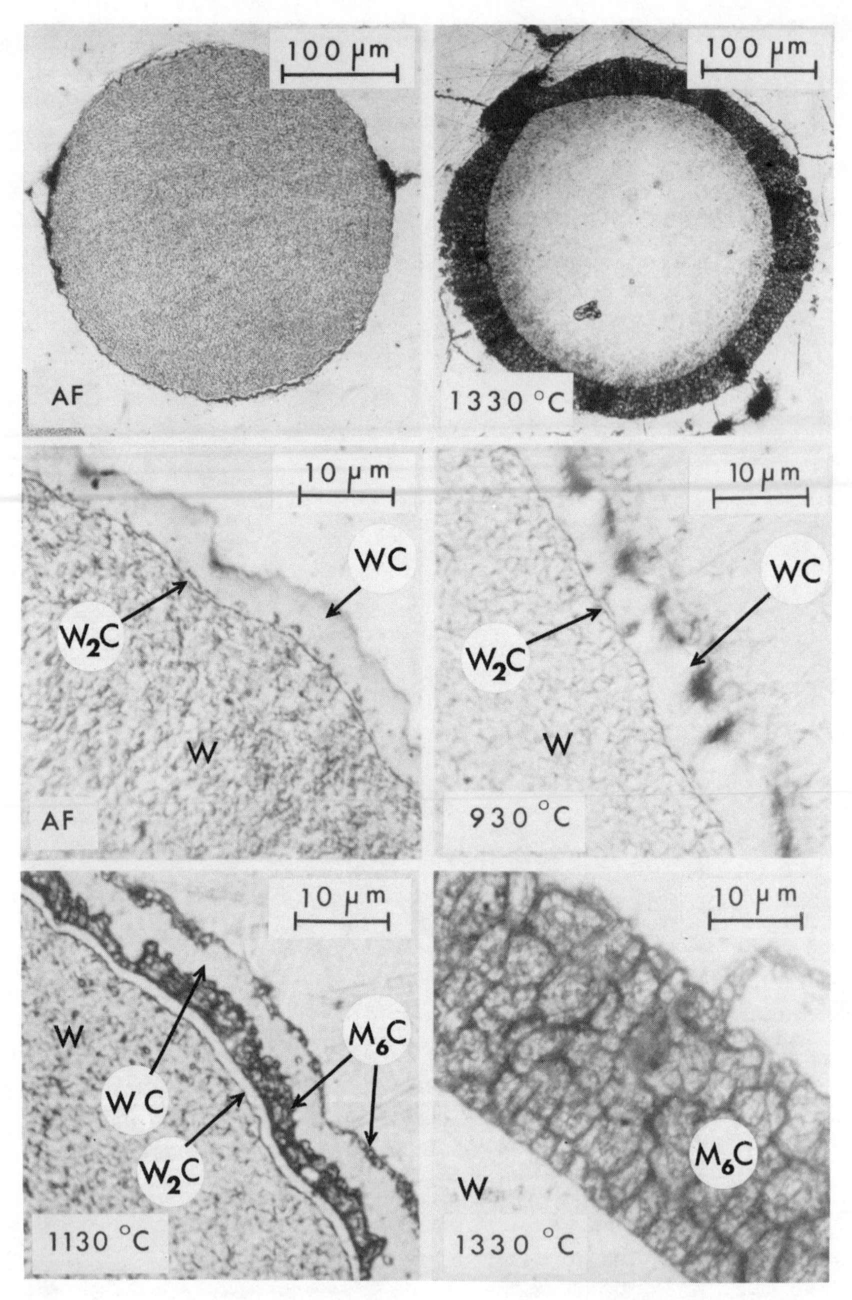

Fig. 9 Optical Photomicrographs of Interfaces Developed in 1095 St/45W-1ThO_2 Composites after Annealing for One Hour at the Indicated Temperatures.

Discussion

Rule-of-Mixture Strength Calculations

Because of the many complicating factors evident in composite material systems, such as void content, fiber misalignment, bonding imperfection, residual stresses, and mechanical and thermo-chemical interactions, the simple linear rule-of-mixtures (ROM) calculations usually are not sufficient to describe accurately the observed experimental properties of composites. The ROM method has, nevertheless, been a useful engineering guideline for composite material analysis. For ideal, non-reactive model composite systems (i.e., tungsten-copper, silver-sapphire, and some eutectic alloy systems), ROM predictions are fairly accurate, and any deviations from the ROM expectations indicate important areas to investigate for clues to how the material may be improved.

Specifically, the ROM strength calculation based on isostrain conditions predicts the tensile behavior of continuous fiber reinforced metal matrix composites quite well from the additive properties of the individual components for the ideal Cu/W system (6). For the Al/W system, the ROM and experimental values also agree quite well (86 percent here and 99 percent in Ref. 4). For the 1010 St/W system, however, the deviation is as large as 43 percent. The SEM fractograph shown in Fig. 4B suggested that the primary cause of weakening was fiber embrittlement, but the source of embrittlement was not clearly identified (4). Even though it was known that tungsten fibers embrittle with elevated temperature exposure, the differences in fracture modes between Al/W and 1010 St/W systems could not be attributed to the difference in consolidation temperatures. As was shown in Fig. 2, the consolidation temperature of 760C for 1010 St/W could not account for the degradation of this system, because the tungsten fibers showed no thermal degradation up to 800C. Furthermore, the annealing temperature for transition from ductile to brittle room temperature failure mode was above 800C (4).

In fact, both x-ray diffraction measurements and SEM studies (Fig. 6) on tungsten fibers extracted from the 1010 St/W composite revealed the existence and structure of intermetallic compounds. Tungsten is known to be a notch sensitive material, and cracks generated at the brittle interface layer presumably act as stress raisers. Hence, the recovery of strength and ductility after long immersion times in HCl acid was apparently caused by dissolution of the interface layer which thereby reduced the interface crack depth. Thus, the upper plateau shown in Fig. 3 corresponds to a critical thickness of interface reaction layer. Reaction layers thinner than this value do not affect the strength and fracture strain of the fiber. Similar experimental work on a tungsten/nickel system also reported a minimum thickness of the brittle interfacial zone, below which the strength of the fiber was constant and approximately equal to that of the unreacted fiber (7). Reaction layers thicker than this critical value led to a gradual reduction of fiber strength, which occurred only under conditions of strong interfacial bonding. Otherwise, the notch formed by fracture of the brittle zone at an early stage of deformation was unable to extend into the fiber due to effective notch-tip blunting caused by premature interfacial debonding between the reaction zone and the fiber. In the St/W composites, debonding was observed between the matrix and the reaction layer, but never between the reaction layer and the fiber.

Several investigators have attempted to relate the size of the interface layer to composite strength. For example, in titanium/boron composites, the effect of the titanium diboride reaction zone thickness on strength and fracture strain produced two plateaus as predicted by the interaction zone theory (8), which was based on elastic stress concentrations generated at the ends of cracks in the interaction zone. A similar theory was developed on the basis of linear elastic fracture mechanics for the cases of both strongly and weakly bonded fiber/brittle zone interfaces (9). As before, the upper plateau corresponded to a minimum thickness of the reaction layer. However, the lower plateau corresponded to a second critical thickness, above which the fiber exhibits a constant lower strength value, implying that the strength and fracture strain of the reaction zone determine the composite material property. The lower plateau was not observed in 1010 St/W, because the reaction layer was only 0.1 - 0.3 μm thick. Therefore, in order to extrapolate the relationship between strength and immersion time to zero time in the present work (Fig. 3), it is necessary to explore the general trends observed in other systems. The lower plateau values of fracture strength in W/Ni (7) and Ti/B (10) systems have been reported to be about half the upper plateau values. In between the upper and lower plateaus, existing theory (7,8) predicts a parabolic strength decrease with increasing thickness of the reaction zone. However, reported experimental data for the Ti/B system shows a somewhat linear relationship (10). Assuming the thickness decreased linearly with immersion time, and the same strength decrease of 50 percent applied in the 1010 St/W system, then the observed data were still within the transient region between those two plateaus. Hence, the extrapolation to zero immersion time (A in Fig. 3A) being well above the lower plateau for the fiber, provides a reasonable lower bound estimate of in situ fiber strength for this system.

The slight discrepancy between the upper plateau properties of 1010 St/W and the steady value of the unreacted Al/W system was most likely caused by the rapid grain boundary diffusion of iron into the tungsten fiber (2), which degraded both strength and ductility. Microhardness measurements on diffusion reaction zones that had not recrystallized definitely indicated a strength degradation of about 10 percent in a depleted uranium/tungsten system (11). In the 1010 St/W composites, about 5 percent degradation in strength resulted from this diffusion of iron into the tungsten fibers.

For tungsten fibers extracted from 1095 St/W composites, reaction zones were more than 5 μm thick and were considered to be beyond the second critical thickness. The data presented in Fig. 3 represent the lower plateau. Strength of the reaction zone is the factor controlling the extracted fiber strength, because the crack formed in the reaction zone is likely to propagate into the fiber immediately upon formation. The tensile strength of thoriated tungsten fibers extracted from 1095 St/W composites was only 37 percent of that measured for unreacted virgin fibers (830 MPa compared to 2260 MPa). Extracted fibers exhibited a constant strength level regardless of immersion time in HCl, because the tungsten monocarbide (WC) reaction layer was insoluble in this acid. To determine the critical reaction layer thicknesses, the WC could have been dissolved gradually in liquid fluorine. This was not pursued, since the extracted fiber strengths already obtained were considered adequate for refining the ROM estimation of strength in this system.

The ROM strength values were recalculated using the extracted fiber strengths based on extrapolations to zero immersion time in acid. Because extracted fibers are quite representative of fibers in the composite in that they have gone through the actual consolidation processes and suffered the thermal, diffusional, and chemical reactions, such a ROM calculation is more realistic than one based on virgin material properties. These modified ROM strength values are compared with experimental data in Fig. 10. The fact that experimental strengths were still generally lower than ROM strengths was attributed to such factors as voids, fiber misalignment, and imperfect bonding. However, the 1095 St/W (QT) composite exhibited anomalous behavior that required analysis of fiber strengthening effects for a reasonable explanation.

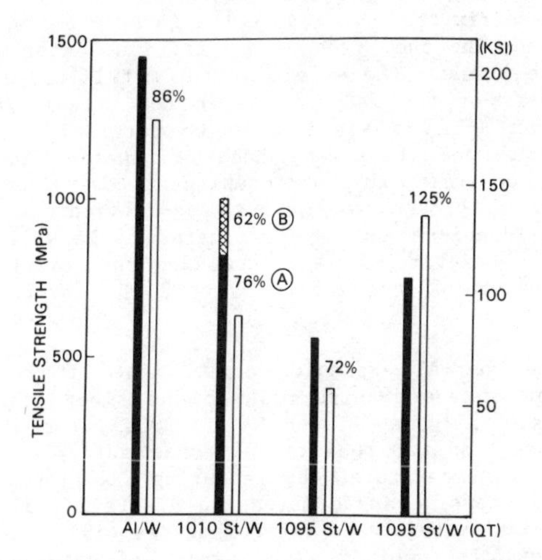

Fig. 10 Comparison of experimental tensile strengths (white), with theoretical (ROM) values modified by using extracted fiber strengths (black). A and B correspond to the extracted fiber strengths as referred to in Fig. 3A.

Fiber Strengthening Effects

One of the important assumptions in predicting the tensile behavior of continuous-fiber reinforced metal matrix composites from additive properties of the individual components is that the constituents are noninteracting (12). The chemical and thermal reaction effects could be accounted for by analyzing, respectively, the properties of extracted fibers and heat-treated fibers. However, the mechanical interactions in the composite should be analyzed in terms of in situ behavior of reacted fibers. The strength anomaly observed in 1095 St/W (QT) composites is not due to the simple matrix strength increase, which was already considered in the ROM computations shown in Fig. 10, and other possible mechanisms had to be considered.

For example, the strengths of notch-insensitive ductile fibers, such as steel and aluminum, may be elevated beyond their original values by the constraining effects of the matrix or the brittle interface zone in multifilament composites (13,14) and single filament composites (14, 15). In both cases, the reported composite strengths were higher than the predicted ROM values primarily due to the arrest of fiber necking by lateral constraining stresses developed by the tightly bonded matrix and/or brittle reaction zone when the composite was subjected to uniaxial longitudinal loads. Similar behavior has been observed at elevated temperatures in a nickel-tungsten composite (16). Another strengthening mechanism consisting of multiple necking of individual fibers has been observed in brass/tungsten composites (17, 18). However, these last two analyses cannot be applied to the current observations, because such necking behavior is limited to ductile fibers. Furthermore, the tungsten fibers extracted from the 1095 St/W system exhibited brittle transgranular failure across the fiber that was accompanied by a drastic decrease of strength, even though the extracted fiber was constrained by a tightly bonded brittle reaction layer.

Another possible strengthening mechanism is that the higher yield strength of the matrix tends to reduce the interface crack face displacement, thereby reducing the stress concentration factor at the fiber end of the interface crack. The effect of the matrix proportional limit on critical thickness of the reaction layer for titanium/boron composites demonstrated that a lower yield strength material required a smaller critical reaction zone thickness, because plastic flow of the matrix eliminates such constraint to crack opening at lower loads (10). However, the reaction zone thicknesses in both as-fabricated and heat-treated composites were identical and were estimated to be well above the lower critical value. Therefore, it was concluded that crack growth retardation due to strength of the matrix was not the governing mechanism in this case.

Because the thermal expansion coefficient of the matrix is much higher than that of the interface layer and fiber (i.e., $23.3 \times 10^{-6} K^{-1}$ for Fe and $5.3 \times 10^{-6} K^{-1}$ for WC and W at 1200K), the 1095 St/W composites developed substantial residual stresses during cooling from their high consolidation temperature. By measuring the proportional limits of the composites in tension and compression (Table 3), it was possible to compute the residual stresses in the composites (19). Knowledge of these residual stresses was combined with results of extensive fractographic analysis of composite failures to construct a possible model of damage evolution and strengthening in these composites. The proposed model is shown schematically in Fig. 11.

Lower residual stresses in the quenched-and-tempered material were thought to be partly the result of the volume expansion associated with the martensitic transformation, which would tend to mitigate the linear thermal expansion effects. Without external loads, the fibers were in compression and the matrix was in tension in both materials. The _in situ_ matrix proportional limits, that is, the matrix yield points in the presence of reinforcing fibers which constrain plastic flow in the matrix, were calculated to be 228 MPa for the as-fabricated 1095 St/W composites and 354 MPa when they were heat treated (QT).

Because of the high residual tensile stress of 157 MPa in the matrix, an applied tensile stress of about 100 MPa was all that was necessary to initiate plastic flow in the as-fabricated 1095 St/W composites in spite of a matrix yield strength of 228 MPa. With the onset of plastic flow in the matrix, the Poisson's ratio increased from its elastic value to the ideal plastic value of 0.5. Hence, appreciable transverse stresses are developed during uniaxial longitudinal loading of the composite as a result of this large difference in the Poisson's ratios of the plastic matrix and elastic fiber. These mechanical interactions have been studied analytically in concentric cylindrical composite models (20-22). As the transverse tensile stress increases in the interface, debonding occurs between the matrix and interface layer in the as-fabricated composite due to necking of the matrix, even though the

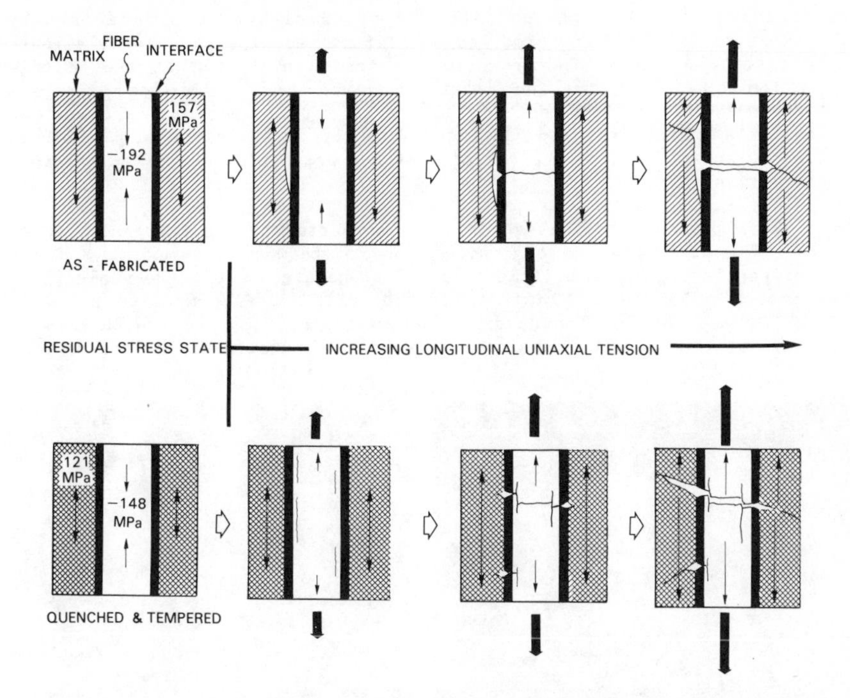

Fig. 11 Schematic of Proposed Damage Evolution in 1095 St/45W-1ThO$_2$(305) Composites during Longitudinal Tensile Loading.

fibers are not yet in a state of tension. As the external load increases further, the brittle reaction zone cracks at its weakest point along this debonded surface, and the crack propagates in brittle fashion across the whole fiber.

On the other hand, in the 1095 St/W (QT) composites, the situation is somewhat different because both the matrix itself and the fiber/matrix bond are stronger. The extracted reaction layer shown in Fig. 7 possesses microscopically rough surface structures on both sides. High strength materials such as the heat-treated matrix and the reinforcing fibers embedded in these surfaces will be strongly interlocked mechanically, in addition to the chemical bond. The Poisson's ratio for 1095 steel is 0.33 compared to 0.27 for tungsten. Because the difference is small, the transverse stress developed at the interface in the elastic-matrix/elastic-fiber regime is only 2-3 percent of the axial stress in the matrix (20), but with the onset of plastic flow in the matrix, the transverse stress quickly rises to about 40 percent of the matrix axial stress (21). In this plastic-matrix/elastic-fiber region, the large radial stress induces delamination of the fibrous elongated grains in the fiber instead of debonding at the interface. This is particularly true along the weak grain boundaries of the diffusion reacted area near the fiber periphery. Tungsten is notorious for its weak grain-boundaries, which have been reported to have as little as 1/10 the tensile strength and 1/1000 the shear strength of the grains themselves (23). Micro-grain boundary delamination in tungsten fibers is always observed in triaxial stress states generated by necking. The combination of thermally-induced triaxial residual stresses with the triaxial stresses developed by uniaxial tensile loading produced sufficient radial stress to delaminate the fibrous grain boundaries prior to crack initiation at the brittle reaction zone. Hence, the interface layer cracks did not propagate through the fiber, but instead the cracks were temporarily arrested at these delaminated grain boundaries. Thus the core regions of the fibers were not as affected by the surface notches and could retain the original fiber strength.

This proposed damage evolution is supported by fractographic analysis. The as-fabricated composite debonded between the matrix and the interface layer, and a brittle fracture propagated across the whole fiber as shown in Fig. 12. In the quenched-and-tempered composite shown in Fig. 13, most of the fibers exhibited multiple, compound fractures with failure at the outer periphery of a fiber being distinctly different from

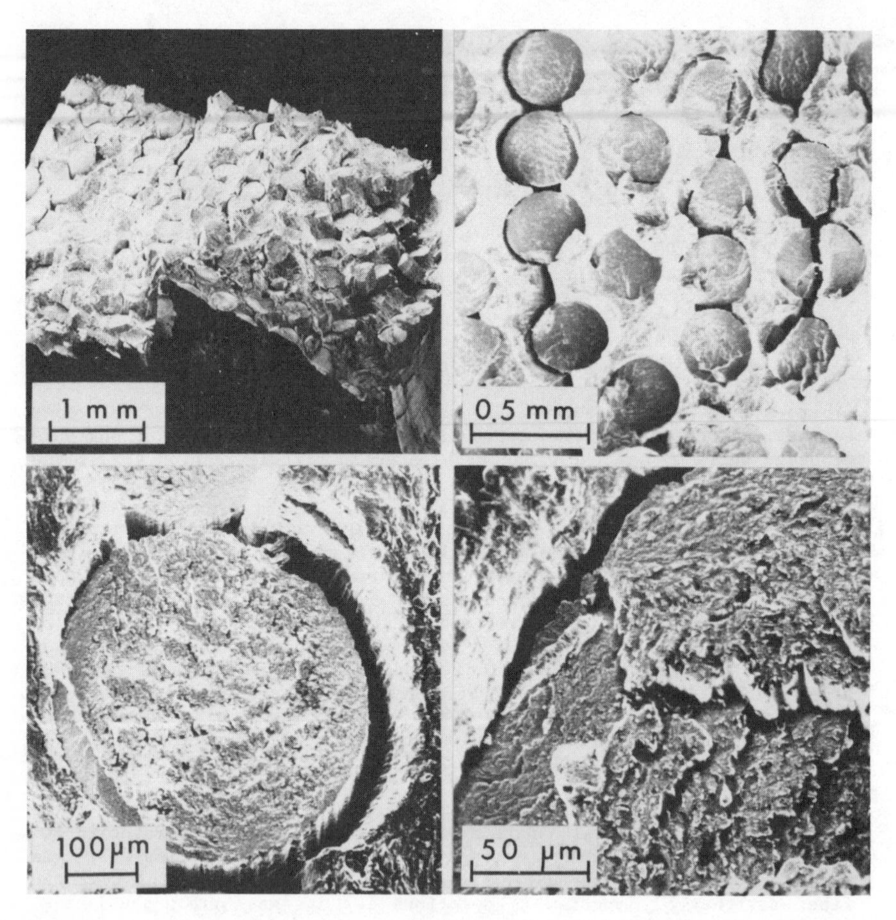

Fig. 12 SEM Fractographs of As-Fabricated 1095 St/45W-1ThO$_2$ Composite Fractured in Longitudinal Tension.

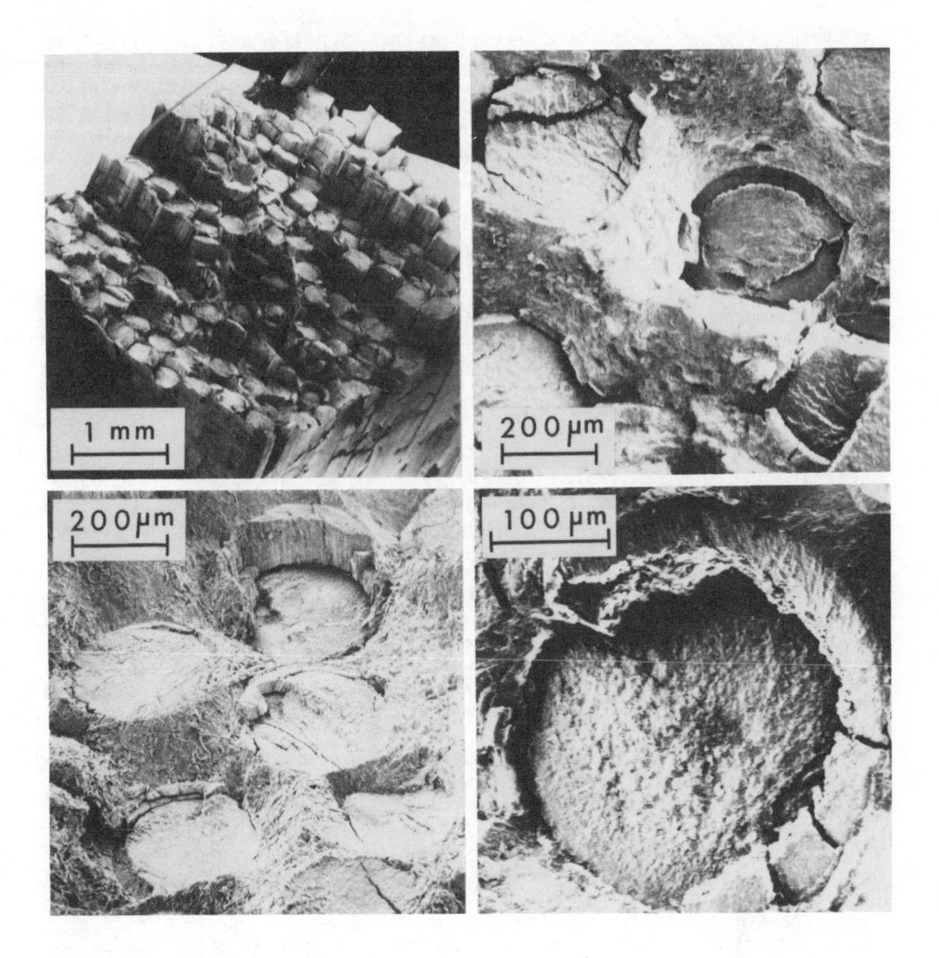

Fig. 13 SEM Fractographs of Quenched-and-Tempered 1095 St/45W-1ThO_2 Composite Fractured in Longitudinal Tension.

the failure in the core region. Occassionally, local conditions were such that nearly symmetrical failures occurred that showed clearly the described sequence of failure behavior in the reaction zone and core region of the fibers. Strong bonding between the matrix and interface layer was also observed.

Reaction Zone Development at Elevated Temperatures

The reaction layer, because it is very thin compared to the irregularity of the tungsten fiber surface in 1010 St/W composites, may easily be overlooked as the source of embrittlement in this system. Previous work showed that iron readily diffused from the matrix into the tungsten fiber at elevated temperatures (3,4). Through electron microprobe and Auger electron spectroscopy analyses, fractographic features in the peripheral regions of the fibers corresponded well with the measured iron contamination profiles and were distinctly different from fracture surfaces in the uncontaminated core regions of the fibers. Thus the embrittling degradation of the tungsten fibers was originally considered to be due solely to this contamination by grain boundary diffusion of iron. However, the present experimental results on fiber strength and ductility recovery (Fig. 3) and x-ray diffraction analyses of interface composition suggest otherwise, and so the interfaces were studied in detail.

Because tungsten is a strong carbide former, the small carbon concentration in 1010 steel (0.1 percent carbon) preferentially formed ditungsten carbide (W_2C) at the interface in equilibrium with $(FeW)_6C$ at the composite consolidation temperature of 760C. Although one study (24) showed that a eutectoidal decomposition of W_2C to W + WC was possible below 1200C, another study (25) found the W_2C to be stable below 1200C, perhaps to as low as 800C. However, no tungsten monocarbide (WC) was observed in this latter system, probably due to insufficient carbon concentration in the matrix. Instead, in the presence of iron, $(FeW)_6C$ double carbide formation is favored by the combination of one weak and one strong carbide former, e.g., Fe and W (25). Formation of $(FeW)_6C$ predominates until the carbon cannot be replenished; then the formation of the Fe_7W_6 binary compound occurs. The Fe_7W_6 layer grows faster into the fiber than into the matrix due to the higher diffusion rate of iron into tungsten. Short circuit diffusion of iron through the grain boundaries of the tungsten fibers further enhances the diffusion rate, leaving behind Kirkendall porosity because the tungsten diffusion does not compensate the material deficit in the matrix. Void formation due to imbalance of the transported species in their diffusion processes (Kirkendall effect) had been observed previously around each tungsten filament in a columbium/tungsten composite (26).

In 1095 St/W composite, the 1.0 percent carbon concentration in the matrix is an order of magnitude greater than that in 1010 steel. As a result of this abundant carbon, the interface reaction proceeded primarily by forming WC with a trace of other carbides such as W_2C and $(FeW)_6C$ at the fabrication temperature of the composite. As the annealing temperature was increased, however, the replenishment of carbon to the interface was not sufficient to maintain the 6.12 percent carbon concentration necessary for WC formation. Instead, the iron diffusion rate controlled to favor $(FeW)_6C$ formation. In other words, the carbon was consumed in formation of eta carbides before it could build up to concentrations sufficient to support formation of WC and/or W_2C at the interface. Hence, at elevated temperatures, the prevailing reaction compound is $(FeW)_6C$. Kirkendall porosity does not occur as long as the carbon diffusion rate controls the reaction rate of $(FeW)_6C$, thus retarding the iron transport rate. Instead, the tungsten diffuses through the grain boundaries of the entire matrix. Bulk diffusion of tungsten in the matrix is evidenced by the formation of islands of $(FeW)_6C$ compound in the matrix regions outside the original fiber/matrix boundary, but the reaction zone growth into the fiber is much greater.

Conclusions

1. The strength degradation in steel/tungsten composites is due to tungsten fiber degradation by thermal, diffusional, and chemical effects, especially the formation of brittle intermetallic reaction layers at the fiber/matrix interface.

2. The interface compounds for the as-fabricated systems were found to be predominantly W_2C in the 1010 St/W composites and WC in the 1095 St/W composites. Enhanced reactions at elevated temperatures developed the Fe_7W_6 binary compound in 1010 St/W composites and the $(FeW)_6C$ ternary compound in 1095 St/W composites.

3. A notch effect in the brittle intermetallic layers in 1010 St/W was demonstrated on the basis of strength and ductility recovery in the extracted tungsten fibers.

4. In the quenched and tempered 1095 St/W, a crack front arrest mechanism based on delamination of fibrous grain boundaries in the tungsten fibers was proposed to account for strength exceeding the expected ROM value.

Acknowledgements

This research program was sponsored by the Defense Advanced Research Projects Agency under DARPA Order No. 3388.

References

1. Weimer, R.J., Kim, C., Pattnaik, A., and Krause, D.J., "Development of Steel/Tungsten Composites for Ballistic Applications," NRL Memorandum Report 4682, Apr. 1982.

2. Davis, G.L., "Embrittlement of Tungsten Wires by Contaminants, "Nature, 181 (1958) p. 1198.

3. Koo, R.C., Gerber, M., Garofolo, J., and Shurgan, J., "A Method to Prevent Iron Contamination During Manufacture of Incandescent Filaments," Journal of the Illumination Engineering Society, 2 (4) (1973) pp. 387-395.

4. Kim, C., Phillips, W.L., and Weimer, R.J., "Fracture of Tungsten Wire in Metal Matrix Composites," ASTM STP 733 (1981) pp. 314-333.

5. Petrasek, D.W., and Weeton, J.W., "Effects of Alloying on Room-Temperature Tensile Properties of Tungsten-Fiber-Reinforced-Copper-Alloy Composites," TMS-AIME, 230 (1964) pp. 977-990.

6. McDanels, D.L., Jech, R.W., and Weeton, J.W., "Stress-Strain Behavior of Tungsten-Fiber-Reinforced Copper Composites," NASA TND 1881 (1963), also Metal Progress, 78 (6) (1960) pp. 118-121.

7. Ochiai, S., Urakawa, S., Ameyama, K., and Murakami, Y., "Experiments on Fracture Behavior of Single Fiber-Brittle Zone Model Composites," TMS-AIME, 11A (1980) pp. 525-530.

8. Metcalfe, A.G. "Interaction and Fracture of Titanium-Boron Composites," Journal of Composite Materials, 1 (1967) pp. 356-365.

9. Ochiai, S., and Murakami, Y., "Tensile Strength of Composites with Brittle Reaction Zones at Interfaces," Journal of Material Science, 14 (1979) pp. 831-840.

10. Metcalfe, A.G., and Klein, M.J., "Effect of Interface on Longitudinal Tensile Properties," pp. 125-168 in Composite Materials Vol. 1, Arthur G. Metcalfe, ed.; Academic Press, New York and London, 1974.

11. Weimer, R.J., Kim, C., Pattnaik, A., Krause, D.J. and Sanday, S.C. "Development of Metal Matrix Composite Penetrators - Final Report, Part III; Material Characterization," NRL Memorandum Report 4686, Apr. 1982.

12. McDanels, D.L., Jech, R.W., and Weeton, J.W., "Analysis of Stress-Strain Behavior of Tungsten-Fiber-Reinforced Copper Composites," TMS-AIME, 233 (1965) pp. 636-642.

13. Piehler, H.R., "Plastic Deformation and Failure of Silver-Steel Filamentary Composites," TMS-AIME, 233 (1965) pp. 12-16.

14. Ochiai, S., and Murakami, Y., "Deformation and Fracture Behavior of Composites with Brittle Zones on Fibre Surfaces," Trans. Japan Inst. Metals, 18 (1977) pp. 384-392.

15. Ochiai, S., and Murakami, Y., "The Strengthening Effects of Brittle Zones on Ductile-Fibre Composites," Metal Science, 10 (1976) pp. 401-408.

16. Mileiko, S.T., "The Tensile Strength and Ductility of Continuous Fibre Composites," Journal of Material Science, 4 (1969) pp. 974-976.

17. Vennett, R.M., Wolf, S.M., and Levitt, A.P., "Multiple Necking of Tungsten Fibers in a Brass-Tungsten Composite," Met. Trans., 1 (1970) pp. 1569-1575.

18. Schoene, C., and Scala, E., "Multiple Necking Phenomena in Metal Composites," Met. Trans., 1 (1970) pp. 3466-3469.

19. Thomas, E.R., Koss, D.A., and Chestnutt, J.C., "Mechanical Behavior of a Carbide Reinforced Co-Cr Eutectic Alloy," Met. Trans., 1 (1970) pp. 2807-2813.

20. Ebert, L.J., Hecker, S.S., and Hamilton, C.H., "The Stress-Strain Behavior of Concentric Composite Cylinders", Journal of Composite Materials, 2 (1968) pp. 458-476.

21. Hecker, S.S., Hamilton, C.H., and Ebert, L.J., "Elastoplastic Analysis of Residual Stresses and Axial Loading in Composite Cylinders, "Journal of Materials, 5 (1970) pp. 868-900.

22. Hamilton, C.H., Hecker, S.S., and Ebert, L.J., "Mechanical Behavior of Uniaxially Loaded Multilayered Cylindrical Composites," Journal of Basic Engineering (Trans. ASME Ser. D), 93 (1971) pp. 661-670.

23. Leber, S., Tavernelli, J., White, D.D., and Hehemann, R.F., "Fracture Modes in Tungsten Wire," Journal of Less-Common Metals, 48 (1976) pp. 119-133.

24. Telegus, V.S., Gladyshevsky, E.I. and Kripyakevich, P.I. Kristallografiya, 12 (1967) pp. 936-939.

25. Bergstron, M., "The Eta-Carbides in the Ternary System Fe-W-C at 1250°C," Material Science and Engineering, 27 (1977) pp. 257-269.

26. Brentnall, W.D., Klein, M.J., and Metcalfe, A.G., "Tungsten Reinforced Oxidation - Resistant Columbium Alloys," NAVAIR Systems Command, First Annual Report on Contract N00019-69-C-0137, Jan. 1970.

MICROMECHANISMS OF CRACK GROWTH IN A FIBER-REINFORCED, TITANIUM-MATRIX COMPOSITE

D. L. Davidson, R. M. Arrowood, J. E. Hack and G. R. Leverant
Southwest Research Institute
6220 Culebra Road
San Antonio, TX 78284

and

S. P. Clough
Perkin-Elmer
Physical Electronics Laboratories
6509 Flying Cloud Drive
Eden Prairie, MN 55344

The behavior of a propagating fatigue crack in a B_4C coated B fiber reinforced titanium alloy was studied as a function of thermal exposure. Crack tip deformation behavior was compared to that found in monolithic titanium. Qualitative observations were made in-situ in an SEM fatigue stage, while quantitative measurements of relative crack tip strain distributions were performed by use of the stereoimaging technique. Failure modes are discussed in terms of changes in the crack tip strain distributions and differences in interfacial surface chemistry due to heat treatment, as measured by Auger Electron Spectroscopy.

Introduction

Although a great deal of work has been performed on fracture mechanisms in fiber reinforced metal matrix composites (1-4), no general predictive capability has been established due to the lack of knowledge of the plastic strain distributions ahead of cracks in these materials (5) and the contradictory nature of the data gathered to date. The inconsistencies arise from the complex interaction of the fiber/matrix interfacial region with various applied states of stress. The presence of the fiber/matrix interface in composites introduces several mechanisms of crack propagation in addition to those observed in monolithic materials (5). Also, standard metallurgical techniques for modifying the properties of the matrix material (i.e., alloying, heat treatment, deformation processing, etc.) affect the metallurgical structure and properties of the interfacial region as well (6,7). Thus, the effects of processing variables on composite behavior through microstructural changes in the matrix are very difficult to separate from the effects of those variables on the fiber/matrix interface.

Early work by Cooper and Kelly (1), Cooper (2) and Kelly (3) showed that the work to fracture of a well bonded brittle fiber-ductile matrix composite was dependent on the work to fracture of the matrix between the fibers. Thus, the more matrix material separating the fibers, the tougher the composite. This result implies that the fibers need to be as large in diameter as possible for maximum toughness at a given volume percent. Another type of particle size effect was observed by Argon and Im (8) and others (9-13) for precipitate particles and inclusions. Interfacial cracks were observed at lower stresses and strains for particles which were larger (8-12), had larger aspect ratios (8-10), and had lower cohesive strengths at the particle/matrix interface (8,9,13). Argon and Im (8) found that the size effect associated with nondeformable particles which decohere from the matrix arose from the interaction between the plastic strain fields of neighboring inclusions which provides a stress concentration at the interface. Goldenberg _et al_ (14) confirmed the plastic interactions and revealed the role of elastic interaction stresses and the onset of weak shear instabilities along characteristic slip lines. The instabilities produce stress concentrations at the interface from matrix flow. Both Argon and Im (8) and Goldenberg _et al_ (14) concluded that a local critical decohesion stress is required at the particle/matrix interface. A similar critical stress or interfacial strength would be expected in the case of fiber reinforced composites.

Since all structural components contain either notches or pre-existing flaws or cracks when put into service, the fatigue and fracture behavior in the presence of a crack is critical. When coupled with the high degree of anisotropy in the mechanical properties of advanced metal matrix composites, the multiaxial stress state induced by a crack oriented normal to the fibers can cause fiber/matrix decohesion at relatively low applied stresses in a unidirectional composite loaded parallel to the interface. In fact, examination of the stress field ahead of a Mode I crack (15) reveals that the transverse tensile stress directly ahead of the crack is equal in magnitude to that perpendicular to the plane of the crack. Thus, the critical stress for interfacial decohesion is important in axial tensile and fatigue loading as well as in direct off-axis loading conditions.

Hancock (6) Hancock and Swanson (16) and Tetelman (5) showed that high strength fibers can help achieve good fatigue resistance and toughness by promoting interfacial splitting without sacrificing transverse tensile strength. Additional work by Hancock and Shaw (17) revealed that resistance to crack propagation could be optimized for fatigue life by altering the degree of bonding at the fiber/matrix interface. In effect this demonstrates that composite properties can be controlled by adjustments in the local interfacial strength.

The present study was performed to determine the effects of thermal exposure on the resistance to fatigue crack propagation in a boron carbide coated boron fiber reinforced Ti-6Al-4V matrix composite (B_4C/B/Ti). Specific goals of the investigation were to 1) quantify the effect of thermal exposure on the resistance to decohesion at the fiber/matrix interface by measuring the local strains which accumulate at the interface prior to decohesion, and 2) identify the microstructural or chemical component which controls the level of the interfacial strength.

Experimental and Materials

Materials

The unidirectional composite material, fabricated by the standard diffusion bonding process at Amercom, Inc., consisted of four layers of 140 μm diameter B_4C-coated boron fibers in a matrix of Ti-6Al-4V (B_4C/B/Ti). This material has been tested in the as-received (AR) condition and after several heat treatments: annealed 4.5 h in vacuum of less than 10^{-6} torr at 899°C (1650°F); annealed seven days in air at 593°C (1100°F); or annealed seven days in vacuum at 500°C (932°F). This last heat treatment was intended to produce a degree of fiber/matrix interface degradation intermediate between the as-received condition and the other two heat treatments. It is identical to the treatment which Mahulikar (18) used to produce intermediate interface weakening on a similar plate of material. For comparison with the composite behavior, additional experiments utilized unreinforced Ti-6Al-4V material which was manufactured by the same diffusion bonding procedure used to make the composite. Incorporation of an additional foil (i.e., six layers of foil instead of the five in the composite) produced an unreinforced sheet with nearly the same thickness as the composite panel. The titanium alloy in both the composite and the unreinforced specimens had a microstructure typical of mill-annealed material: an equiaxocl α structure, with β phase located at the triple points between α grains.

Tensile Testing

Room temperature tensile testing was performed on specimens of material in both the as-received condition and after heat treatment at 899°C for 4.5 hours in vacuum. Specimens 15.2 cm long by 1.3 cm wide were cut from the composite panel with the fibers at either 0 or 90° to the length direction. Ti-6Al-4V tabs 5.1 cm long by 1.3 cm wide by 0.08 cm thick were epoxy bonded to both sides of the ends of the specimens to protect the grip sections of the composite during testing. Strain gages were placed in the middle of the gage section to monitor specimen elongation. A strain rate of 0.005 cm/m/min was used for all tests. Fractography was performed on failed specimens by optical and scanning electron microscopy (SEM).

Fatigue Crack Growth

The fatigue crack growth specimens used in most of the experiments contained a single edge notch produced by electrical discharge machining, as shown in Figure 1. However, two specimens of the composite heat-treated at

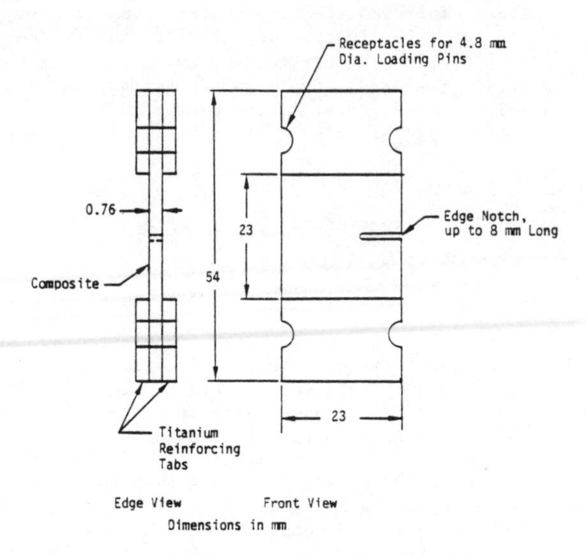

Fig. 1 - Fatigue Crack Growth Specimen.

500°C contained double edge notches, one on each edge of the coupon. This modification was part of a continuing effort to assess the influence of loading geometry on crack path. Present evidence indicates that the single- and double-notch geometries produce similar amounts of delamination and crack branching. Prior to testing, the surface of the specimen was carefully hand polished, removing just enough metal to reveal the outer surface of the interfacial reaction zones of the outermost fiber layer. This procedure made it possible to relate surface crack tip behavior to the distance from the nearest intact fiber. The specimens were cycled under load control at frequencies of 5-10 Hz, and crack advance was monitored by intermittent metallographic inspection. As the crack extended, the load amplitude was periodically reduced to keep the stress intensity factor range (ΔK) nearly constant. The minimum to maximum load ratio (R) was 0.1 in all tests. Unless otherwise noted, fatigue cycling was performed in air at room temperature.

Crack Tip Strain Analysis

In order to examine the details of the crack propagation processes, the specimen was transferred to a cyclic loading stage, which operates in-situ in an Etec scanning electron microscope (SEM), for further load cycling. Micrographs and/or videotape were used to monitor the crack advance and permitted use of the stereoimaging technique developed by Davidson (19). This technique applies stereographic methods developed for topographic mapping to the precise measurement of local strains. When a pair of micrographs, one of an unloaded crack and the other of the same crack under load, are viewed as a stereopair, in-plane particle displacements image as "topography," or vertical displacements. With modern photogrammetric equipment, very small displacements can be measured, and from these displacements the strains around the crack are computed. When the stereopair consists of the loaded and unloaded crack, the resulting strain values are the cyclic strain ranges around the crack. If, instead, the stereopair consists of the unloaded crack and the same field of view prior to arrival of the crack, the resulting strains are the cumulative strains associated with the introduction of the crack. A detailed description of the stereoimaging technique for crack tip strain analysis can be found in reference (20).

Auger Electron Spectroscopy

Specimens for Auger Electron Spectroscopy were sheared from coupons of material in the as-received and all heat treated conditions. Shearing was performed so that the longitudinal axes of the specimens were perpendicular to the fiber axis. The samples were then fractured parallel to the fiber/matrix interface in-situ in the ultra-high vacuum chamber of a Physical Electronics Model 595 high resolution Scanning Auger system. The vacuum level in the system was maintained at 1.3×10^{-10} torr throughout the fracture and analysis steps.

SEM fractographs, Auger point spectra, depth profiles, elemental scanning maps and high magnification line scans and point spectra were taken for all specimens. Beam conditions used for the various analysis modes are given in Table I. Data was collected, stored and analyzed with an on-line minicomputer system. Quantitative Auger analysis was conducted based on published and generally accepted sensitivity factors (21). All maps and line scans were corrected for topological variations in the specimens, unless otherwise noted. Depth profiles were obtained by sputtering with Ar ions accelerated at a potential of 4 KeV with the current being dependent on the selected sputtering rate.

Table I. Beam Conditions For Auger Analysis

Analysis Mode	Accelerating Potential (KeV)	Beam Current (nA)	Beam Diameter (nm)
SEM	5	0.1	50
Low Magnification Point Spectra and Depth Profiles	5	1000	1000
Elemental Scanning Maps	5	250	700
High Magnification Line Scans and Point Spectra	5	0.1	50

Results

Tensile Testing

The results of the tensile tests are presented in Table II. As can be seen from the data, the heat treatment degraded the tensile strength

Table II. Tensile Test Results

Fiber Orientation	Condition		Ultimate Tensile Strength (MPa)	Modulus (GPa)	Total Strain to Failure
Longitudinal	As-Received		1070	-	0.49%
			1061	213	0.51%
			1150	239	0.50%
			980	221	0.45%
		Mean	1065	224	0.49%
	Heat-Treated 899°C for 4.5 hours in vacuum		892	223	0.44%
			878	231	0.42%
		Mean	885	227	0.43%
Transverse	As-Received		316	170	0.39%
			319	172	0.52%
			372	158	9.54%
			355	157	0.54%
		Mean	341	164	0.50%
	Heat-Treated 899°C for 4.5 hours in vacuum		279	-	0.50%
			302	-	0.56%
			291	-	0.53%

of the composite in both the longitudinal and transverse orientation. The modulus and strain to failure values were unaffected by the thermal exposure. Fractography was performed on typical fracture surfaces and the results are presented in Figures 2-5. As can be seen from the figures, there is no apparent difference in failure mode between the longitudinal samples of as-received and heat treated material (Figures 2 and 3, respectively). Both samples demonstrated relatively flat fracture with no significant amount of fiber pullout. Higher magnification photographs [Figures 3(b) and 3(c)] revealed a brittle region immediately adjacent to the B_4C coating on the fiber, which has separated from the fiber in most cases, with an area of dimpled rupture in the matrix. Some evidence of brittle and/or crystallographic cracking behavior is also evident in the matrix material.

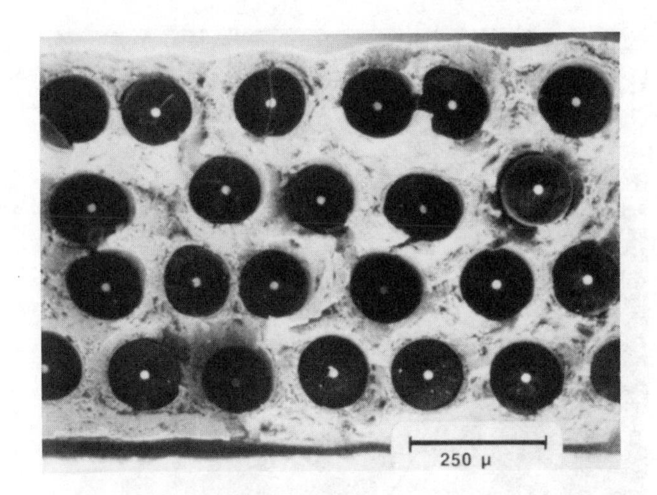

Fig. 2 - Fracture Surface of As-Received Composite Longitudinal Tensile Specimen.

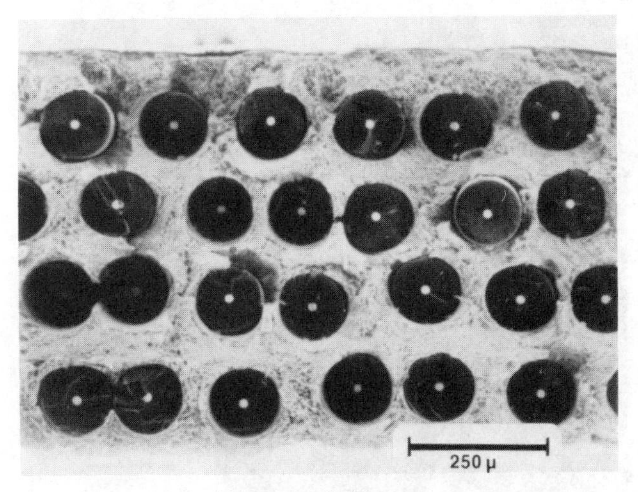

(a) General View

Fig. 3 - Fracture Surface of 899°C Heat-Treated Composite Longitudinal Tensile Specimen.

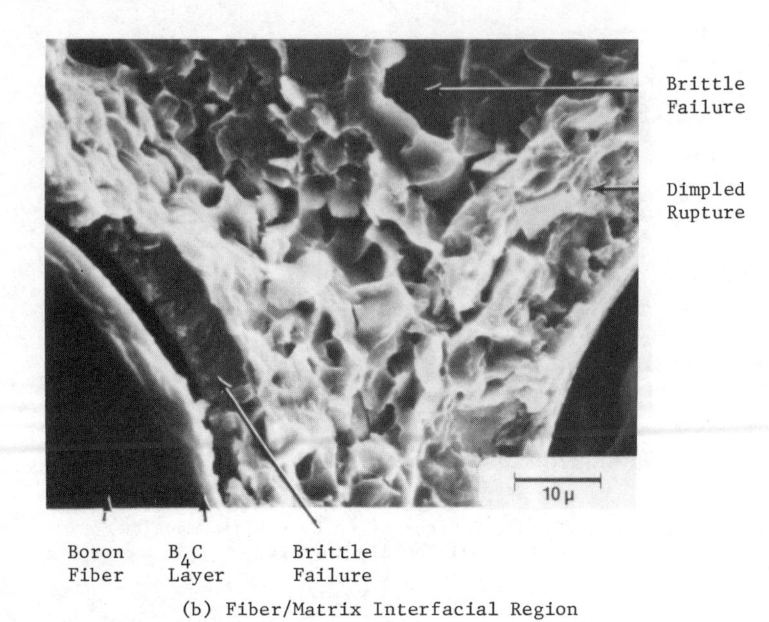

(b) Fiber/Matrix Interfacial Region

Dimpled Rupture

Brittle Failure

5 μ

Boron Fiber

B_4C Layer

(c) Fiber/Matrix Interfacial Region

Fig. 3(continued) - Fracture Surface of 899°C Heat-Treated Composite Longitudinal Tensile Specimen.

Fig. 4 - Fracture Surface of As-Received Composite Transverse Tensile Specimen.

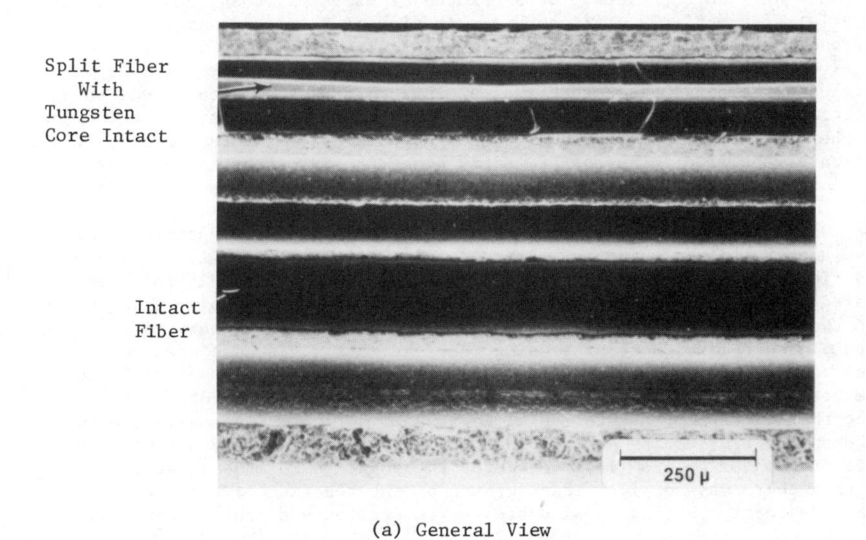

(a) General View

Fig. 5 - Fracture Surface of 899°C Heat-Treated Composite Transverse Tensile Specimen.

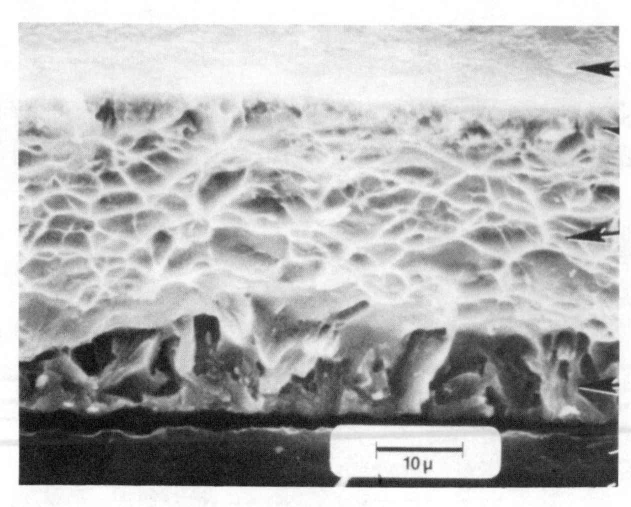

(b) Fiber/Matrix Interfacial Region

Fig. 5 (Continued) - Fracture Surface of 899°C Heat-Treated Composite Transverse Tensile Specimen.

Figures 4 and 5 show that there is also no apparent difference in failure mode between the transverse specimens of as-received and heat-treated material, respectively. Although examination of the figures may yield the impression that there is a higher degree of fiber splitting in the as-received material, this has not been confirmed on a sufficient number of samples to be conclusive at this time. Figure 5(b) shows that a region of brittle fracture lies adjacent to the fibers with ductile failure occurring in the matrix, as was the case for the longitudinal failures. Again, separation of the fiber occurred primarily along the interface between the B_4C coating on the fiber and the area of brittle fracture in the matrix.

Observations on polished transverse sections of the composite revealed a volume fraction of fiber of approximately 0.44. A close examination of the fiber/matrix interfacial region in the two composites, Figure 6, revealed a reaction zone between the B_4C coating and the titanium alloy matrix. The two-phase nature of this region, a sheath-like coating next to the fiber with a needle-like phase protruding into the matrix, is consistent with the observations of others (22,23). This phase is apparently responsible for the region of brittle fracture surrounding the fibers in Figures 1-4. The fact that fiber/matrix interfacial separation during failure always occurs between the B_4C coating and the reaction zone is also consistent with independent observations (23). An important point to be made about Figure 6 is that there was no significant growth of the reaction zone due to the heat treatment, yet the tensile properties were only slightly degraded.

Fatigue Crack Growth

Although the transverse tensile strength was not drastically affected by the 899°C heat treatment (≃15% reduction), fatigue tests revealed a

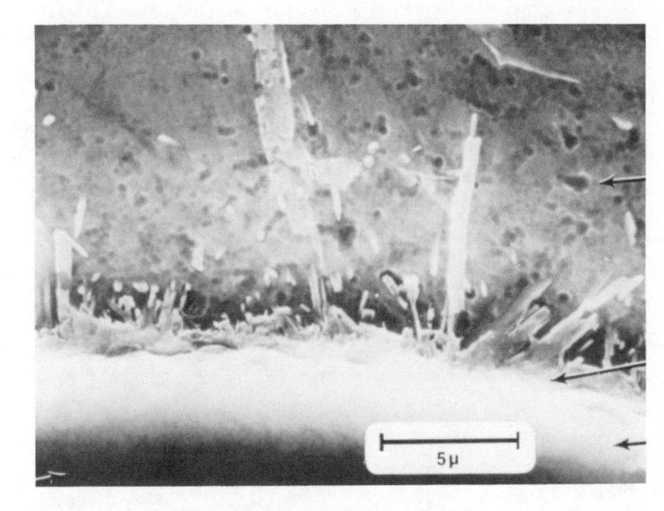

(a) As-Received

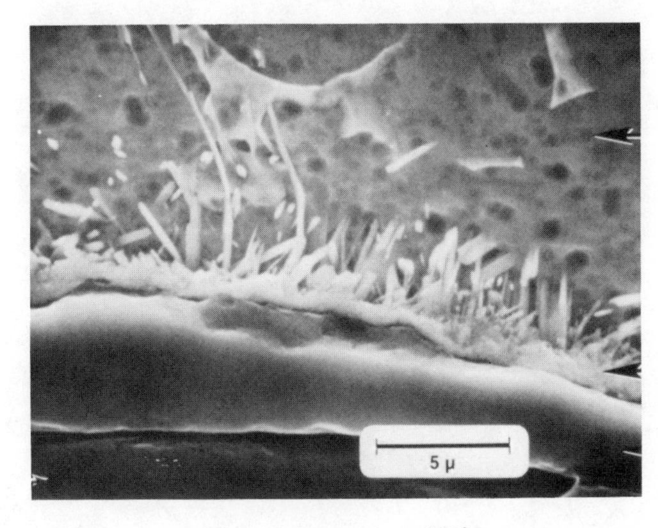

(b) Heat-Treated at 899°C

Fig. 6 - Comparison of Fiber/Matrix Interfacial Region.

distinct change in the crack propagation behavior in the three heat-treated materials as contrasted to the as-received composite. In the as-received composite, the crack propagated perpendicular to the fibers, with little deflection at intersections with fiber/matrix interfaces. In the heat-treated samples, however, crack branching was typical of the interaction between the propagating cracks and the interfaces (Figure 7). This deflection of the transverse crack by interfacial splitting resulted in

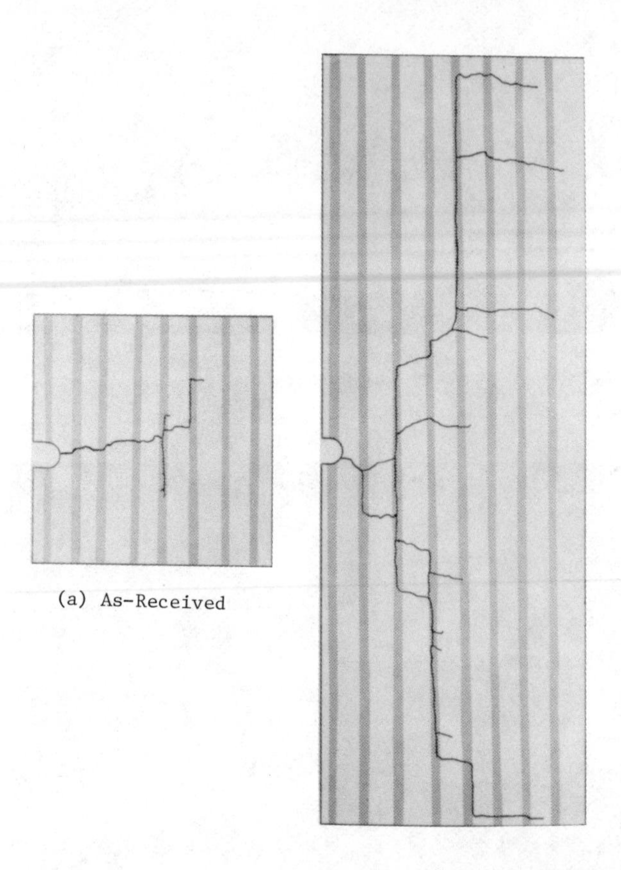

(a) As-Received

(b) Heat-Treated at 899°C

Fig. 7 - Comparison of Fatigue Crack Paths.

reduced crack propagation rates. For example, in one test of material heat-treated at 899°C, 1.8×10^6 cycles at a ΔK of 22-26 MPa $\sqrt{m}$ were necessary to propagate the crack across four successive fibers. In contrast, only 1.2×10^5 cycles were required to propagate a crack across for fibers in the as-received material at a ΔK of 20-22 MPa $\sqrt{m}$.

Figure 8 is a graph of crack length (plus notch length) versus number of cycles for as-received and 899°C heat-treated specimens. The crack

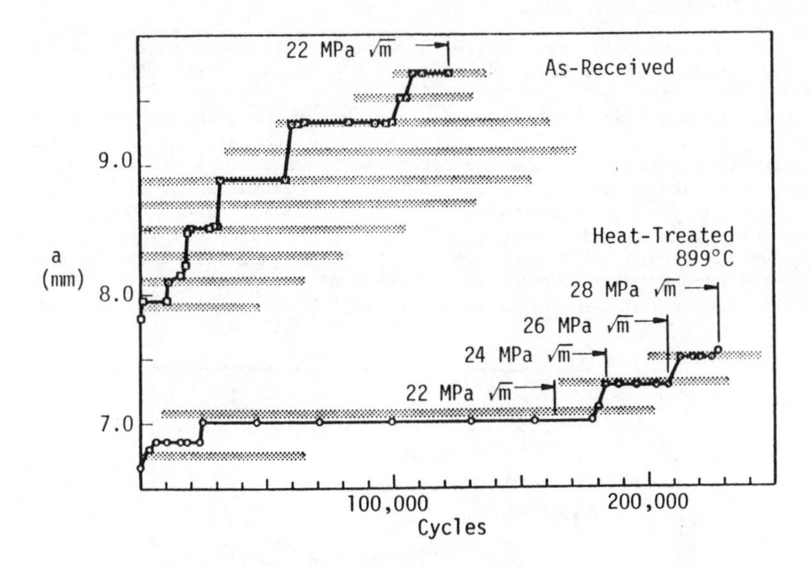

Fig. 8 - Transverse Crack Propagation During Fatigue Test. (Parameter a is the sum of the notch length and the transverse crack propagation distance. Stippled bands indicate the exposed portion of a fiber-matrix interface.)

length plotted here is the projection of the foremost crack tip on the plane of the notch, i.e., the crack propagation transverse to the fibers. Transverse propagation occurs in steps, separated by induction periods in which the crack usually propagates along the fiber/matrix interface. In Figure 8 the intersections of interfacial reaction zones with the specimen surface are indicated schematically by stippled bands across the graph. As the figure shows, the lower average rate of transverse crack propagation in the heat-treated material is primarily due to arrest of the crack by interfacial debonding. In this particular experiment on heat-treated material, the crack was induced to resume transverse propagation by increasing the load amplitude. In another experiment, using the material heat treated at 500°C the load amplitude was not increased and the interfacial crack reached the reinforcing grip tabs, ending the experiment.

It is evident that fiber/matrix interface splitting was easier in the heat-treated materials, and that this had a profound effect on the crack path and the transverse crack propagation rate. The effect of all three heat treatments used was similar, and the effect was so strong that is posed experimental problems. The crack branched so profusely that it became impractical to monitor events at all the individual crack tips, and the stress intensity experienced by any one crack tip could no longer be calculated. However, the local strains around the crack tips could still be measured by the stereoimaging technique.

Crack Tip Strain Analysis

It is instructive to compare the behavior of cracks in the composite to the behavior of cracked, unreinforced Ti-6Al-4V. Experimental data were obtained on samples of this alloy which were produced by hot-pressing of foils to duplicate the processing history of the titanium matrix in the composites. Fatigue tests were run in the SEM fatigue stage at various ΔK values, and the crack opening displacements and crack-tip strains were measured by stereoimaging. Figure 9 shows the variation of crack opening displacement (COD) with distance from the crack tip. Included are data from the matrix material (i.e., diffusion bonded Ti-6Al-4V fails with no reinforcing fibers) at three different ΔK levels and data from the as-received composite at 22 MPa $\sqrt{m}$. A sketch of the crack path (Figure 10) in the composite shows the relation of the COD sets to the position of the

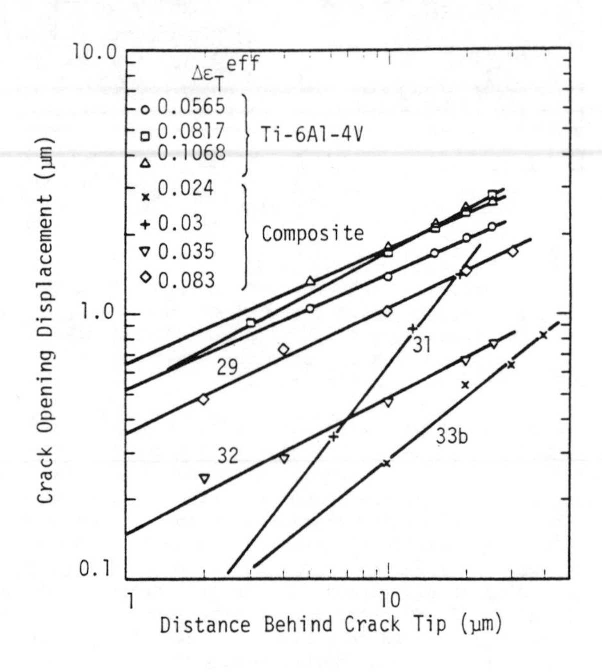

Fig. 9 - Variation of Crack Opening Displacement With Distance Behind the Crack Tip.

crack tip. The matrix material data fell on three straight, nearly parallel lines. The crack opening displacements of cracks in the composite, when the crack tips were not near a fiber, also fell on straight lines with slopes similar to that of the matrix material data. However, the displacements were reduced in the composite, so that these lines were below those of the matrix material. When the crack tip was near a fiber/matrix interface [Data Sets 31 and 33(b)], the slope of the COD line was much steeper. This was particularly evident in the case where the crack tip was adjacent to a debonded fiber matrix interface (Data Set 31).

When the crack tip effective strain range data for the unreinforced matrix material were plotted as a function of ΔK (Figure 11), the resultant

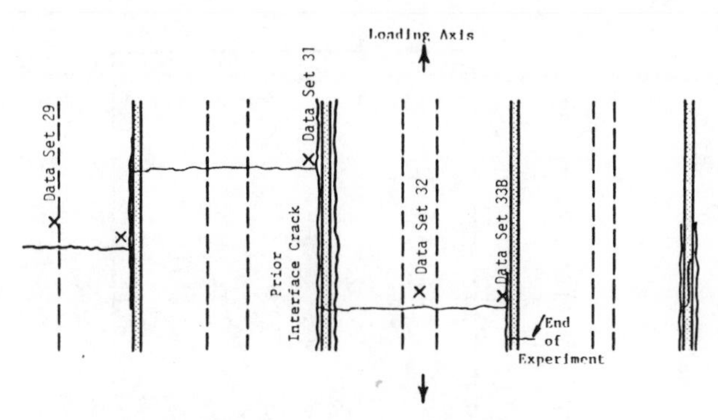

(a) Location of Cracks at Various
Points Where Data Was Obtained

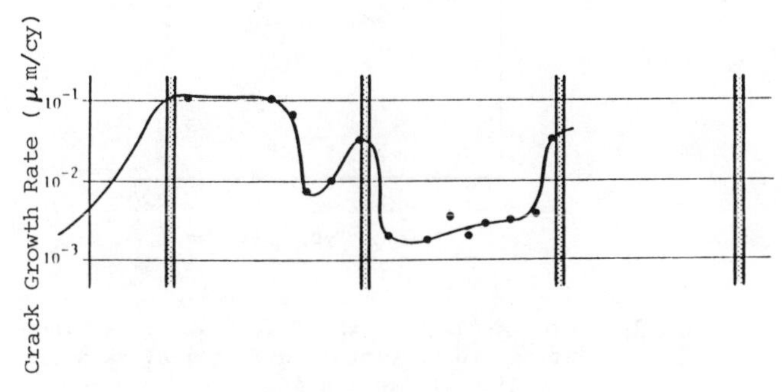

(b) Measured Position Dependent
Crack Growth Rates

Fig. 10 - Schematic of Crack Growth Through the As-Received Composite (Dotted lines indicate the outermost position of the buried portion of the fiber and stippled bands indicate the exposed portion of a fiber/matrix interface).

curve indicated that the crack tip strain range was proportional to $\Delta K^{2.8}$. By using this graph, the observed crack tip strain ranges for the as-received composite specimen were converted to effective ΔK levels experienced by cracks in the composite at an applied ΔK of 22 MPa $\sqrt{m}$. Results of this procedure (Table III) show that the material ahead of the crack experiences only 5.8 - 9.4 MPa $\sqrt{m}$. This is over a 50% reduction as compared to the applied ΔK. Using the crack propagation data of Yuen, et al (24), the effective ΔK levels were used to predict crack propagation rates in the composite. As shown in Table III, the results are in excellent agreement with observed rates for three of the four points. The rate was consistently lower than that predicted on the basis of the increased elastic modulus for the composite. The only inconsistent point was associated with the data set where the crack tip was located near a debonded fiber/matrix interface.

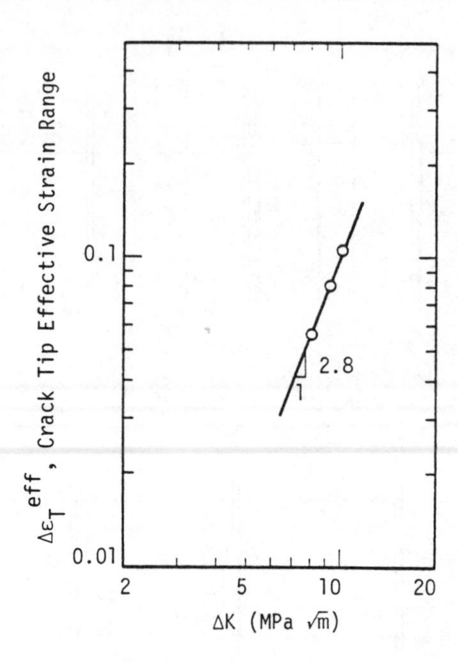

Fig. 11 - Cyclic Range of the Effective Crack Tip Strain in Unreinforced Ti-6Al-4V as a Function of Applied ΔK.

Figures 12(a)-12(e) display the maximum shear strain ranges around crack tips, as determined by stereoimaging. The strains around a crack tip in unreinforced Ti-6Al-4V are shown in Figure 12(a). There is a strong concentration of strain at the crack tip, as expected. The strain range distribution around a crack in the composite, when the crack tip was remote from a fiber/matrix interface, was similar to the distribution in unreinforced material [Figure 12(b)]. Figure 12(c) shows the maximum shear strain ranges for a fatigue crack tip which was approaching a previously debonded interface. The cracked interface has clearly modified the strain distribution and reduced the degree of strain concentration (note the expanded vertical scale in this figure). Figure 12(d) reveals that the strain range distribution was also modified when the crack was near an unbroken interface. Figure 12(e) related to the same crack position as in Figure 12(d); however, it displays the cumulative, rather than cyclic, strains due to the propagation of the crack into the area. The accumulated strains build up along the fiber/matrix interface and are comparable in magnitude to the cyclic strain ranges. Figure 12(f) shows the accumulated strains normal to the interface for the same analysis as shown in Figure 12(e). Along the interface the strains vary from tensile (positive) to compressive (negative). Debonding initiated in the central region of tensile strain, near a local maximum in the strain plot, within ten cycles after the strain measurements in Figures 12(e) and 12(f) were obtained. The normal tensile strain at this point was approximately 0.02% and the maximum shear strain was in the order of 2.5%.

Table III. Effective ΔK Experienced by Cracks in Composite at Applied ΔK of 22 MPa $\sqrt{m}$

Data Set	Crack Location	$\Delta\varepsilon_t^{eff}$	ΔK_{eff} (MPa$\sqrt{m}$)	Predicted da/dN (μm/cycle)	Observed* da/dN (μm/cycle)
29	Remote from Fiber	.0832	9.4	$2x10^{-2}$	$2x10^{-2}$
32	Remote from Fiber	.0350	6.6	$4.5x10^{-3}$	$4.5x10^{-3}$
31	At Debonded Interface	.0299	6.4	$4.5x10^{-3}$	$3.3x10^{-2}$
33B	At Sound Interface	.0240	5.8	$3x10^{-3}$	$3x10^{-3}$

*The modulus corrected growth rate using the method of Pearson (25) is $2.66x10^{-2}$ μm/cycle.

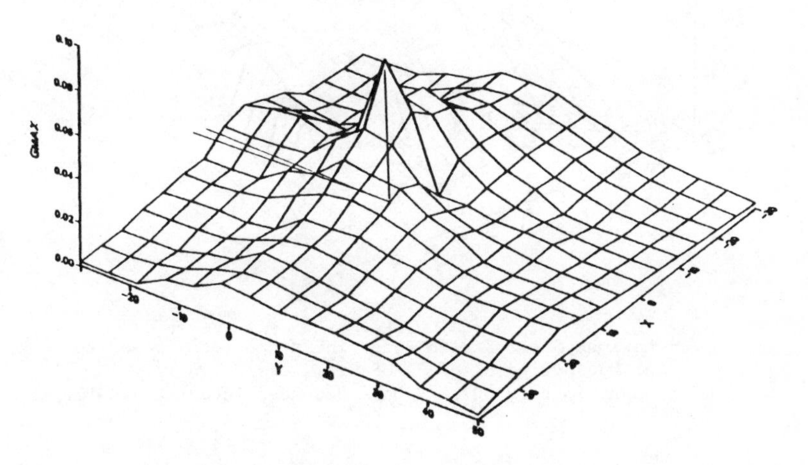

(a) Maximum Shear Strain Range in Unreinforced Ti-6Al-4V at a ΔK of 8 MPa $\sqrt{m}$

Fig. 12 - Typical Strain Range Distributions For Cracks Studied (Cracks are shown schematically on the zero strain surface, with the tip at x = 0, y = 0; x and y are in μm).

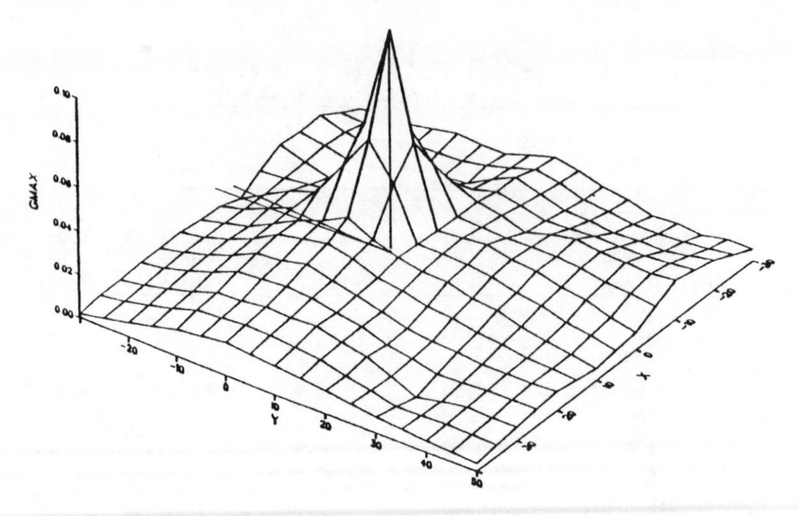

(b) Maximum Shear Strain Range in As-Received Composite at an Applied ΔK of 22 MPa $\sqrt{m}$. Crack Tip is Approximately 170 μm From the Next Fiber.

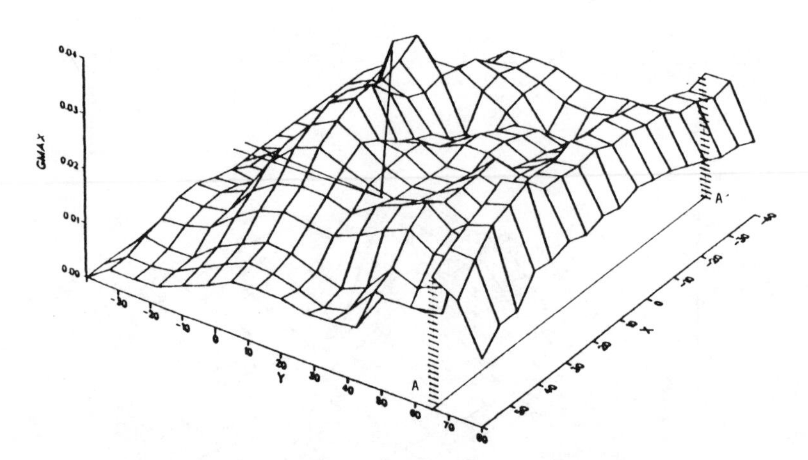

(c) Maximum Shear Strain Range in As-Received Composite at an Applied ΔK of 22 MPa $\sqrt{m}$. Crack Tip is Approximately 60 μm From a Debonded Interface (AA').

Fig. 12 (Continued) - Typical Strain Range Distributions For Cracks Studied (Cracks are shown schematically on the zero strain surface, with the tip at x = 0, y = 0; x and y are in μm).

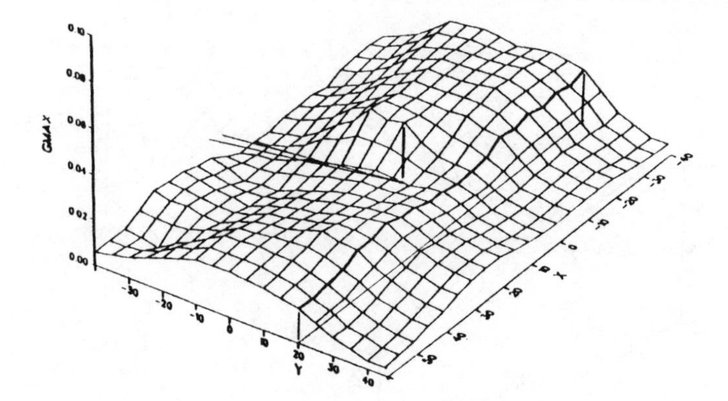

(d) Maximum Shear Strain Range in As-Received Composite at an Applied ΔK of 22 MPa $\sqrt{m}$. Crack Tip is Approximately 20 μm From an Uncracked Fiber/Matrix Interface (Darkened Line).

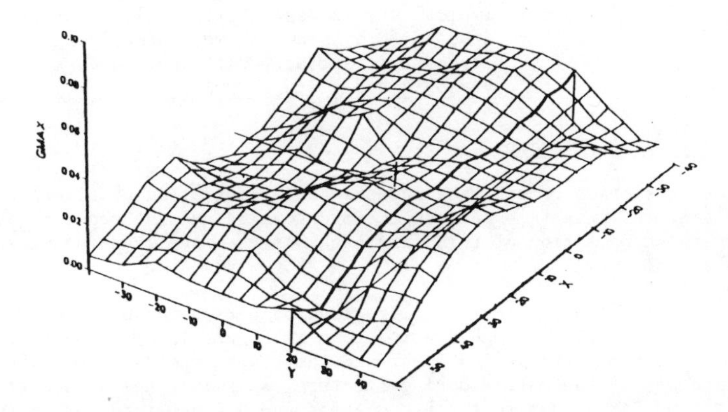

(e) Cumulative Shear Strain in As-Received Composite in the Same Region as 12(d).

Fig. 12 (Continued) - Typical Strain Range Distributions For Cracks Studied (Cracks are shown schematically on the zero strain surface, with the tip at x = 0, y = 0; x and y are in μm).

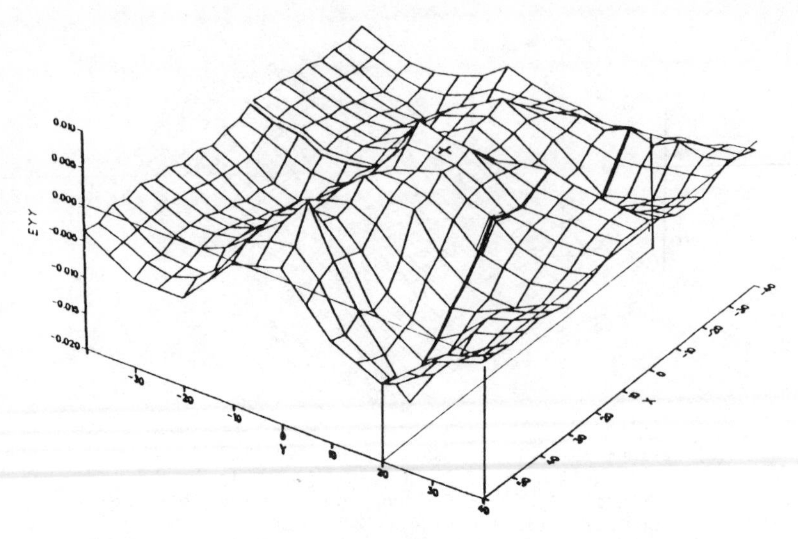

(f) Cumulative Normal Strain in As-Received Composite in the Same Region as 12(d)

Fig. 12 (Continued) - Typical Strain Range Distributions For Cracks Studied (Cracks are shown schematically on the zero strain surface, with the tip at x = 0, y = 0; x and y are in μm).

Auger Electron Spectroscopy

Since the crack growth rate appears to be controlled by fiber/matrix debonding at the interface between the B_4C coating and the reaction zone, the chemical composition of this interface was examined as a function of heat treatment.

A typical scanning electron micrograph of a transverse fracture surface of the as-received composite is shown in Figure 13(a). The field of view includes the surface of a fiber and the trough where an adjacent fiber has pulled out of the matrix during fracture. A sliver has broken from the fiber, giving SAM access to the B_4C coating and B fiber interior as well as the surface. Figures 13(b)-13(d) show SAM maps of calcium, boron and titanium, respectively, for the same field of view.

The boron and titanium maps show excellent correlation with the B_4C coating, boron fiber and metal matrix. The most interesting result, however, is that the calcium map shows a significant concentration of calcium on the fiber and trough surfaces along which fracture occurred. Similar calcium distributions were found in three heat-treated specimens. Table IV shows the results of quantitative Auger analyses of points on the fiber surface and trough for each of the four material conditions. The table also lists the sputter time for the reduction of the calcium concentration to one-fourth of its surface value and the corresponding approximate thickness of the Ca-enriched layer. In general, the calcium concentration at the fiber and trough surfaces is higher in the heat-treated specimens than in the as-received specimen by a total of about 3-5%.

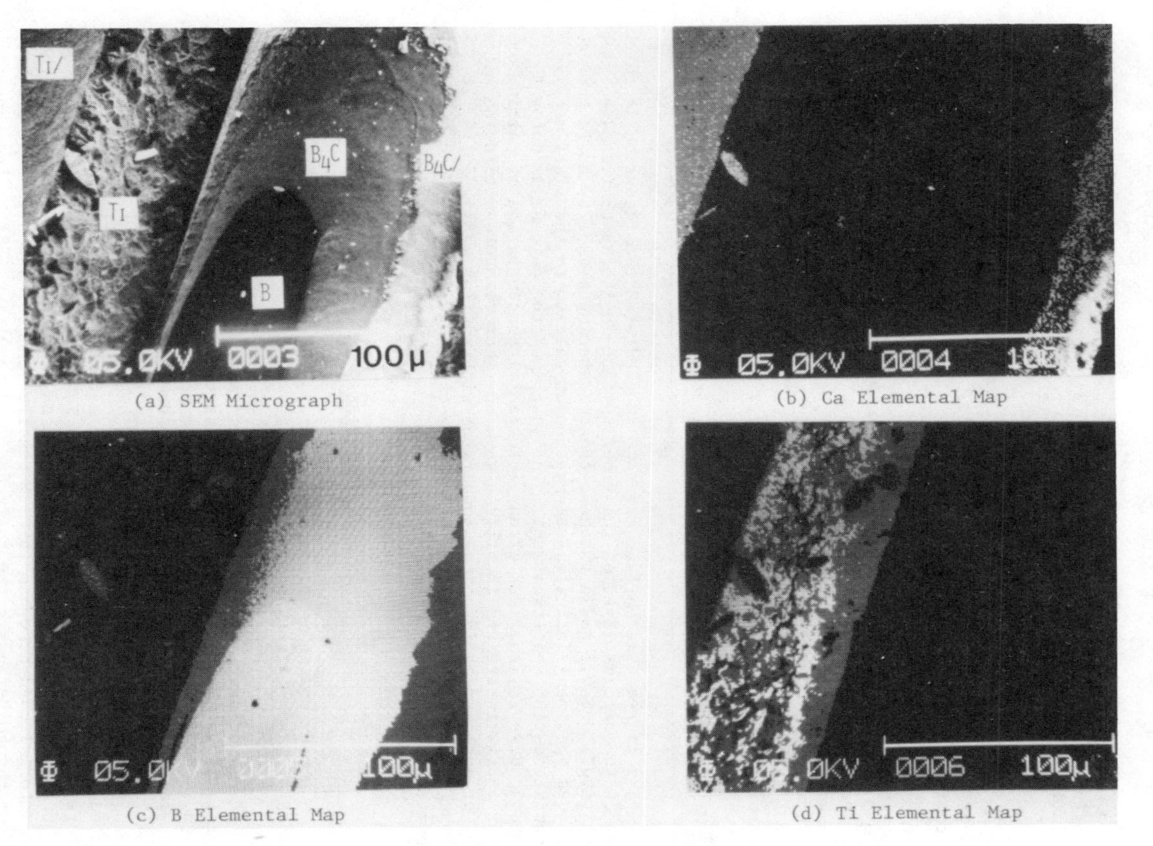

(a) SEM Micrograph

(b) Ca Elemental Map

(c) B Elemental Map

(d) Ti Elemental Map

Fig. 13 - Region of Transverse Fracture Used in Auger Analysis.

Table IV. Auger Electron Spectroscopy Results

		As-Received	Heat-Treated 500°C	Heat-Treated 593°C	Heat-Treated 899°C
Fiber Surface (atomic %)	Ca	6.3	7.8	11.1	10.5
	B	65.5	59.1	59.7	58.2
	Ti	1.7	1.2	1.1	0.8
Fiber Trough (atomic %)	Ca	8.7	11.3	9.5	7.8
	B	40.1	36.3	29.4	50.6
	Ti	29.3	35.3	33.6	23.5
Sputter Time for 75% Ca Reduction on Fiber (min.)		5.4	1.1	1.3	1.3
Approximate Ca Segregation Depth on Fiber Surface (nm)		2.7	0.6	0.7	0.7

This increase in surface coverage coupled with the reduction in thickness of the calcium layer on the fiber surface in the heat-treated specimens (e.g., 0.6-0.7 nm in the heat-treated specimens vs. 2.7 nm in the as-received specimen) indicates segregation of calcium to the fiber/matrix interface during heat-treatment. The source of the calcium is not yet certain; however, the lack of any interfacial concentration of calcium on fiber/matrix interfaces in silicon carbide fiber reinforced titanium material produced under identical conditions with similar matrix foils (26) strongly indicates that it is originally contained in the B_4C coated boron fiber.

Discussion

Effect of Thermal Exposure

It is evident from the results given above that for the material used in this investigation thermal exposure at temperatures in excess of 500°C did not have a drastic effect on tensile properties or reaction zone growth. Upon initial consideration, this would seem to conflict with the results of others (27,28). However, nominal longitudinal and transverse tensile strengths in excess of 1400 MPa and 500 MPa, respectively, have been reported for similar as-received B_4C/B/Ti material (29). After thermal exposure these values generally drop to a low of approximately 900 MPa and 350 MPa, respectively (28). From the data in Table I, it can be seen that the material used here had initial tensile behavior which is only slightly better than material in the fully degraded condition. After the thermal exposure at 899°C, the tensile strengths fall to those found for the fully degraded condition.

The microstructural evidence which indicates no growth in reaction zone thickness after the 899°C heat-treatment is also contrary to the results of others (28). Close examination of the reaction zones in Figures 6(a) and 6(b) reveals that the total reaction zone thickness is on the order of 5 µm, even in the as-received material. The data of Smith et al (28) show that, in the as-received condition, the reaction zone thickness of material with a longitudinal tensile strength on the order of 1400 MPa is only 0.5 µm thick. In fact, their parabolic growth rate curves for reaction zone thickness as a function of time and temperature indicate that quite long times, even at temperatures in excess of 850°C, would be required to develop a reaction zone thickness on the order of 5 µm. Thus, the tensile and microstructural evidence provide strong support to the belief that the material used in this study was almost fully degraded by thermal exposure in the as-received condition.

Although the above discussion suggests that tensile behavior, both in the longitudinal and transverse direction, is quite sensitive to the growth of the reaction zone between the fiber and matrix, fatigue crack propagation may be relatively insensitive to the extent of the reaction zone. The drastic change in fatigue crack growth mode, from relatively flat fracture in the as-received material to severe interfacial debonding in all of the heat-treated conditions, and resultant drop in fatigue crack growth rate occurred with no commensurate loss in tensile strength or increase in reaction zone thickness. This implies that reaction zone growth, in and of itself, may not affect the interfacial strength and that processes which do alter interfacial strength can act independently from simple growth of the reaction zone.

One possible explanation of the onset of interfacial decohesion in the heat-treated crack growth specimens is weakening or actual breaking of the fiber/matrix bond during the heat-up and/or cool-down portions of the thermal exposure cycle. It is conceivable that the stresses generated due to the thermal expansion mismatch between the fiber and matrix could reach a high enough level to break the bond, at least intermittently, along the fiber/matrix interface. This is supported by the results of Marcus (30) that oxygen penetration along fiber/matrix interfaces in SiC/Ti composites occurred to a much higher extent in specimens which were thermally cycled to 500°C compared to those held isothermally at 500°C for the same overall time at temperature. In addition, Bhatt and Grimes (31) have shown that damage accumulated during thermal fatigue of Al_2O_3 fiber reinforced magnesium alloys is more dependent on the temperature range of the thermal cycle rather than the total number of cycles. If the temperature range was above a critical value, most of the damage due to thermal fatigue occurred in the first few cycles.

A more probable source of the thermally induced change in fatigue failure mode may be the observed increase in segregation of calcium to the fiber/matrix interface during heat treatment. Data on grain boundary embrittlement due to segregation of a tramp element in a number of alloys (32-34) would suggest that the significant coverage of the fiber/matrix interface by calcium should play a major role in interfacial decohesion resistance and thus fatigue crack growth resistance. The positive identification of the source of the calcium and crack growth experiments run under controlled levels of segregation would be required to provide a definite link between the enhanced calcium levels and the observed decrease in resistance to interfacial decohesion.

Comparison Between the As-Received Composite and Monolithic Material

The local crack growth rates shown in Figure 10 are one to two orders of magnitude lower than those observed in monolithic Ti-6Al-4V at a ΔK of 22 MPa $\sqrt{m}$ (24). Even when the values are adjusted for the higher modulus of the composite (25) there is a significant decrease in crack growth rate dur to the presence of the fibers. The reason for this is apparent in the data in Table III. The data clearly shows that there has been a significant reduction in crack tip plasticity in the composite due to the large constraint imposed by the fibers. Except for the case where the constraint was lost due to fiber/matrix debonding, the effective ΔK ranges derived from the crack tip strain data show excellent correlation with the local crack growth data. These results explain why attempts at understanding crack growth behavior in metal matrix composites on the basis of the applied stress intensity have not been successful, even if fiber/matrix interfacial debonding was not prevalent (35). Any attempt to model the crack growth process, under fatigue or static loading, must take into account the reduced but variable level of effective ΔK range experienced by the material ahead of the crack tip prior to debonding and the deflection of the crack after interfacial decohesion has occurred.

Conclusions

(1) Trends in tensile behavior and microstructural development during high temperature thermal exposure of B_4C/B/Ti composites do not necessarily relate to the mode of failure to be expected during fatigue.

(2) The decrease in resistance to interfacial decohesion induced by high temperature thermal exposure is most likely due to the enhancement of calcium concentration at the fiber/matrix interface due to segregation. An additional increment of damage may be induced by the thermal expansion mismatch between the fiber and matrix.

(3) The degradation of resistance to interfacial decohesion due to thermal exposure causes a precipitous drop in overall fatigue crack growth rate by deflecting the propagating crack.

(4) The constraint produced by the fibers significantly reduces the effective ΔK range at the crack tip, over and above any effect of elastic modulus, by directly reducing the extent of crack tip plasticity.

(5) The results obtained by stereoimaging for the crack tip strain fields are encouraging in that it has been shown that decohesion occurs in a region of high cumulative shear strain and maximum cumulative normal strain.

Acknowledgements

The authors would like to express their appreciation for the encouragement and support of Dr. Alan H. Rosenstein of the Air Force Office of Scientific Research under Contract F49620-78-C-0022.

References

(1) G. A. Cooper and A. Kelly, Journal of the Mechanics and Physics of Solids, Vol. 15, 1967, p. 279.

(2) R. E. Cooper, Journal of the Mechanics and Physics of Solids, Vol. 18, 1970, p. 279.

(3) A. Kelly, Proceedings of the Royal Society of London, Vol. 319, 1970, p. 95.

(4) D. L. McDanels, R. W. Jech and J. W. Weeton, Metallurgical Transactions, Vol. 233, 1965, p. 636.

(5) A. S. Tetelman, Composite Materials: Testing and Design, STP 460, American Society for Testing and Materials, 1969, p. 473.

(6) J. R. Hancock, Composite Materials: Testing and Design (Second Conference), STP 497, American Society for Testing and Materials, 1972, p. 483.

(7) G. D. Swanson and J. R. Hancock, ibid, p. 469.

(8) A. S. Argon and J. Im, Metallurgical Transactions, Vol. 6A, 1975, p. 839.

(9) A. S. Argon, J. Im and R. Safoglu, Metallurgical Transactions, Vol. 6A, 1975, p. 825.

(10) T. B. Cox and J. R. Low, Jr., Metallurgical Transactions, Vol. 5, 1974, p. 1457.

(11) A. Gangulee and J. Gurland, Transactions of TMS-AIME, Vol. 239, 1967, p. 269.

(12) I. G. Palmer and G. C. Smith, Proceedings of the Second Bolton Landing Conference on Dispersion Strengthening, Gordon and Breach, New York, 1968, p. 253.

(13) G. T. Hahn, M. F. Kanninen and A. R. Rosenfield, Annual Review of Materials Science, Vol. 2, 1972, p. 381.

(14) T. Goldenberg, T. D. Lee and J. P. Hirth, Metallurgical Transactions A, Vol. 9A, 1978, p. 1663.

(15) P. C. Paris and G. C. Sih, Fracture Toughness Testing and Its Applications, ASTM STP 381, American Society for Testing and Materials, 1965, p. 30.

(16) J. R. Hancock and G. D. Swanson, Composite Materials: Testing and Design (Second Conference), ASTM STP 497, American Society for Testing and Materials, 1972, p. 299.

(17) J. R. Hancock and G. G. Shaw, Composite Materials: Testing and Design (Third Conference), ASTM STP 546, American Society for Testing and Materials, 1974, p. 497.

(18) D. S. Mahulikar, Private Communication, University of Texas at Austin, 1981.

(19) D. L. Davidson, SEM 1979, Vol. 2, SEM Inc., AMF O'Hare, Illinois, 1979, p. 79.

(20) D. R. Williams, D. L. Davidson and J. Lankford, Experimental Mechanics, Vol. 20, 1980, p. 134.

(21) L. E. Davis, N. C. MacDonald, P. W. Palmberg, G. Riach and R. E. Weber, Handbook of Auger Electron Spectroscopy, Perkin-Elmer Corporation, Minnesota, 1976, p. 13.

(22) J. Thebault, R. Pailler, G. Bontemps-Moley, M. Bourdeau and R. Naslain, Journal of the Less Common Metals, Vol. 47, 1976, p. 221.

(23) A. G. Metcalfe, Composite Materials, Vol. I, L. J. Broutman and R. H. Krock, ed., Academic Press, 1974, p. 67.

(24) A. Yuen, S. W. Hopkins, G. R. Leverant and C. A. Rau, Metallurgical Transactions, Vol. 5, 1974, p. 1833.

(25) S. Pearson, Nature, Vol. 211, 1966, p. 1077.

(26) D. S. Mahulikar, Y. H. Park and H. L. Marcus, Proceedings of the Greece/USA Mixed Mode Fracture Conference, Athens, Greece, 1980.

(27) R. Naslain, J. Thebault and R. Pailler, Proceedings of the 1975 International Conference on Composite Materials, E. Scala, E. Anderson, I. Toth, and B. R. Noton, ed., AIME, 1976, p. 116.

(28) P. R. Smith, F. H. Froes and J. T. Carnmett, Mechanical Behavior of Composite Materials, J. E. Hack and M. F. Amateau, ed., AIME, 1983, p. 141.

(29) J. A. Cornie, Private Communication, Avco Corporation, 1982.

(30) H. L. Marcus, Titanium Matrix/Continuous Fiber Composite Interface Interactions and Their Influence on Mechanical Properties, AFOSR Report 80-0052, University of Texas at Austin.

(31) R. T. Bhatt and H. H. Grimes, Mechanical Behavior of Composite Materials, J. E. Hack and M. F. Amateau, ed., AIME, 1983, p. 50.

(32) A. Joshi and D. F. Stein, Metallurgical Transactions, Vol. 1, 1970, p. 2543.

(33) B. D. Powell and H. Mykura, Acta Metallurgica, Vol. 21, 1973, p. 1151.

(34) W. C. Johnson and D. F. Stein, Metallurgical Transactions, Vol. 5, 1974, p. 549.

(35) W. R. Hoover, Failure Modes in Composites III, T. T. Chiao and D. M. Schuston, ed., AIME, 1976, p. 304.

CORRELATION OF FRACTURE CHARACTERISTICS AND MECHANICAL PROPERTIES FOR TITANIUM-MATRIX COMPOSITES

P. R. Smith*, F. H. Froes* and J. T. Cammett**

* Metals and Ceramics Division
Materials Laboratory, AFWAL
WPAFB OH 45433

** Metcut Materials Research Group
P. O. Box 33511
WPAFB OH 45433

INTRODUCTION

In recent years material requirements for advanced aerospace applications have increased dramatically as performance demands have escalated. As a result, mechanical properties of monolithic metallic materials such as titanium often have been insufficient to meet these demands. Attempts have been made to enhance the performance of titanium by reinforcement with high strength/high stiffness filaments.

Titanium matrix composites have for quite some time exhibited enhanced stiffness properties which approach rule-of-mixtures (ROM) values. However until quite recently, with few exceptions, both tensile and fatigue strengths were well below ROM levels and were generally very inconsistent. With the advent of more recently developed fibers improvements in these latter properties have been made although inconsistencies still exist.

The present study was conducted to determine which material characteristics most influence the tensile and fatigue failure mechanisms and hence the levels of these properties. The three major areas investigated included: reinforcing fibers, various titanium alloy matrix materials, and fabricated composites. The fiber evaluation covered those fibers which were commercially available at the time of the study and were being considered for use in advanced aerospace structures with major emphasis on the effect of fiber structure on fiber and composite properties. The matrix materials investigated were the workhorse alpha-beta titanium alloy, Ti-6Al-4V, and three richer alloys. The latter were studied to define effects of microstructure and alloying constituents on reaction zone growth. Composite characteristics examined included: fiber/reaction zone interface bonding, reaction zone microcracking, reaction zone thickness, and matrix residual stresses.

MATERIAL AND EXPERIMENTAL PROCEDURE

Materials

The fibers evaluated were the four commercially available fibers being considered for use in titanium: Silicon Carbide (SiC), SiC coated Boron (Borsic), Boron Carbide coated Boron (B_4C/B) and SCS-6. All filament was continuous and approximately 5.6 mil in diameter. The recently developed SCS-6 fiber, which is essentially SiC with the the addition of a silicon-rich layer to the surface [1] only became available towards the end of this program and thus was not studied in as much detail as the other fibers. The fibers were consolidated for evaluation by diffusion bonding into various titanium matrices utilizing specific combinations of parameters (ie. time, temperature, pressure) depending on matrix flow characteristics (Table I). The predominant matrix of interest was the alpha-beta titanium alloy Ti-6Al-4V. Attention was focused on this alloy because of its wide usage in the aerospace field as well as for ease of comparison with the data base of mechanical properties previously established for this alloy, both with and without reinforcement. Other titanium matrices investigated included: CP-Ti, Ti-10Mo, Ti-4.5Al-5Mo-1.5Cr (CORONA 5), [2] and Ti-11.5Mo-6Zr-4.5Sn (Beta III). [3,4] These alloys were selected in an attempt to reduce the amount of reaction product formed between the fiber and matrix during composite fabrication (with the exception of CP-Ti which was used as a control for the Ti-10Mo study) since previous work [5] had suggested that the extent of this product critically affects mechanical properties. Earlier studies [6] with Boron filaments had shown that alloying titanium with molybdenum significantly reduced the amount of reaction product. The Beta III [7,8] and CORONA 5 [9] alloys were selected as earlier studies had demonstrated that careful processing control resulted in a fine grain size and thus should allow lower temperature fabrication. Table I shows that composite fabrication was accomplished at much reduced temperatures with these alloys (when compared to Ti-6Al-4V). Borsic filament was selected as the reinforcement in the CP-Ti, Ti-10Mo, CORONA 5, and Beta III matrices because it had previously exhibited composite properties (in Ti-6Al-4V) intermediate to the SiC and B_4C/B fibers. [10,11] This would allow composite property improvement or degradation to be more easily detected.

TABLE I. COMPOSITE FABRICATION PARAMETERS

Fiber/Matrix	Temperature $^{o}C(^{o}F)$	Time (hr)	Pressure MPa(ksi)
SCS-6/Ti-6Al-4V	925 (1700)	0.50	70 (10)
SiC/Ti-6Al-4V	925 (1700)	0.50	70 (10)
Borsic/Ti-6Al-4V	925 (1700)	0.50	70 (10)
B_4C/B/Ti-6Al-4V	925 (1700)	0.50	70 (10)
Borsic/Ti-10Mo	870 (1600)	0.75	55 (8)
Borsic/CP-Ti	870 (1600)	0.75	55 (8)
Borsic/CORONA 5	850 (1565)	0.75	55 (8)
Borsic/CORONA 5	905 (1665)	0.75	55 (8)
Borsic/Beta III	730 (1350)	24	70 (10)

Experimental Procedure

Fiber Evaluation. In an attempt to distinguish between effects that thermal exposure and/or chemical reaction with the matrix may have on the fibers during composite fabrication, the loop strength and tensile strength were determined for each of the fibers (with the exception of SCS-6*) in four conditions: 1) as-received, 2) heat treated (HT), 3) digested (matrix dissolved away with acid) and 4) acid treated (AT). The loop test involves bending the filaments around cylinders of decreasing diameter until fiber fracture occurs. The loop stress σ_L, is then given by:

$$\sigma_L = E_f d_f / d_L$$

where:

E_f = fiber modulus

d_f = fiber diameter

d_L = loop diameter (at fracture)

Fiber tensile testing was conducted by Composite Technologies, Inc. on a computer-automated high-speed testing machine. The rationale for conducting both types of tests is that the loop strength should be more sensitive to surface flaws while tensile testing should be sensitive to both internal and surface flaws, thus allowing effects from the two types of defects to be distinguished.

* The SCS-6 filament was evaluated in the as-received (loop and tensile) and heat treated (loop) conditions only.

The as-received test condition was the filament condition as supplied by the manufacturer. The heat treated condition (HT) corresponded to 925°C (1700°F)/30 minutes in a vacuum of 10^{-7} Torr which was used to simulate the composite fabrication thermal cycle. Fibers were also examined for effects of chemical reaction during composite fabrication by loop testing after digestion. This was accomplished by chemically etching away the titanium matrix (Ti-6Al-4V) using a 90% HNO_3 + 10% HF solution. The acid treated condition (AT) was investigated to determine the effect that the etchant used in extracting the filament had on the fibers.

Reaction Kinetics. Small coupons 0.6 cm x 1.3 cm (0.25 in. w 0.5 in. 1) containing each of the four types of fibers in a Ti-6Al-4V matrix were encapsulated in Vycor at a vacuum of 10^{-5} Torr and thermally exposed as shown in Table II. The range shown represents at the lower extreme, 595°C (1100°F), the maximum service temperature, and at the upper extreme, 955°C (1750°F), the maximum processing (e.g. SPF/DB) temperature, to which these composites might be exposed. After thermal exposure the samples were metallographically prepared and high magnification SEM micrographs were taken of the reaction zone (up to x10,000). From each of four micrographs, eight measurements of

TABLE II. THERMAL EXPOSURE RANGE FOR REACTION KINETICS STUDY

Temperature ^{o}C (^{o}F)	Time (hr)		
595 (1100)	32	128	512
760 (1400)	4	16	64
870 (1600)	1	4	16
955 (1750)	0.5	2	8

the reaction zone thickness were made giving a total of thirty-two counts per thermal exposure. The measurement for the B_4C/B/Ti-6Al-4V reaction product was the average extent of the TiB needles as illustrated in Figure 1.

Residual Stress Measurements. Matrix residual stresses were measured using the technique recommended by the Society of Automotive Engineers [12] which differs from the more common SAE technique in two respects. First, the diffraction peak used for stress measurements was located using a five-point parabolic regression procedure rather than the three-point algebraic procedure and second, the intensities measured at each of the five points were corrected for background intensity. These modifications improve the repeatability of stress measurements. Details of the technique and diffraction fixturing

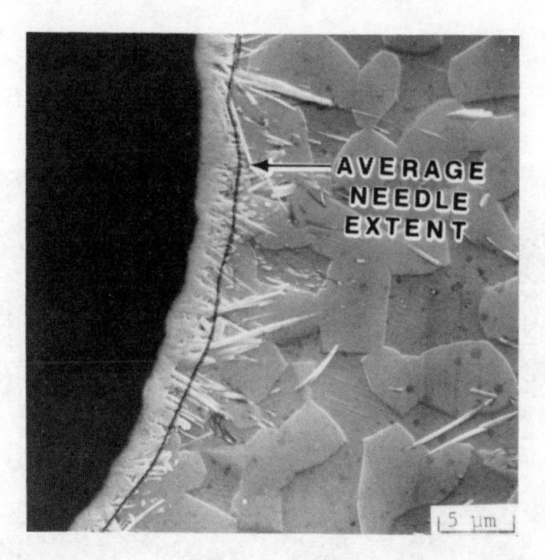

Figure 1. Measurement of the B_4C/B/Ti-6Al-4V Reaction Zone

for residual stress measurements on the Ti-6Al-4V matrix materials were:

Crystallographic Plane (hkl): (213)

Radiation: Cuk_α

Incident Beam Divergence: 1°, 3°

Detector Slit Divergence: 1°, 0.5^o

Filter: 18 μm (0.0007 in.) Nickel

ψ Rotation: 0.0^o and 45.0^o

$E/(1 + \nu)$: 84,100 MPa (12,200 ksi)

The value of the elastic constant, $E/(1 + \nu)$, in the direction normal to the (213) planes was taken to be the same as that previously determined in calibration experiments on Ti-6Al-4V.

In order to produce sufficient diffracted beam intensity from the Ti-6Al-4V matrix, it was necessary to irradiate an area on each specimen of from 0.3cm (0.125 in.) square to about 0.6cm (0.25 in.) square. Because the irradiated area is large relative to the inter-fiber spacing, the measured stress is essentially the average matrix residual stress within the entire composite.

For subsurface residual stress measurements, specimens were electropolished in a 5% H_2SO_4 - methanol solution to expose the subsurface material. Material was removed uniformly layerwise from a 1.7cm (0.5 in.) square area on each specimen.

Mechanical Testing. Longitudinal and transverse tensile and fatigue test specimens 10 cm x 1.7 cm (4 in. x 0.5 in.) with a 2.5 cm (1 in.) gage length were sheared from the composite panels (Figure 2). Aluminum tabs, 0.3 cm (0.125 in.) thick x 2.5 cm (1 in.) long, were adhesively bonded to each end and tapered to meet the gage section. Fatigue specimen edges were prepared in a manner described elsewhere [13] before testing.

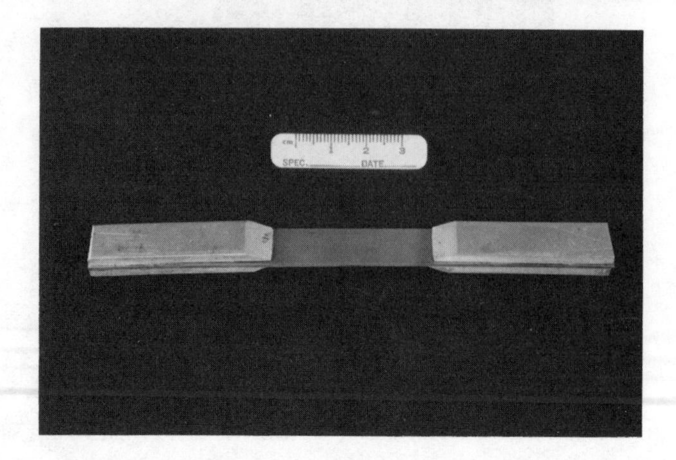

Figure 2. Specimen used for Tensile and Fatigue Testing

Tensile testing was carried out in an Instron machine with strain rates of approximately 0.013 cm(.005 in.)/cm/min. while fatigue testing was conducted on a Schenck machine, in a tension-tension mode, R = +0.1, using a sinusoidal waveform in the frequency range of 2200-2400 cpm. All testing was done at room temperature in ambient air at a maximum stress level of 515 MPa (75 ksi).

Metallography. Metallographic examination included both optical microscopy and scanning electron microscopy (SEM) of polished specimens as well as SEM of failed tensile and fatigue samples.

RESULTS AND DISCUSSION

Reaction Zone Evaluation

Ti-6Al-4V Matrix. Figures 3-7 illustrate the as-consolidated reaction product for each of the composite systems evaluated. In Figures 3a, 3b, and 3d it can be seen that the reaction zone formed between the Ti-6Al-4V matrix and the SiC, Borsic, and SCS-6 fibers, respectively, consists of a uniform layer of intermetallic compounds approximately 0.5 μm thick. In each of these instances the reaction zone is formed by the interdiffusion of silicon and carbon atoms with titanium atoms. Brewer et al [14] have shown by X-ray diffraction that the compounds formed for a similar system (Borsic/Ti-3Al-2.5V) include: TiSi, Ti_5Si_3, TiC, TiB and TiB_2.

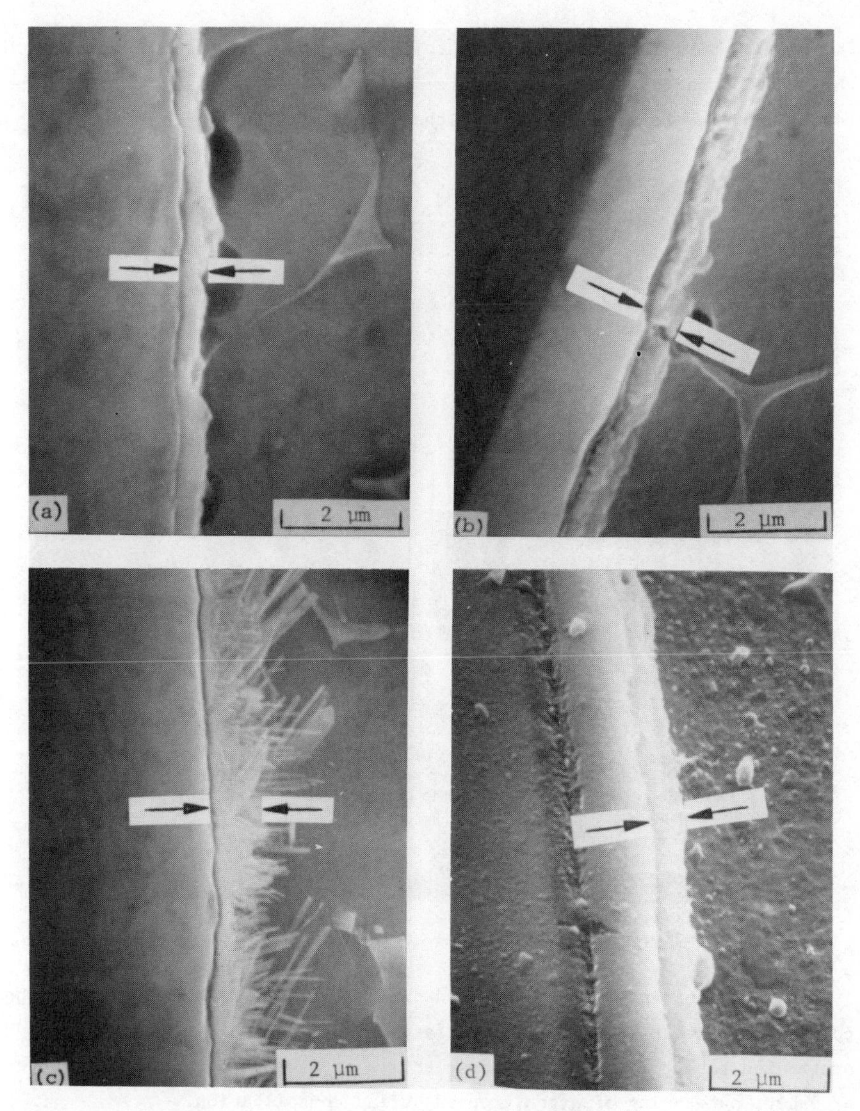

Figure 3. As Fabricated Reaction Zones for (a) SiC, (b) Borsic, (c) B_4C/B and (d) SCS-6 Fibers in a Ti-6Al-4V Matrix.

The reaction zone formed for the B_4C/B/Ti-6Al-4V system (Figure 3c) consists of a uniform inner zone, and a needle-like outer region. The diffusion mechanism enabling the reaction to occur is a one-way diffusion of Boron atoms into the titanium matrix. Pailler et al [15] using microprobe analysis have identified the uniform inner layer as TiB_2 while the outer needle-like layer is TiB (These regions can be more distinctly seen in the fracture sample of

Figure 4). Both compounds are non-stoichiometric with vanadium entering into the reaction while aluminum is rejected into the matrix ahead of the advancing reaction zone. It is also probable that TiC is present in the reaction zone.

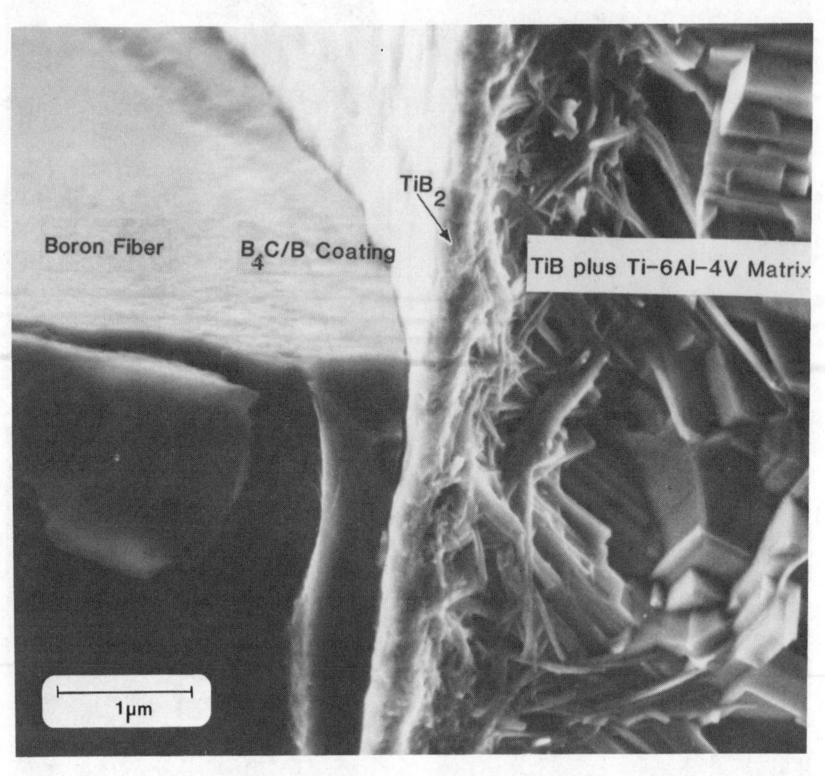

Figure 4. B_4C/B/Ti-6Al-4V Fracture Surface Depicting Elements of Composites

In many instances it was noticed that small, half-spherical voids appeared at the reaction zone/matrix interface (Figure 3a, 3b). These were attributed to either concentrated selective attack by the etchant or, more probably, outgassing of argon trapped during consolidation.

Other Matrices. Figure 5-7 depict the fiber spacing, degree of bonding, and reaction zone morphology for Borsic filament in CORONA 5, Beta III, and the Ti-Mo alloys, respectively. In Figures 5c and 5d it can be seen that the Borsic/CORONA 5 composites have completely consolidated at 905°C (1665°F) (although some debonding has occurred at the matrix/matrix interface most probably due to foil surface contamination). The uniform reaction layer formed has been reduced below that formed with a Ti-6Al-4V matrix at higher temperatures. At 850°C (1565°F) the composite is completely consolidated

(Figure 5a and 5b) and the reaction layer thickness is only one-half (0.25 μm) that typically exhibited with a Ti-6Al-4V matrix (here no debonding due to foil surface contamination is evident).

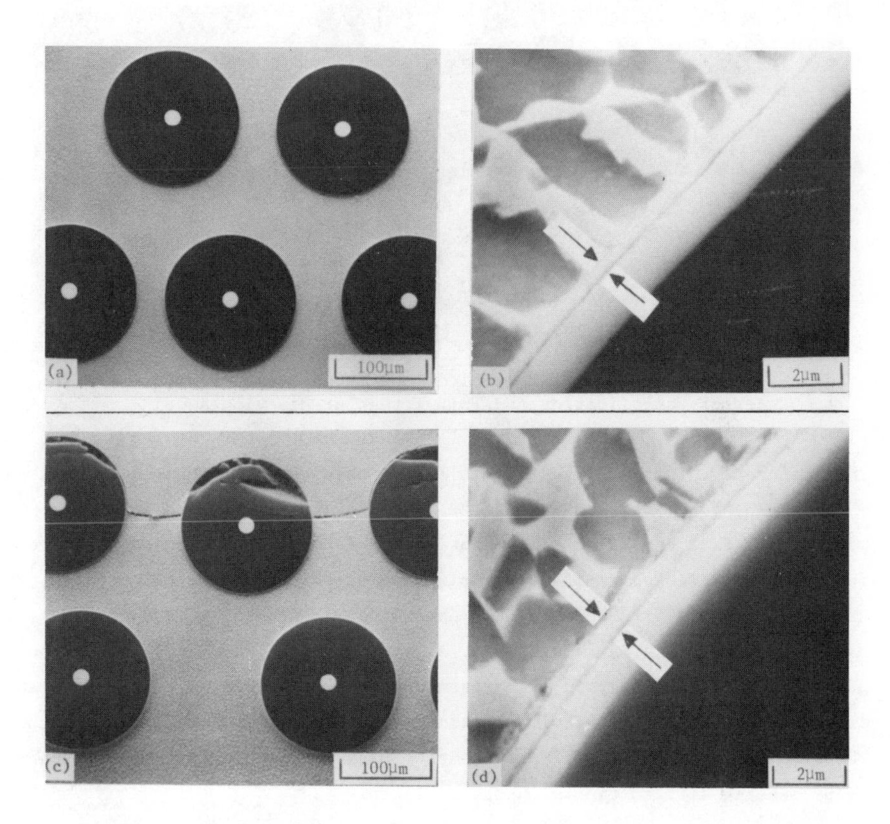

Figure 5. Borsic/CORONA 5 Composites fabricated at (a&b) 850°C (1565°F) and (c&d) 905°C (1665°F) illustrating the degree of consolidation and resulting reaction zone.

Since the beta transus of Beta III occurs at 745°C (1375°F) it was desirable to compact below this temperature to avoid excessive grain growth.[7,8] Preliminary tensile work indicated that with the maximum available platen pressure, and using a temperature of 730°C (1350°F)full compaction should take about 16 hours. Consequently a 24 hour pressing cycle was used. However the compacted material was slightly less than 100% compacted. The reaction product in this case was very thin and irregular (Figure 6b).

Figure 7a-f show the results of the Ti-Mo evaluation. In Figures 7a, 7b it can be seen that the control sample (Borsic/CP-Ti) was completely consolidated (except where close fiber spacing prohibited) and that the uniform

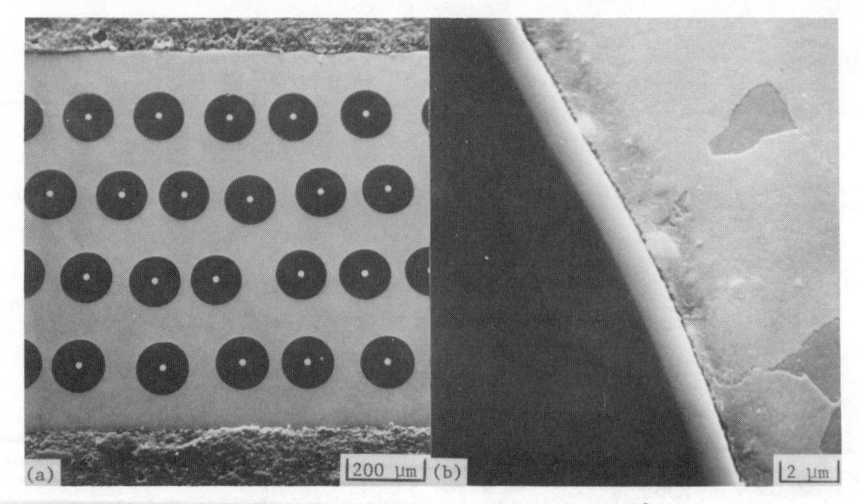

Figure 6. Borsic/Beta III Composite Fabricated at 730°F) Illustrating the Degree of Consolidation and Thin Irregular Reaction Zone

reaction zone thickness (~0.7 μm) is larger than that for a Ti-6Al-4V matrix which is in agreement with Schmitz et al [16]. Figures 7c, 7d are photo micrographs from the Ti-Mo alloy showing a reaction zone which has been reduced below CP-Ti and Ti-6Al-4V matrix levels (to ~0.35 μm). Figures 7e, 7f show that for the Ti-20Mo alloy the composite was not fully consolidated (because of an increase in flow stress, a result of both the increased Mo content per se and the larger grain size often observed in alloys of this beta stability [17]) but that the uniform reaction layer has been reduced still further. The results from the higher Mo contents (30% and 40%) are not reported because of processing difficulties with these alloys.

Reaction Kinetics. Figures 8a-d show the reaction zone growth of all four fibers in a Ti-6Al-4V matrix over the temperature range 595°C - 955°C (1100°F - 1750°F). At both 595°C (1100°F) and 760°C (1400°F) the curve asymptotically approach a limit with the SCS-6 fiber yielding a slightly smaller reaction product. At 870°C (1600°F) and 955°C (1750°F) the reaction zone thickness increased rapidly, with the SCS-6 reaction zone eventually becoming the largest.

The standard deviation in reaction zone thickness for each fiber at the corresponding times and temperatures is indicated in the enclosure. The largest standard deviation predominantly occurred for the SiC fiber (especially at the higher temperatures) indicating significant variation in reaction zone thickness for this system. This variation was both a function fiber location and circumferential position (around a particular fiber) and has been

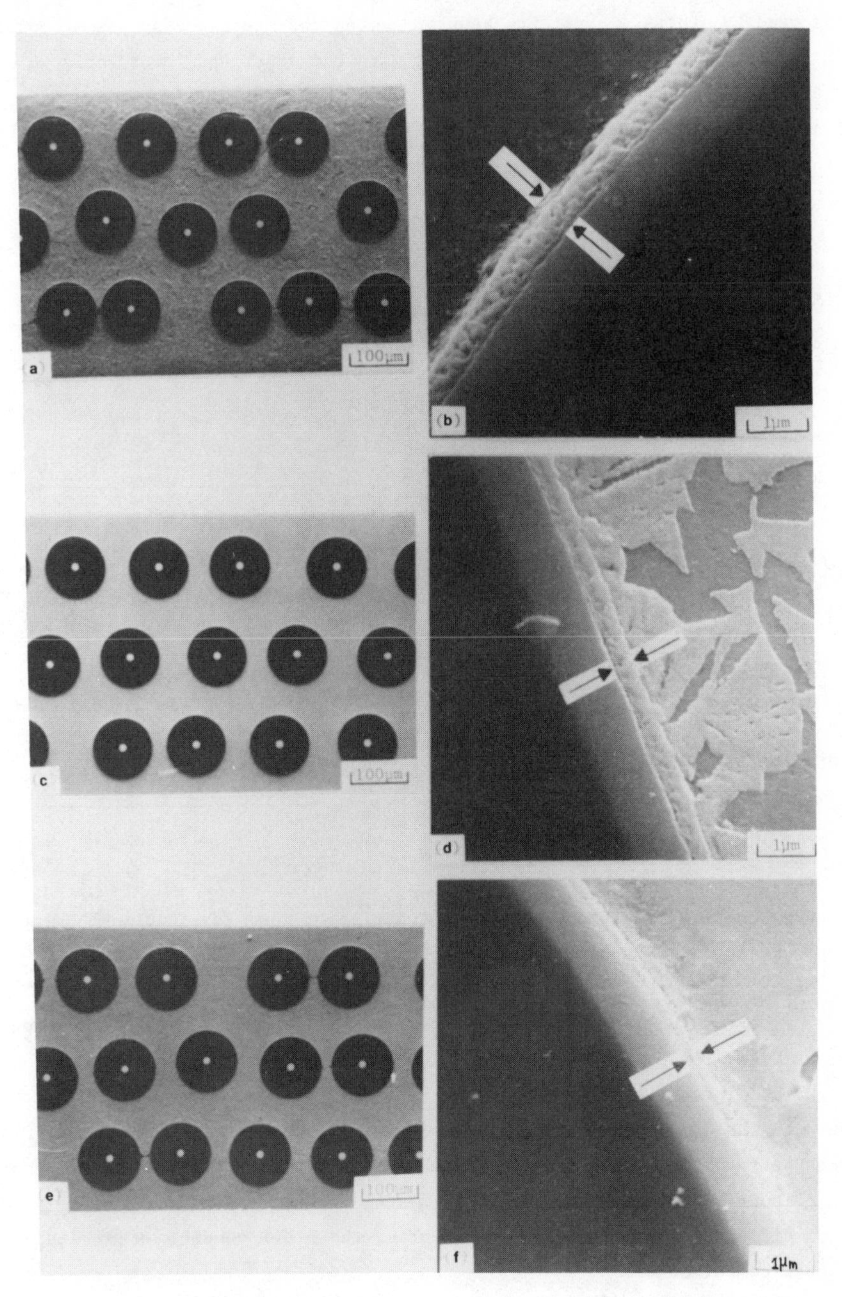

Figure 7. As fabricated (a&b) CP-Ti, (c&d) Ti10Mo, and (e&f)Ti2oMo composites showing the degree of consolidation and resulting reaction zone.

observed by others.[18] It has been suggested that this may be due at least in part to the matrix being in contact with the upper and lower surfaces of the fiber longer during consolidation since they are the first to come in contact with the matrix. However, this does not explain the increased variation for the SiC system relative to the other fibers.

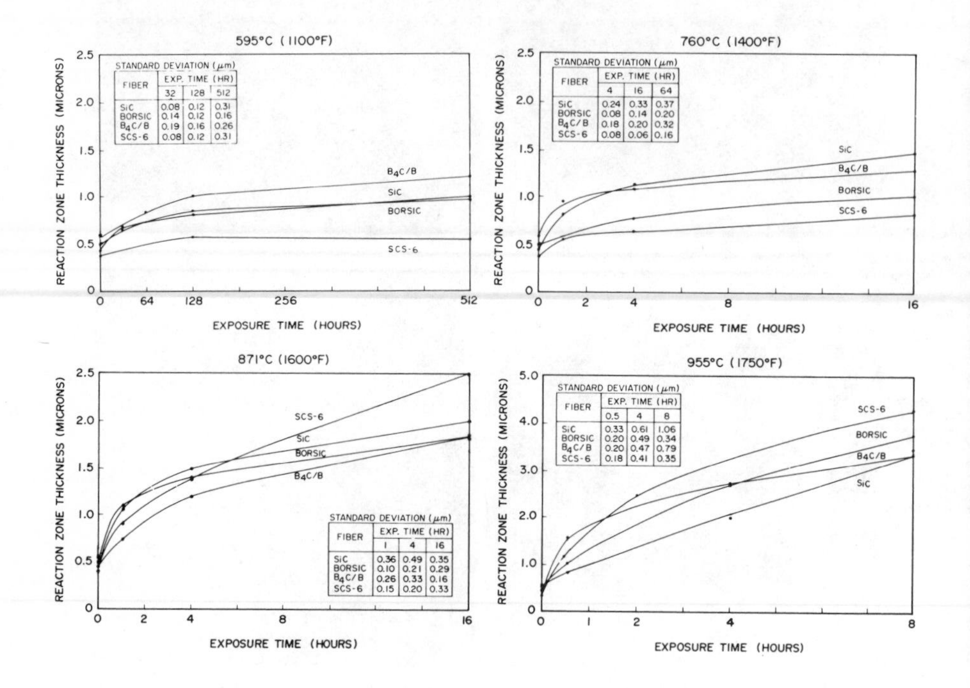

Figure 8. Reaction Zone Growth as a Function of Post-Fabrication Thermal Exposure for SiC, Borsic, B_4C/B and SCS-6 Fibers in a Ti-6Al-4V Matrix.

<u>Fiber Evaluation</u>. Figures 9a-d and 10a-d illustrate the loop and tensile properties, respectively, for each of the fibers in various conditions. As previously noted, these conditions were chosen to follow the fiber properties through the composite fabrication process. Two points should be noted: 1) Because the SCS-6 only recently became available, this fiber was tested only in selected conditions and 2) fibers tested in the digested conditions were from different lots than those representing the other conditions precluding direct comparison.

Since each condition corresponds to only 25-50 tests per fiber type, and the results showed considerable scatter, statistically significant conclusions cannot be drawn. However the following trends were observed:

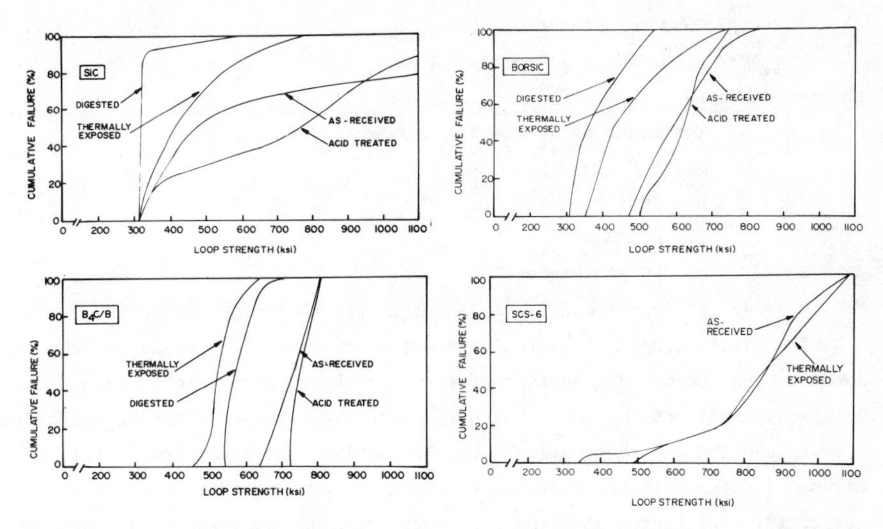

Figure 9. Loop Strength of SiC, Borsic, B_4C/B and SCS-6 Fibers in Various Conditions.

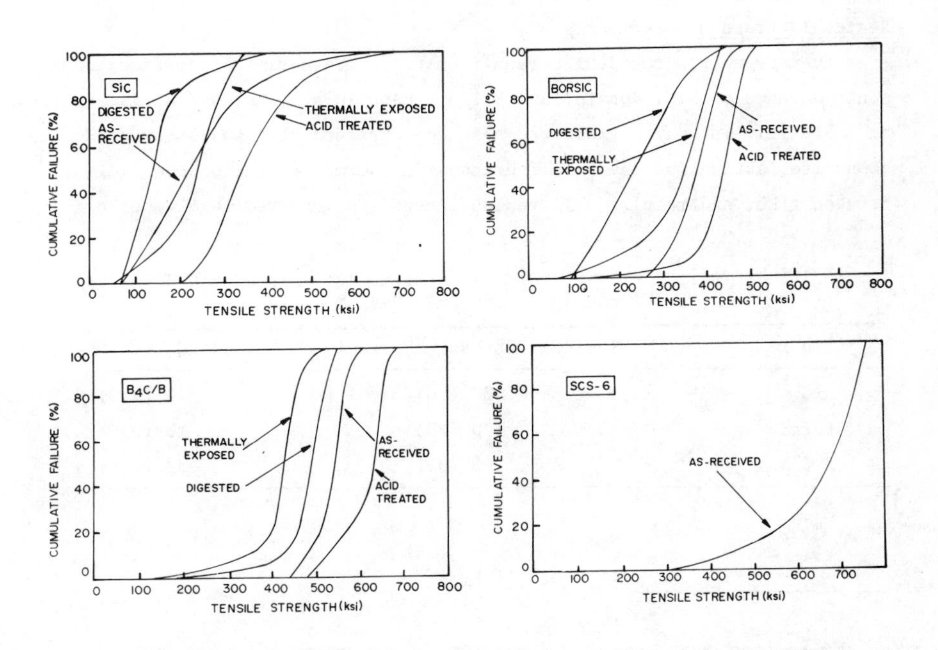

Figure 10. Tensile Strength of SiC, Borsic, B_4C/B and SCS-6 Fibers in Various Conditions.

1) Good correlation between loop and tensile strengths.

2) Boron-based filaments (Borsic, B_4C/B) degrade with fabrication thermal exposure.

3) The order of increasing filament strength is:

a) SiC (lowest)

b) Borsic

c) B_4C/B

d) SCS-6 (highest)

4) An improvement in filament strength with acid exposure.

The first trend appears real since the two sets of data vary in the same manner. The thermal degradation of the Boron-based materials is significant in that it indicates a problem with the stability of this filament aside from any degradation due to reaction with the matrix. The third trend will be discussed further in the section on mechanical properties of the composite materials. The fourth of these trends was somewhat unexpected in that this condition was tested to determine if the acid used for digestion had any degrading effects on the filaments. The increase in strength was apparently due to removal of surface flaws by the etchant.

Residual Stress Measurement

Average matrix residual stress results from a series of surface* measurements on each of SiC, Borsic, and B_4C/B fibers in a Ti-6Al-4V matrix are shown in Table III. It may be inferred that the observed stresses occurred during composite fabrication due to differences in thermal expansion coefficients between fiber and matrix. All residual stresses measured were tensile in

Table III. Surface Residual Stresses for SiC, Borsic and B_4C/B Fibers in a Ti-6Al-4V Matrix

Fiber	Residual Stress, MPa (ksi)	Standard Dev, MPa (ksi)
SiC	480 (70)	30 (4.0)
Borsic	170 (25)	18 (2.6)
B_4C/B	265 (39)	32 (4.6)

* Subsurface measurements were made to ensure that surface residual stresses did not merely reflect some surface finishing treatment. Subsurface residual stresses were of the same order of magnitude as surface values.

nature with the SiC/Ti-6Al-4V composites exhibiting the highest values, the Borsic/Ti-6Al-4V composite had the lowest values, while the B_4C/B/Ti-6Al-4V composite had intermediate values. These relative magnitudes are consistent with thermal expansion coefficient differences (Table IV). Residual stresses of these magnitudes are considered significant and might be expected to adadversely influence fatigue crack initiation and propagation resistance in the matrix (see tensile and fatigue properties).

Table IV. Coefficients of Thermal Expansion

Material	Coefficients of Theramal Exp (K^{-1})
Ti-6Al-4V	1.1×10^{-5}
SiC	4.8×10^{-6}
Boron	8.3×10^{-6}
B_4C	4.8×10^{-6}

Toth [19] had previously shown that the fatigue strength of SiC/Ti-6Al-4V could be improved by exposing specimens to liquid nitrogen prior to testing. This improved fatigue performance was rationalized by calculations which indicated that matrix residual stress levels would be favorably influenced by such a treatment. In the current investigation, a series of residual stress measurements were made after liquid nitrogen treatment to determine the validity of these concepts. Results for each of the materials after liquid nitrogen treatment are shown in Table V. The treatment resulted in a residual stress reduction of about 140 MPa (20 ksi) for the SiC/Ti-6Al-4V composite which was statistically significant at the 95% probability level. The liquid nitrogen treatment resulted in only a 40 MPa (6 ksi) residual stress reduction for the Borsic/Ti-6Al-4V system. This difference was small and of the same order of magnitude as the standard deviation and is therefore of questionable significance. Results for the B_4C/B/Ti-6Al-4V composite indicated essentially no change in residual stress level with liquid nitrogen treatment.

Table V. Surface Residual Stresses for SiC, Borsic, and B_4C/B Fibers in a Ti-6Al-4V Matrix After Liquid Nitrogen Treatment

Fiber	Residual Stress, MPa (ksi)	Standard Dev, MPa (ksi)
SiC	350 (51)	28 (4.1)
Borsic	130 (19)	33 (4.8)
B_4C/B	260 (38)	31 (4.5)

Based on these residual stress measurements it is reasonable to conclude that matrix residual stresses of relatively high magnitude relative to matrix yield strength do exist in as-fabricated composites. In combination with results reported by Toth [19] it may also be concluded that these residual stresses negatively influence composite fatigue properties. Attempts to improve mechanical properties by liquid nitrogen treatment would not be expected to be successful in either the B_4C/B/Ti-6Al-4V or Borsic/Ti-6Al-4V composite materials because the observed reduction in matrix residual stresses was minimal.

Mechanical Properties

Tensile Properties. Table VI lists the tensile properties of all composite materials evaluated. The Ti-6Al-4V material was used as control and had the same thickness and had been given the same thermal treatment as all as-fabricated composites. Testing was conducted on each fiber/Ti-6Al-4V combination in the as-fabricated condition and after thermal exposure. These exposures were chosen to parallel those previously used in the reaction kinetics evaluation.

The longitudinal tensile properties of the SiC/Ti-6Al-4V and Borsic/Ti-6Al-4V composites were equivalent to or less than those of the monolithic Ti-6Al-4V. Both of these composite materials were relatively insensitive to additional thermal exposure. The B_4C/B/Ti-6Al-4V composites had longitudinal tensile strengths very similar to or slightly greater than the monolithic Ti-6Al-4V. The DWA produced materials displayed better thermal stability than that produced by AVCO in the present work. The SCS-6/Ti-6Al-4V composite produced by far the highest as-fabricated longitudinal tensile properties of all of the composite materials evaluated. Unfortunately, these properties degraded even with the more moderate thermal exposures. It should be noted that other researchers have not seen this degradation in composites produced from another batch of this fiber [20]. The longitudinal tensile properties of both the Borsic/Beta III and Borsic/Ti-Mo composites were relatively low.

The transverse tensile properties of all the composites utilizing a Ti-6Al-4V matrix were very similar and were virtually unaffected by subsequent thermal exposure. Somewhat surprisingly the Borsic/Ti-10Mo combination yielded a high transverse strength (approaching 550 MPa (78 ksi)) relative to the other materials.

Longitudinal Young's modulus values were approximately the same for all composites evaluated, approaching rule-of-mixture (ROM) values although both the SiC/Ti-6Al-4V and SCS-6/Ti-6Al-4V composites showed a slight degradation with thermal exposure.

Table VI. Room Temperature Tensile Properties

1 ksi = 6.89 MPa/1 Msi= 6.89 GPa

MATERIAL	CONDITION	UTS (LONG) (Ksi)	UTS (TRANS) (Ksi)	E (LONG) (Msi)
Ti-6Al-4V*	As Rec'd	129	129	-
"	595°C(1100°F)/512 Hr	142	135	17.5
"	760°C(1400°F)/64 Hr	145	138	15.5
"	900°C(1650°F)/5 Hr	148	144	18.5
"	870°C(1600°F)/16 Hr	140	136	16.2
"	955°C(1750°F)/8 Hr	129	127	15.3
Borsic/Ti-6Al-4V**	As Rec'd	130	-	-
"	595°C(1100°F)/512 Hr	133	53	30.0
"	760°C(1400°F)/64 Hr	118	47	30.0
"	900°C(1650°F)/5 Hr	120	42	29.1
"	870°C(1600°F)/16 Hr	108	49	29.5
"	955°C(1750°F)/8 Hr	110	46	-
SiC/Ti-6Al-4V**	As Rec'd	119	-	-
"	595°C(1100°F)/512 Hr	109	55	32.6
"	760°C(1400°F)/64 Hr	113	42	27.5
"	900°C(1650°F)/5 Hr	112	55	28.5
"	870°C(1600°F)/16 Hr	104	45	29.0
"	955°C(1750°F)/8 Hr	102	45	-
B_4C/B/Ti-6Al-4V**	As-Rec'd	142	44	-
"	595°C(1100°F)/512 Hr	143	51	28.9
"	760°C(1400°F)/64 Hr	152	45	29.4
"	900°C(1650°F)/5 Hr	155	50	31.2
"	870°C(1600°F)/16 Hr	137	49	30.3
"	955°C(1750°F)/8 Hr	133	42	-
B_4C/B/Ti-6Al-4V*	As-Rec'd	153	45	-
"	595°C(1100°F)/512 Hr	150	47	30.0
"	760°C(1400°F)/64 Hr	127	45	31.7
"	900°C(1650°F)/5 Hr	128	46	33.2
"	870°C(1600°F)/16 Hr	105	42	30.0
"	955°C(1750°F)/8 Hr	98	44	30.7
SCS-6/Ti-6Al-4V*	As-Rec'd	211	47	-
"	595°C (1100°F)/512 Hr	151	49	34.8
"	760°C(1400°F)/65 Hr	132	44	33.8
"	900°C(1650°F)/5 Hr	121	53	30.9
"	870°C(1600°F)/16 Hr	98	48	32.8
"	955°C(1750°F)/8 Hr	125	49	28.5
Borsic/Beta III ***	As-Rec'd	124	-	-
Borsic/CP Ti ***	As-Rec'd	90	47	-
Borsic/Ti-10Mo ***	As-Rec'd	106	48	-

* Fabricated by Avco
** Fabricated by DWA
*** Fabricated by Amercom

The two most likely possibilities for the degradation of the composite longitudinal tensile properties below ROM levels are: (1) the brittle intermetallic reaction zone exceeds some critical size and the resulting stress intensity produced on the adjoining fiber by a crack in the reaction zone is high enough to fracture it at low strength levels, or (2) the fiber strength is degraded during fabrication by either pitting of the filament surface due to chemical reaction with the matrix or due to a structural change of the filament surface from amorphous to crystalline during the fabrication thermal exposure.

The first of these mechanisms discussed by Ochiai et al [21] requires that the reaction product behave as though it were a coating on the filament surface (i.e., the reaction zone must adhere strongly to the fiber). All fractographic evidence obtained during this investigation indicates that this is not the case and in fact during fracture the reaction zone separates from the fiber and remains well bonded to the matrix (Figure 5). Additionally, the as-fabricated reaction zone of the SCS-6/Ti-6Al-4V composite is nearly equivalent in size to the other systems (some researchers have even found it to be larger [22]) yet its as-fabricated longitudinal tensile strength approaches ROM values while those values of the other composites are well below ROM. Therefore, it appears that the critical reaction zone size mechanism is not the primary mechanism causing degradation in these composites.

The second mechanism, that of fiber strength degradation, is supported by the evidence presented in Figures 11 and 12. All of the filaments were degraded in strength after being digested from the composite (SCS-6 was not evaluated in this condition) suggesting that the filament surface was damaged

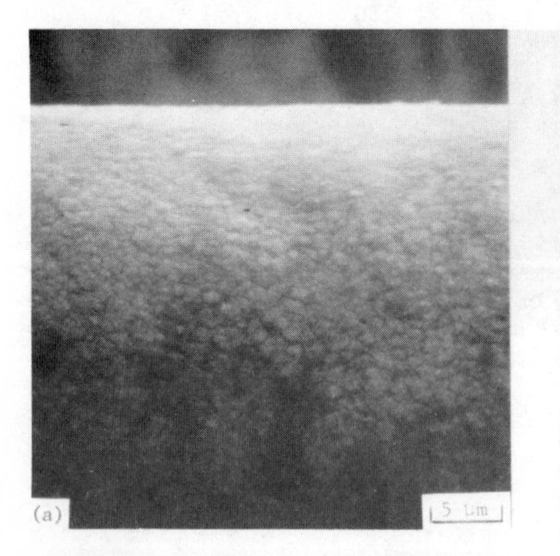

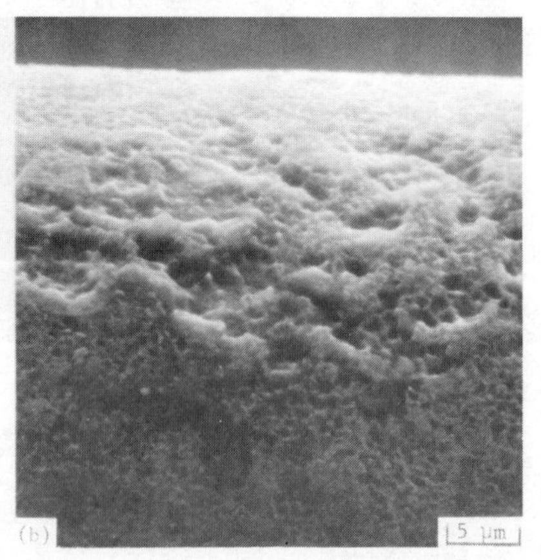

Figure 11. Surface of B_4C/B Fiber a) As-Received and b) After Digestion Indicating Surface Damage Caused by Chemical Reaction with Ti-6Al-4V Matrix.

(Figure 11). Further, the Boron-based filaments (Borsic and B_4C/B) were degraded even when exposed to the composite fabrication thermal cycle. It has been suggested that this latter effect could be due to the change in the Boron surface from an amorphous to a crystalline structure [23]. It has previously been shown that an amorphous surface can act to heal crystalline subsurfaces and improve filament strength [24]. The SiC filaments evaluated degraded to very low tensile values (860 MPa) (~125 ksi) when the protective amorphous C-rich layer is consumed during fabrication, exposing the more surface sensitive β-crystalline SiC below. These extremely low fiber strengths after fabrication indicate why the SiC/Ti-6Al-4V composite does not degrade significantly in strength after additional thermal exposure. The Borsic filament had the next lower strength level. This fiber has a crystalline SiC coating with a nodular surface (Figure 12) which can be very sensitive to pitting during reaction with the matrix and also could be degraded by the fabrication thermal cycle. The strength of this filament after digestion from the as-fabricated composite was also so low (690-1380 MPa (100-200 ksi)) that very little degradation in fiber strength with additional thermal exposure would be expected. The B_4C/B fiber was the next highest in strength level ((2760 MPa) (400 ksi) in the digested condition) and although, as in the Borsic fiber, degradation occurred due to thermal exposure, unlike the Borsic fiber which has a crystalline SiC coating, the B_4C/B fiber has an amorphous B_4C coating which may not be as sensitive to surface pitting. The SCS-6 fiber exhibited the highest as-received tensile properties of all the filaments (4135-4825 MPa (600-700 ksi)). Although this fiber was not evaluated in the digested condition, other researchers [22] have reported very little fiber strength degradation during

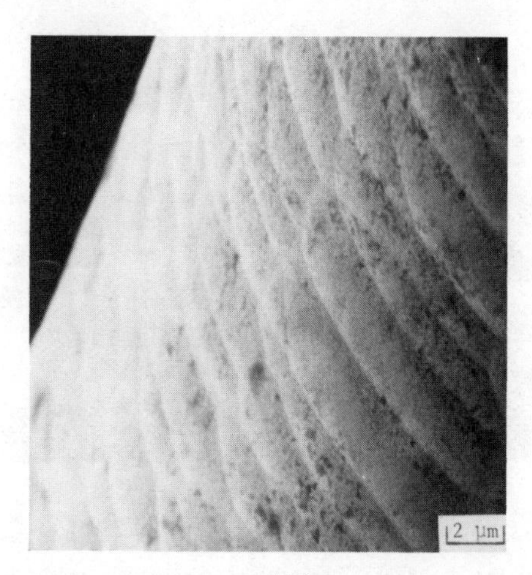

Figure 12. Surface of Borsic Fiber Showing the Nodular Topography of the Crystalling SiC Coating.

composite fabrication. Based on loop test results this fiber, which has a 2 μm amorphous silicon-rich layer that acts as a protective buffer for the crystalline SiC below, does not appear to degrade with thermal exposure. The fact that the composite strengths follow the same trend as the in-situ fiber strengths indicates that this second mechanism may predominate and that in particular, both the fiber strength and hence composite strengths are dependent on the structural make-up of the fiber.

As previously noted the transverse tensile properties of most of the composite evaluated were very similar in magnitude (310-380 MPa (45-55 ksi)) and relatively insensitive to post-fabrication thermal exposure. This insensitivity may be due to the lack of bonding at the reaction zone/fiber interface which can be most readily detected in the transverse fracture mode (Figure 13a). Only the Borsic/Ti-10Mo composite demonstrated enhanced transverse strengths (535 MPa (78 ksi)). These higher values were apparently due to an improvement in interfacial bonding which resulted in more filament splitting (Figure 13b) during fracture indicating increased support from the fiber in this case compared to the other system investigated where fracture occurs around the fiber.

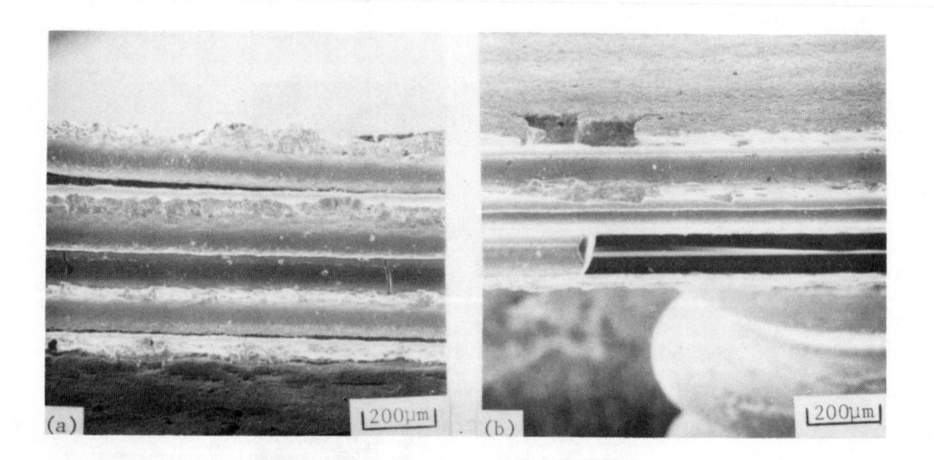

Figure 13. Transverse Fracture of a) Borsic/CP-Ti and b) Borsic/Ti-10Mo Composite Illustrating Failure Around Fibers and Fiber Splitting, Respectively.

Fatigue Properties. The smooth axial fatigue properties are shown in Figure 14 for all of the composites evaluated. Composites of B_4C/B, SiC, and SCS-6 fibers in a Ti-6Al-4V matrix all had fatigue lives greater than that of the control monolithic Ti-6Al-4V material with B_4C/B composites consistently exhibiting the longest life. Composites of Borsic fibers in matrices other than Ti-6Al-4V exhibited fatigue properties equivalent to or less than monolithic Ti-6Al-4V depending on the particular matrix material. The large amount of scatter in the SiC/Ti-6Al-4V properties (particularly the low values) may be attributed to poor composite consolidation (Figure 15). This was also the case for the Borsic/Beta III material although poor consolidation was in evidence on a much smaller scale. In all composite materials evaluated, fatigue life was degraded significantly by the 595°C (1750°F)/hr exposure.

As mentioned in the experimental procedure, the lateral edges of fatigue specimens were finished by special procedures. In previous work Cooke et al [13] found that for the B_4C/B/Ti-6Al-4V material, specimens in which

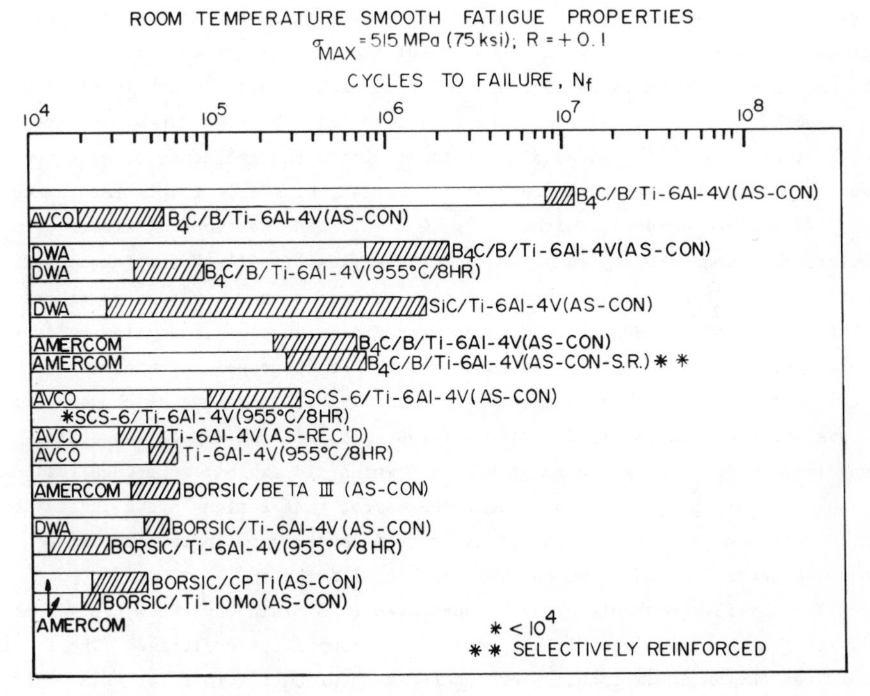

Figure 14. Room Temperature Fatigue Properties for Composites Fabricated by Various Manufacturers (The Cross-Hatched Areas Indicate Range of Values).

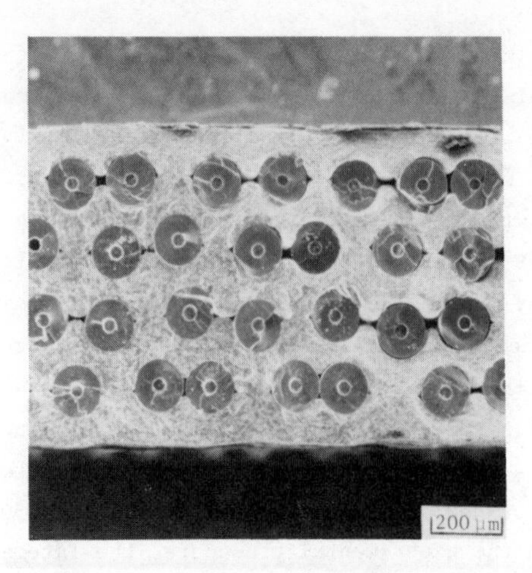

Figure 15. Fracture Surface of SiC/Ti-6Al-4V Composite Showing Incomplete Consolidation.

edges were ground and metallographically polished after sectioning by shearing exhibited fatigue lives more than five times greater than specimens with as-sheared edges and more than twenty-five times greater than specimens with ground and shot peened edges. These results suggest that mechanical damage to specimen edges degrades fatigue life by promoting early crack initiation at exposed fibers and vacated fiber locations near specimen edges, in agreement with other work [25]. In an attempt to eliminate the influence of specimen edges on fatigue life in the current investigation, a selectively reinforced B_4C/B/Ti-6Al-4V composite with 1.2 cm (0.5 in.) wide sections of fiber reinforcement separated by equal widths of base-line Ti-6Al-4V was fabricated. Fatigue specimens were machined from the panel in such a way that no fibers were exposed at the edges. The edges were ground and longitudinally polished to a 600 grit finish before fatigue testing. Fractographic examination after fatigue testing revealed that no fatigue cracks had initiated at the specimen edges yet the observed fatigue lives (Figure 14) did not exceed those of specimens which were fully reinforced and which did exhibit some edge fatigue crack initiation. It was concluded, therefore, that fatigue crack initiation at specimen edges per se does not limit fatigue strength in the composite once mechanical damage from cutting has been eliminated.

Two possible mechanisms may be suggested to explain why the B_4C/B/Ti-6Al-4V results were consistently the best of all composites evaluated. The first involves the unique morphology of the reaction product formed for this system. As previously noted the B_4C/B reaction product consists of two distinct regions a uniform inner layer and an acicular outer layer (Figure 4) while the reaction zones for all other systems consist of a single uniform layer. Fractographic analysis suggests that initiation in all of these systems occurs in the immediate proximity of the fibers probably in the brittle reaction zone which has been shown to be mechanically well-bonded to the matrix. The stress intensity associated with any crack formed in the reaction zone would be a function of the reaction zone width. Therefore, if fatigue initiation occurs in the inner uniform layer adjacent to B_4C/B fibers one would expect that the crack would propagate more slowly than in other composites which exhibit wider uniform reaction zones.

A second possible mechanism relates to the structure of the fiber, itself, and its resistance to fracture. The Borsic and B_4C/B fibers are essentially the same except that the former has a crystalline 2 μm thick SiC coating while the latter has a 6 μm thick amorphous B_4C coating. Figure 16, 17 show fatigue fracture surfaces for the Borsic and B_4C/B systems, respectively. It can be seen that for the Borsic system the primary mode of fiber fracture is from the fiber edge while for the B_4C/B system, fracture is from the fiber core. It seems reasonable to assume that when a propagating matrix fatigue crack impinges on the nodular crystalline SiC coating of a Borsic fiber, the fiber fractures immediately. When a similar crack impinges on the thick amorphous coating of a B_4C/B fiber, fracture does not occur immediately. Apparently the crack must surround the fiber before fiber fracture occurs, until such an event occurs the fiber retards crack propagation within the matrix. This could also account for the intermediate fatigue values obtained for the SiC/Ti-6Al-4V and SCS-6/Ti-6Al-4V systems both of which have small amorphous layers on top of crystalline substrates. Either of these two mechanisms could be acting independently or as a combined effect during composite fatigue.

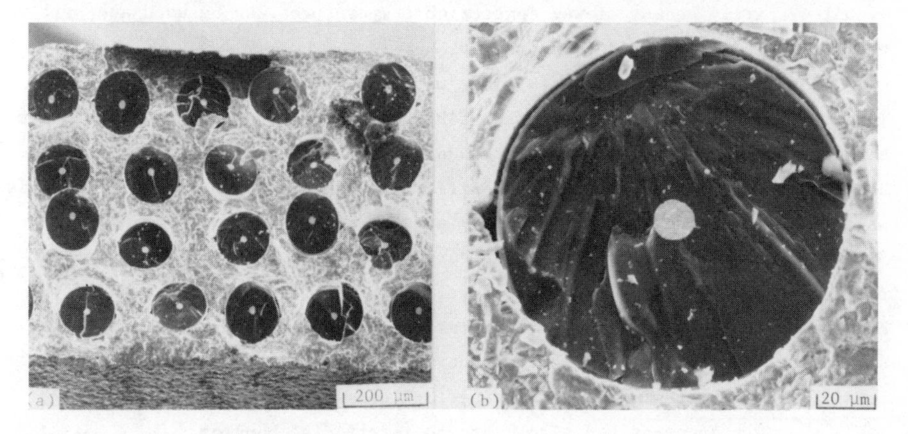

Figure 16. Borsic/Ti-6Al-4V Composite Showing a) Failed Fatigue Surface and b) Fiber Edge Initiation.

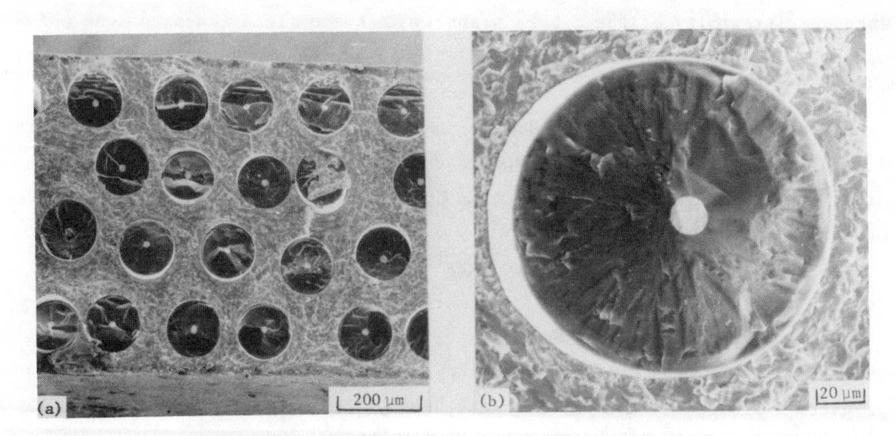

Figure 17. B_4C/B/Ti-6Al-4V Composite Showing a) Failed Fatigue Surface and b) Internal Fiber Initiation.

CONCLUSIONS

1. Of the filaments investigated SCS-6 demonstrated the most consistent and highest fiber tensile and loop strengths followed by B_4C/B, Borsic, and SiC.
2. Both the B_4C/B and Borsic fibers were degraded in tensile and loop strength by exposure to a simulated fabrication thermal cycle, probably because the amorphous surface layer became crystalline during this exposure.
3. A reduction in the reaction zone width can be achieved with other matrices (Ti-10Mo, Beta III, and CORONA 5) compared to that formed in fiber reinforced Ti-6Al-4V partly because lower compaction temperatures can be used and partly because of the alloy additions present.
4. SCS-6/Ti-6Al-4V was the only fiber/matrix combination to yield near ROM longitudinal tensile strengths after fabrication. However degradation to lower strength occurred even with moderate additional thermal exposures. It should be noted that other workers using a different batch of SCS-6 did not find this same level of degradation on secondary exposure.
5. Longitudinal tensile strength appears to be mainly a function of in-situ fiber strength and not a function of critical reaction zone size.
6. Transverse tensile strength is relatively constant from system-to-system and insensitive to post-fabrication thermal exposure with the exception of the Borsic/Ti-10Mo system which had a superior transverse strength (535 MPa (78 ksi)) that was apparently due to improved interfacial bonding.
7. Longitudinal smooth axial fatigue strength was influenced by specimen edge preparation; with grinding followed by metallographic polishing being the recommended procedure to avoid edge damage.

8. The SiC/Ti-6Al-4V system exhibited significant matrix tensile residual stresses (480 MPa (70 Ksi)) after fabrication which might adversely affect fatigue properties, however, this stress could be reduced to 350 MPa (50 ksi)) by subjection to cryogenic temperatures.

9. B_4C/B/Ti-6Al-4V demonstrated the best fatigue life, while reinforcement of any matrix with Borsic filament resulted in the worst.

10. Fatigue life appears to be a function of reaction zone width and/or fiber structure.

ACKNOWLEDGEMENTS

The authors would like to express their appreciation to R. Bacon, R. Brodecki, C. Cooke, A. Houston and G. Lovell for their help with the experimental work. The assistance of Dr. C.H. Hamilton in developing the tensile flow stress data for the Beta III alloy is gratefully acknowledged. In addition helpful discussions with D. Beeler, T. Cordell, Dr. J. Cornie, and T. Steelman are appreciated.

REFERENCES

1. Private Communication, J. Cornie, AVCO, Feb. 1982.

2. F. H. Froes and W. T. Highberger, "Synthesis of CORONA 5 (Ti-4.5Al-5Mo-1.5 Cr)," JOM, Vol. 32, No. 5, May 1980, pp. 57-64.

3. V. C. Petersen, F. H. Froes and R. F. Malone, "Metallurgical Characteristics and Mechanical Properties of Beta III, A Heat-Treatable Beta Titanium Alloy, "Proceedings of the Second International Titanium Conference, Cambridge, MA, 2-5 May 1972, pp. 1969-1980.

4. F. H. Froes, C. F. Yolton, J. C. Chesnutt and C. H. Hamilton, "Microstructural Control in Titanium Alloys for Superplastic Behavior," Conference on Forging and Properties of Aerospace Materials, Leeds, England, 5-7 Jan. 1977, Proceedings pp. 371-398.

5. A. G. Metcalfe, "Interaction and Fracture of Ti-B Composites," Journal of Composite Materials, Vol. 1, pp. 356-365 (October 1967).

6. M. J. Klein, M. L. Reid and A. G. Metcalfe, "Compatibility Studies for Viable Titanium Matrix Composites," AFML-TR-69-242 (October 1969).

7. B. B. Rath, F. J. Lederich, C. F. Yolton and F. H. Froes, "Recrystallization and Grain Growth in Metastable Beta III Titanium Alloy," Met. Trans. A, Vol. 10A, Aug. 1979, pp. 1013-1019.

8. F. H. Froes and C. F. Yolton, "Effect of Chemistry Variations on the Recrystallization and Grain Growth Behavior of Beta III (Ti-11.5Mo-6Zr-4.5Sn)," To be published.

9. F. H. Froes, J. C. Chesnutt, C. F. Yolton, C. H. Hamilton and M. E. Rosenblum, "Superplastic Forming Behavior of CORONA 5 (Ti-4.5Al-5Mo-1.5Cr)," Proceedings of the Fourth International Conference on Titanium, Kyoto, Japan, 19-22 May, 1980, pp. 1025-1031.

10. J. F. Dolowy, B. A. Webb and W. C. Harrigan, "Fiber Reinforced Titanium Composite Materials, "Enigma of the 80's: Environment, Economics, and Energy, Vol. 24, Book 2, 1979, published by SAMPE, pp. 1443-1450.

11. F. H. Lorenz, T. E. Steelman, and W. D. Padian, "Selective Reinforcement of Low-Cost Titanium Components," AFWAL-TR-80-4014 (March 1980).

12. SAE J784A, "Residual Stress Measurements by X-ray Diffraction," (August 1971).

13. C. M. Cooke, J. T. Cammett, D. E. Eylon and P. R. Smith, "Edge Preparation of Titanium Matrix Composite Specimens for Mechanical Testing," to be published as proceedings of the 14th technical meeting of the IMS, July 19-22, 1981.

14. S. D. Brewer, D. R. Tenney, V. B. Rao, C. R. Houska and J. Unnam, "Interfacial Reactions in Borsic/Ti-6Al-4V Composite," proceedings of the AIME symposium entitled "Physical Metallurgy and Composites," Oct 1978, pp. 347-349.

15. R. Pailler, M. Lahaye, J. Thebault and R. Naslain, "Chemical Interaction Phenomenon at High Temperature Between Boron Fibers and Titanium Metal (or TA6V alloy)," proceedings of TMS-AIME symposium entitled "Failure Modes of Composites IV". Chicago, ILL, Oct. 1977.

16. G. K. Schmitz, J. J. Klein, M. L. Reid and A. G. Metcalfe, "Compatibility Studies for Viable Titanium Matrix Composites," AFML-TR-72-34, (January 1972).

17. F. H. Froes, D. Moracz, C. F. Yolton, J. P. Hirth and R. H. Ondercin, "Recrystallization and Grain Growth in Rich Metastable Beta Titanium Alloys," To be published.

18. Private Communication, C. H. Rhodes, Rockwell International, June 1981.

19. Private Communication, I. J. Toth, TRW Inc., March 1981.

20. Private Communication, T. E. Steelman, Rockwell International, February 1982.

21. S. Ochiai and Y. Murakami, "Tensile Strength of Composites with Brittle Reaction Zones at Interfaces," Journal of Material Science, Vol. 14, pp. 831-840, (1979).

22. Private Communication, T. E. Steelman, Rockwell International, November 1981.

23. Private Communication, F. Wawner, University of Virginia, June 1981.

24. Private Communication, T. E. Steelman, Rockwell International, December 1980.

25. R.T. Bhatt and H.H. Grimes "Fatigue Behavior of SiC-Reinforced Titanium Composites." A paper presented at the ASTM Conference, Fatigue of Fibrous Composite Materials." San Francisco, California, 22-23 May 1979.

THE EFFECTS OF HOT ROLLING ON THE MECHANICAL PROPERTIES OF SiC-REINFORCED 6061 ALUMINUM*

W.C. Harrigan, Jr., G. Gaebler, E. Davis and E.J. Levin

DWA COMPOSITE SPECIALTIES, INC.
21119 Superior Street
Chatsworth, California 91311
USA

Abstract

Hot rolling of silicon carbide particulate reinforced 6061 Al is a desirable technique for producing sheets of this composite. This study concentrated on determining the mechanical properties of this composite after large reductions in thickness by hot rolling. Composites containing 15, 18, 25 and 30 volume percent SiC were rolled to reductions of 70 to 96%. Tensile properties of these composites were measured before and after rolling in an as-rolled (F) and heat-treated (T-6) condition. This study demonstrates that the strength and strain to failure of these composites are improved by hot rolling at levels greater than 80% reduction.

*Research supported by DARPA Task 3979 administered by Naval Sea Systems Command, Contract N00024-80-C-5637.

<u>Introduction</u>

During the past few years there has been a renewed interest in discontinuously reinforced metal-matrix composites. The primary reinforcement has been silicon carbide. In a recent study by Harrigan, et al, Ref. 1, the influence of the reinforcement level on the mechanical properties of several alloys was investigated. In general, the strength values were found to increase as the reinforcement was increased, from 20 volume percent to 40 volume percent for all six matrix alloys studied. In particular, the strength levels found for the 6061 Al matrix systems are shown in Figures 1 and 2 for as-produced and heat-treated specimens. Ultimate strengths of 70 ksi were demonstrated for samples containing 20 v/o SiC and as high as 80 ksi for samples containing 40 v/o SiC after a standard T-6 heat treatment. These properties were determined for samples taken from un-worked sheet material.

It is common practice for producers of aluminum powder metal products to hot- or cold-work the products to improve the mechanical properties (Ref. 2). In an effort to improve the properties of SiC reinforced aluminum alloys, the present investigation was initiated. This study concentrated on hot rolling of 6061 Al reinforced systems and determining the influence of

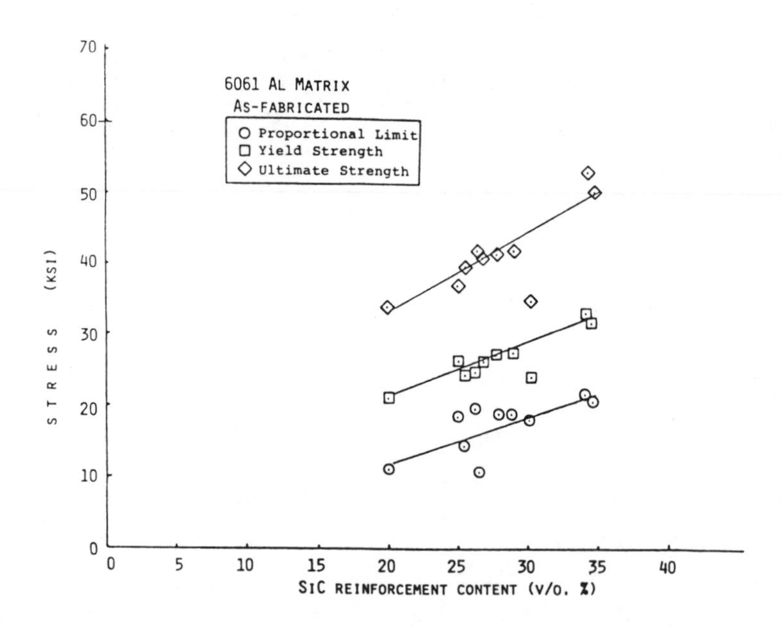

Figure 1 - Mean Strength Data for As-fabricated 6061 Al Matrix Composites as a Function of SiC Content.

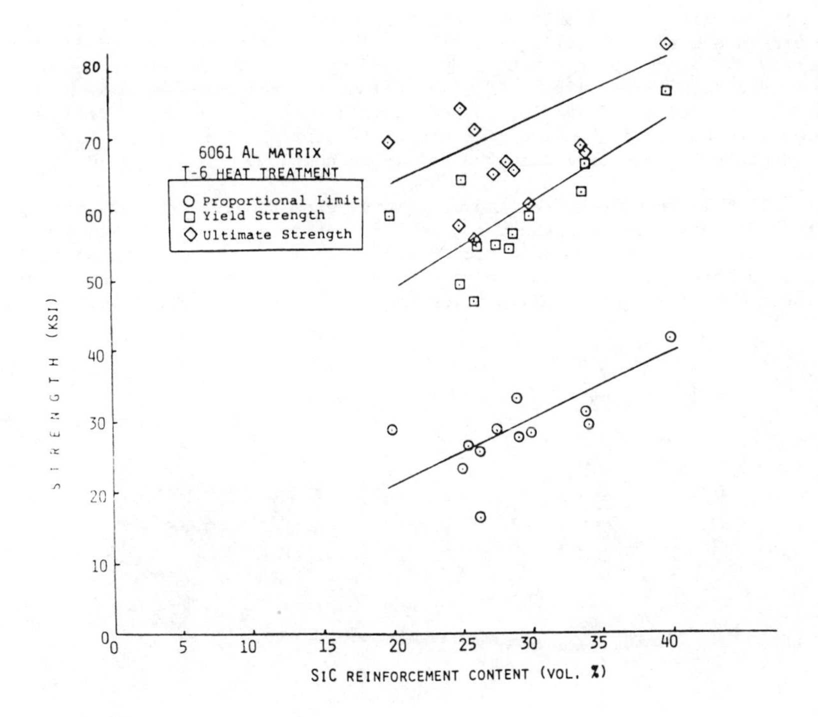

Figure 2 - Mean Strength Data for Heat-treated 6061 Al Matrix Composites as a Function of SiC Content.

hot rolling on the room-temperature tensile properties of these composites. Prior work in this area, Newborn (Ref. 3) and Harrigan (Ref. 1) investigated roll reductions between 10 and 50 percent and concluded that no improvements in strength were caused by hot rolling to a 50% thickness reduction. This study therefore emphasized the behavior of composite after greater than 50% reduction in thickness.

Experimental Procedure

Commercially available 6061 aluminum powder and silicon carbide particulate were used as starting materials. These powders were blended, compacted and sintered in a manner developed by DWA. These composites were made in the form of a billet, 10" x 12" x 1-1/2". Billets were made with reinforcement levels of 15, 18, 25 and 30 volume percent of SiC. Tensile samples were cut from the edge of these billets prior to rolling. Tests were conducted in the as-fabricated and T-6 conditions.

All billets were hot-rolled at temperatures ranging from 800 to 950°F. The roll diameter was 36 inches, and roll reductions per pass ranged from 10% to 50%. A standard roll lubricant was employed. A typical example of the rolled sheets is shown in Figure 3. Tensile coupons were sheared from the sheet material at various locations and in several orientations.

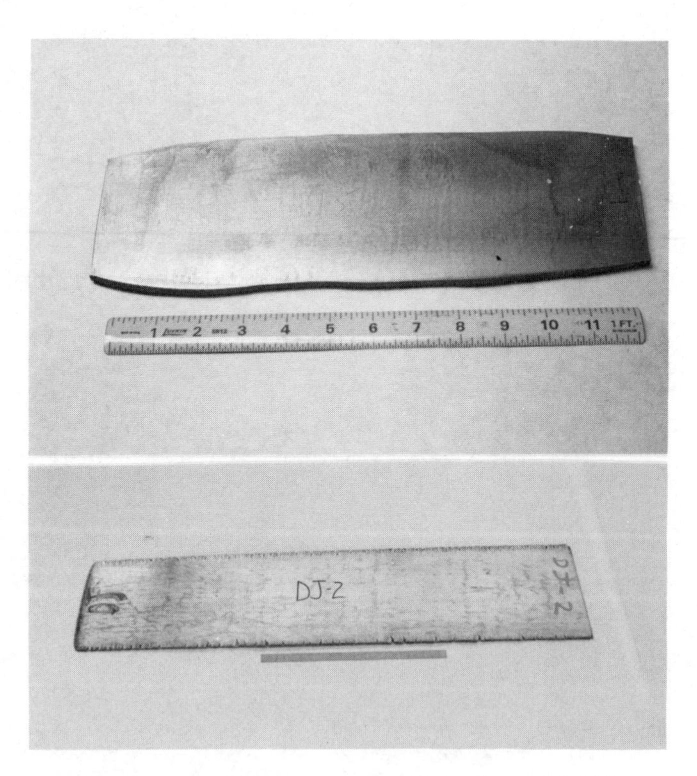

Figure 3 - Photographs of as-rolled Plates of SiC-reinforced Aluminum.

Reduced-gauge sections were machined into the coupons as shown in Figure 4.

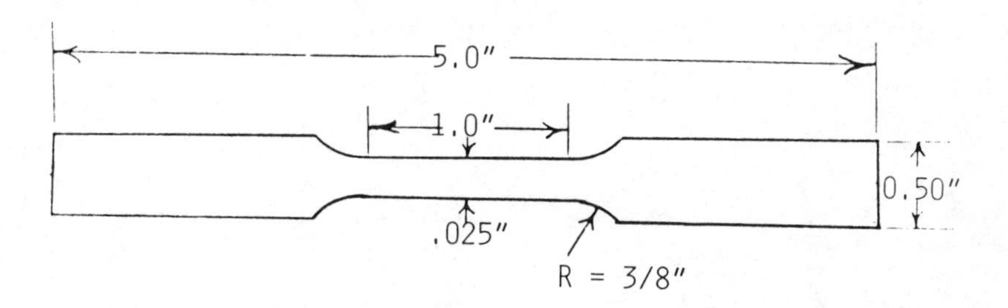

Figure 4 - Tensile Sample Configuration

Tensile tests were conducted on the samples in the as-rolled condition (F), as well, after a standard T-6 heat-treated cycle defined in Table 1.

Table I. Heat-Treat Cycle for 6061 Al-Matrix Composite

Solution Treat:	980°±10°F, 2 hours
Water Quench	
Age:	325°±10°F, 18 hours

Results and Discussion

Summaries of the tensile data for non-heat-treated (F) and T-6 heat-treated samples are listed in Tables 2 and 3. The hot rolling at all levels investigated in this study increases the yield strength and ultimate strength of non-heat-treated samples. This is shown in Figure 5, where the percent of original strength is plotted as a function of true strain. Up to 60% increase in yield strength was found after 94% reduction for both 15 and 30 v/o composites. It is difficult to fully assess the strength values for as-fabricated and as-rolled composites. This is especially true for very thin sheets that were produced after 80 to 90% reduction in thickness. The amount of cooling prior to rolling of these sheets is unknown; and consequently, the level of cold work retained is difficult to estimate. The amount of cold work will markedly influence the yield strength and have a small effect on the ultimate strength.

Table II. Mean Tensile Data for Rolled SiC-Reinforced 6061 Aluminum: F Heat-Treat Condition

SiC Content (v/o)	(Billet I.D.)	Prior to Rolling E (10^6psi)	σ_Y (ksi)	σ_{ULT} (ksi)	ϵ_F (%)	After Rolling Thickness Reduction (%)	E (10^6psi)	σ_Y (ksi)	σ_{ULT} (ksi)	ϵ_F (%)
15	(1661)	13.5	14.9	30.0	*	93.8	14.7	23.6	37.5	*
18	(1616)	12.6	17.6	37.1	*	91.0	16.2	23.3	37.6	*
25	(1601)	-	-	-	-	95.3	16.8	37.4	46.2	2.86
	(1655)	15.8	24.2	39.2	3.24	92.0	17.7	27.3	42.3	4.67
30	(1664)	16.7	22.1	36.2	*	71.8	15.9	25.1	39.1	5.23
	(1587)	16.9	26.9	43.8	3.10	64.4	17.1	28.2	46.0	3.25
						91.5	18.3	34.8	51.6	*
	(1666)	15.0	21.6	37.5	*	83.1	16.8	23.6	41.0	6.03
						96.1	18.0	33.5	43.1	4.90

*STRAIN GREATER THAN 3.2%, STRAIN TRANSDUCER SATURATED AMPLIFIER CIRCUIT.

Table III. Mean Tensile Data for Rolled SiC-Reinforced 6061 Aluminum: T-6 Heat-treated Condition

SiC Content (v/o)	(Billet I.D.)	Prior to Rolling E (10^6psi)	σ_Y (ksi)	σ_{ULT} (ksi)	ϵ_F (%)	After Rolling Thickness Reduction (%)	E (10^6psi)	σ_Y (ksi)	σ_{ULT} (ksi)	ϵ_F (%)
15	(1661)	13.0	56.4	64.2	2.24	93.8	13.8	56.0	65.9	7.56
18	(1616)	14.1	52.3	59.0	2.64	91.0	15.4	59.4	67.9	4.50
25	(1601)	-	-	-	-	95.3	16.6	60.0	69.3	2.58
	(1655)	15.8	61.8	70.2	1.82	92.0	18.2	64.8	74.5	2.80
30	(1664)	16.9	64.9	72.2	1.16	71.8	17.4	64.9	72.7	1.61
	(1587)	15.7	64.5	76.2	1.71	64.4	16.4	60.3	71.6	2.03
						91.5	18.7	66.2	80.0	2.70
	(1666)	15.0	64.5	68.6	1.39	83.1	16.8	65.3	74.5	2.21
						96.1	17.8	64.8	73.6	2.86

The major improvement found after rolling in the heat-treated samples is an increase in strain at failure.

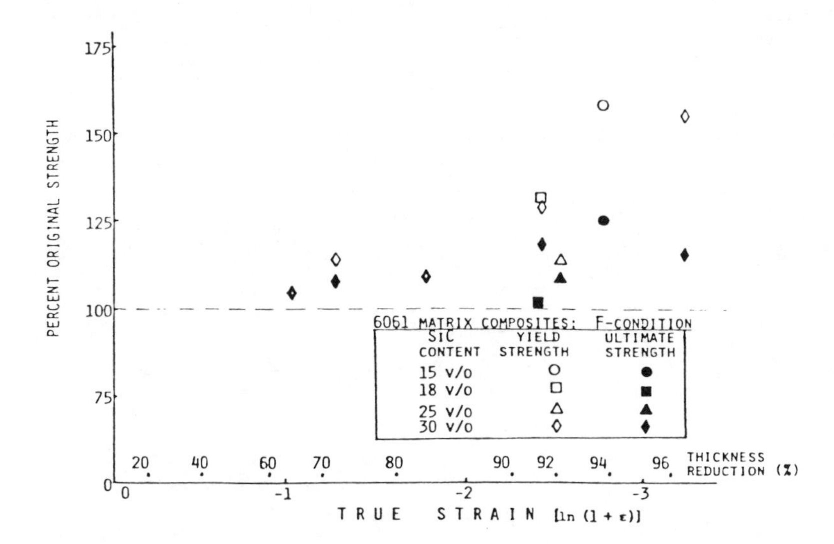

Figure 5 - Percent of Original Strength as a Function of True Strain for Hot-rolled 6061 Al Composites in As-rolled Condition.

The strain at failure level is shown to be a function of SiC content and amount of hot work, Figure 6.

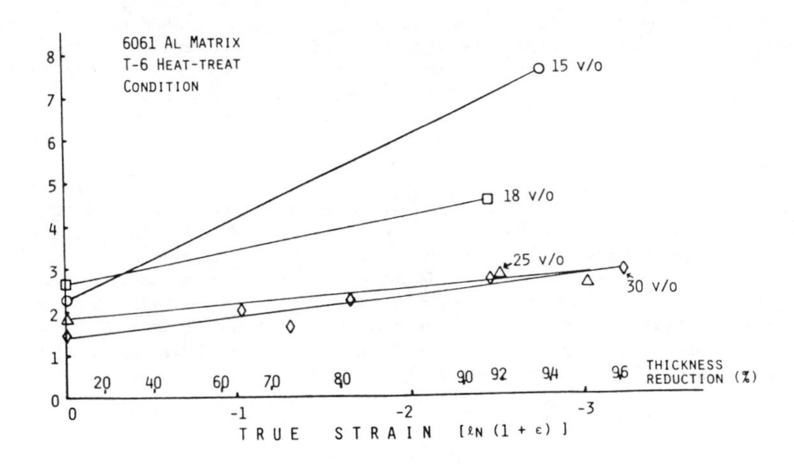

Figure 6 - Strain at Fracture as a Function of True Strain for Hot-rolled 6061 Al Matrix Composites after T-6 Heat Treatment.

Strain levels of 7.5% were found in the heat-treated 15 v/o composite after 93.8% thickness reduction and as high as 3% were found in heat-treated 30 v/o composites after 96.1% thickness reduction. The strength values of heat-treated composites were found to increase only after 80% thickness reduction, Figure 7.

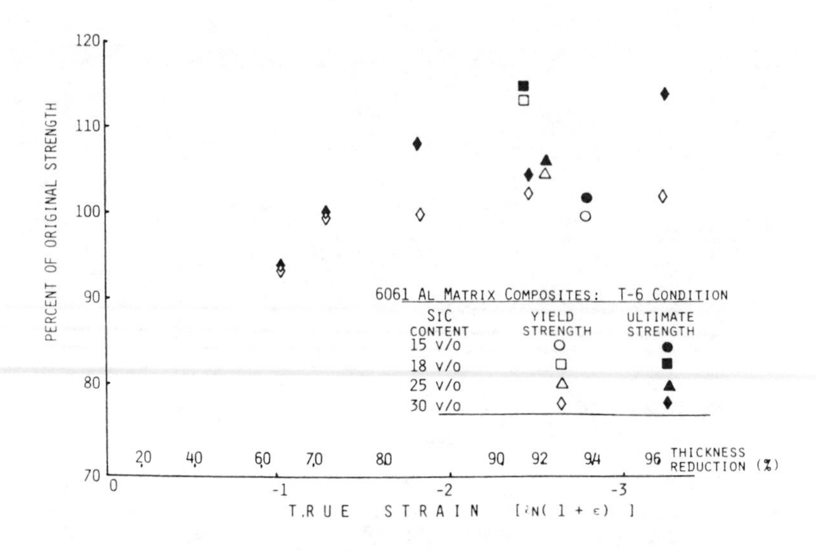

Figure 7 - Strength of Composites as a Percentage of Original Strength as a Function of True Strain for Rolled 6061 Al Matrix Composites in T-6 Heat-treated Condition.

The strength level of the heat-treated composites is strongly dependent upon work hardening during the straining. This can be seen by comparing the yield strength improvement with the ultimate strength improvement, Figure 7. In each case, the ultimate strength increases more than yield strength and the improvement is greater at larger thickness reductions where the strain to failure is the highest. This implies that the strength improvements seen with these composites is due to the increased strain at fracture and work-hardening of the matrix during this straining. The authors think that the increased strain capability is due to improved bonding of the metal particles and the matrix behaving more like a wrought metal. In order to compare the present results with prior work (Ref. 1), the yield strength and ultimate strength of current composites are plotted as a function of SiC content in Figure 8. Mechanical properties are shown before and after a 90% or greater thickness reduction. In all cases, individual test specimens were heat-treated for the T-6 condition prior to testing. The data show an increase in strength caused by hot work of these composites at all reinforcement levels.

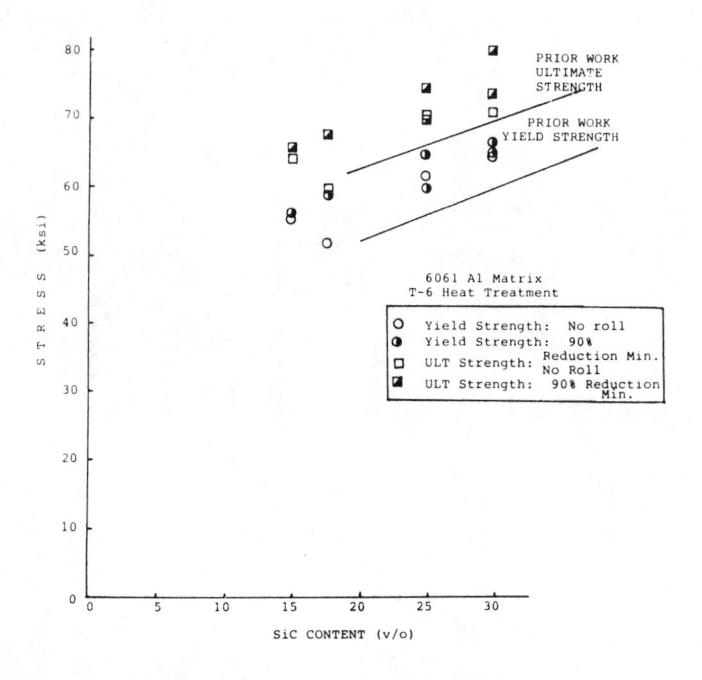

Figure 8 - Mean Strength for 6061 Al Matrix Composites before and after Rolling in the T-6 Heat-treat Condition as a Function of SiC Content.

The microstructure of 18 v/o and 30 v/o SiC composite, before and after hot rolling, is shown in Figures 9 and 10. Prior to rolling, the large particles of aluminum that are present in the aluminum powder appear as SiC-free areas in the composites. After rolling, both of the composites have a very uniform distribution of aluminum and SiC. Some evidence of aluminum lamination caused by the rolling can be seen, but this is minor. The consequences of this microstructure are not known at the present time.

The results of this study indicate that the mechanical property improvement of SiC reinforced 6061 Al matrix composites after hot rolling is related to the improvement in metal-to-metal bonding. Earlier work by Skibo (Ref. 4) showed that solution treating 2024 Al matrix composites at 950°F. for 72 hours lead to improved response to heat treatment and hence higher strengths, most likely due to homogenization of the matrix. After hot rolling, the 6061 composites have almost the same response to heat treatment that they had prior to rolling. If matrix homogenization were important for this system, the large roll reductions should cause an improvement in heat-treatment response; however, the strength increases found are a result of the increase in strain to failure and work hardening of the matrix. This is most clearly seen in Figure 7, where very little improvement in yield strength is found when compared with as-fabricated properties; while greater gains are found in the ultimate strengths.

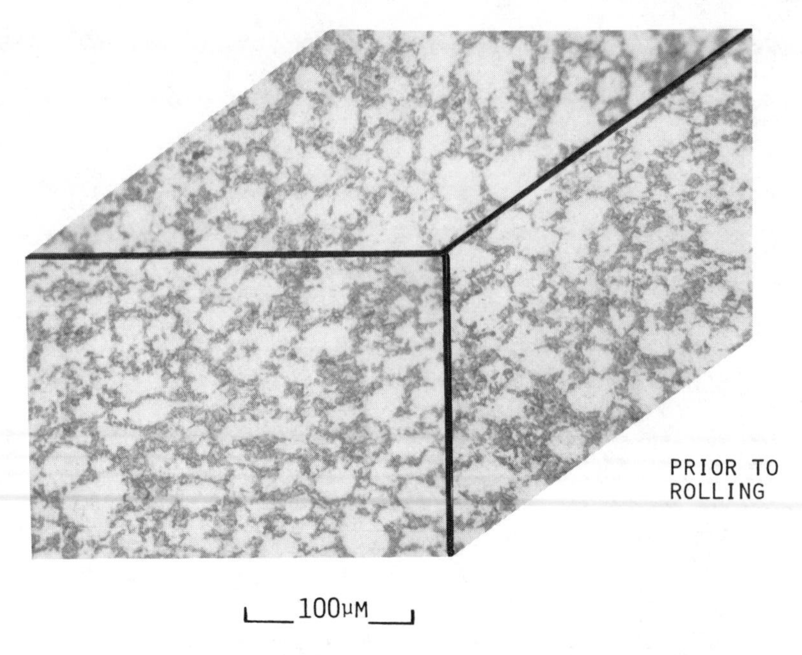

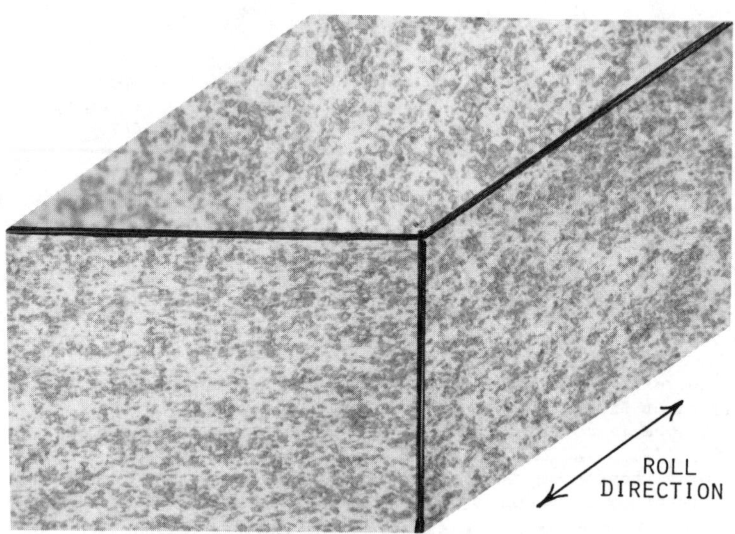

Figure 9 - 18 v/o SiC/6061 Al before and after 91. %
Reduction by Hot Rolling.

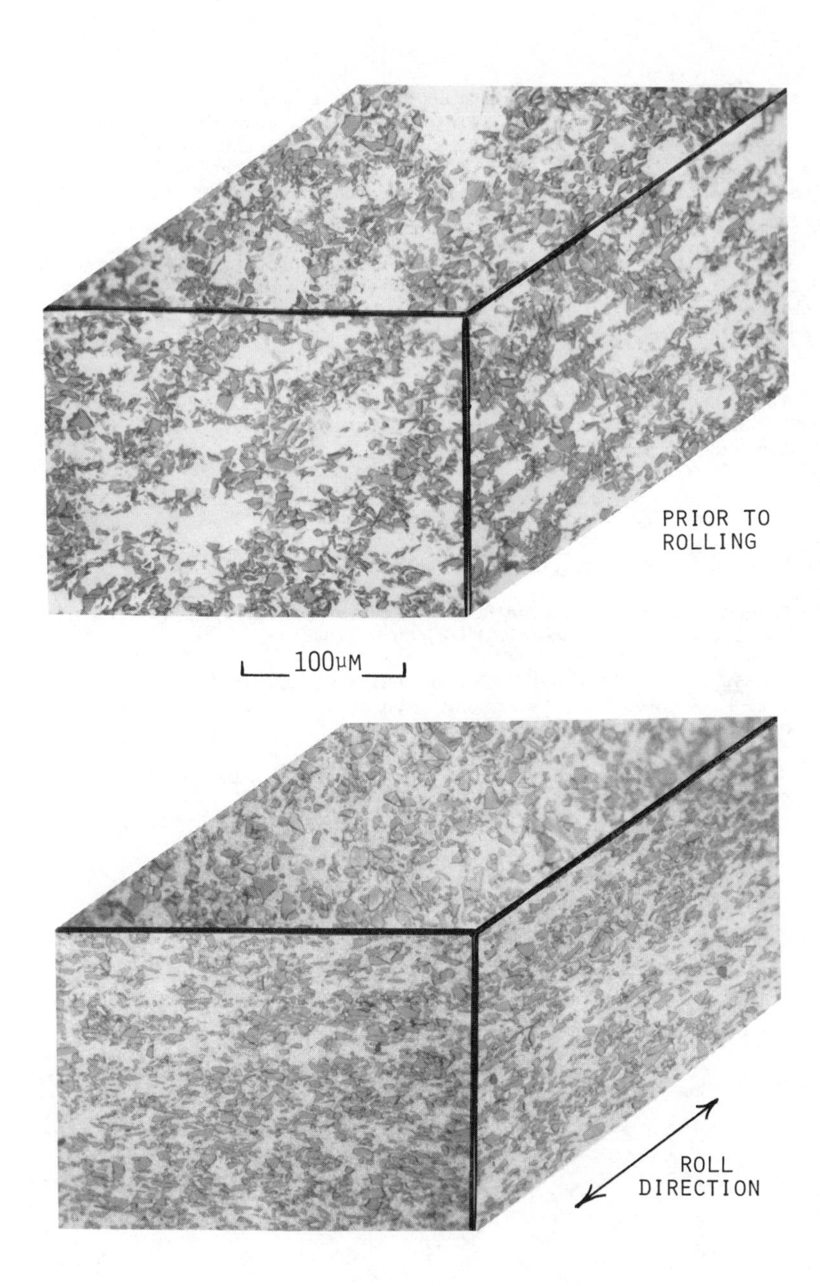

Figure 10 - 30 v/o SiC/6061 Al before and after 71.2% Reduction by Hot Rolling.

Conclusions

The results of this study demonstrate that hot rolling substantially increases the ductility of SiC reinforced 6061 Al composites. This improvement takes place both in the as-fabricated (F) and heat-treated (T-6) conditions for composites containing reinforcement levels between 15 and 20 volume percent. Strength increases after heat treatment were found for roll reductions greater than 80%.

In order to take advantage of the material-property capabilities of the SiC/6061 composites, hot working to a level equal to 80% thickness is recommended.

Acknowledgements

The authors wish to thank Mr. J.P. Gudas of NSRDC, Mr. M.A. Kinna of NAVSEA, and Dr. E.C. Van Reuth and Lt. Col. Jacobson of DARPA for their guidance and support of this research.

References

1. Webb, B.A. and Supan, E.C., unpublished research.

2. Bradbury, Samuel, "Source Book on Powder Metallurgy." American Society for Metals, Metals Park, Ohio (1979).

3. Newborn, H.A., "Strengthening of 6061 Aluminum by Fiber and Particulate Silicon Carbide." Presentation at 1981 AIME Fall Meeting, October 1981, Louisville, Kentucky.

4. Skibo, M.D., "Stiffness and Strength of SiC-Al Composites." Report No. SAND81-8212, Sandia Laboratories, 1981.

THE IMPACT TOLERANCE OF FIBER-AND PARTICULATE-REINFORCED METAL-MATRIX COMPOSITES

Karl M. Prewo

United Technologies Research Center
East Hartford, Connecticut 06108
USA

The impact energy dissipation capabilities of fiber and particulate reinforced aluminum and titanium alloys are presented using a format which permits comparison over a range of stress states. It is shown that material impact performance is a function of both material properties and imposed stresses. Using the format of representation chosen, it is shown that SiC particulate and FP alumina fiber reinforced composites are far less impact tolerant than their parent aluminum matrices. By comparing with boron reinforced aluminum and titanium matrix composites it is shown that energy dissipation capability can be related to fiber, matrix and interfacial properties.

Introduction

Fiber, whisker and particulate reinforced metals have been available to the scientist and engineer ever since the mid 1960's. Up until now the applications for which they have been considered have been primarily in the high technology aerospace industry. Because of the very high cost of systems such as boron fiber (B) reinforced aluminum (Al) and graphite fiber (C) reinforced Al, consideration of more industrial applications has seemed unreasonable. More recently, however, the potential for significantly lower cost systems such as alumina fiber (FP) and silicon carbide particulate (SiC_p) and whisker (SiC_w) reinforced Al has caused a major surge of industrial interest. While still more expensive than unreinforced metals, these and other metal matrix composites are suggested for use over their resin matrix counterparts because of their greater environmental stability and in many cases higher strength. It is also frequently stated that, because they have ductile metals as matrices, these composites will exhibit superior levels of impact and damage tolerance. While sometimes true it will be shown in the following article that this need not be the case and that, in fact, it can be very far from the truth.

As metal matrix composites attempt to step out of the relatively pampered aerospace industry and into the harsher industrial arena it is important that this subject of damage tolerance be understood. Unlike material fracture toughness which at least has an established format for quantifying material performance, the resistance of a material to damage due to impact is always described in much less specific terms. This is in large measure because material response to imposed impact loads frequently involves large strain plasticity and also because the impact event is frequently not well enough understood to permit rigorous analysis. In the case of metals, successful ranking based on damage resistance is as much a function of many years of engineering experience as it is of data base. This does not provide a satisfactory format for composite materials whose structural complexity and diversity require a very specific format.

In this paper a format is proposed based on the use of the well known instrumented pendulum impact test. Involving relatively simple equipment available in a large number of laboratories, it will be shown that this test can be used to rank material impact tolerance for a wide range of possible applications. A demonstration of the use of this test will be provided through the testing of several different metal matrix composites. It will be shown that, to form a complete picture of composite performance, two different graphs must be generated.

- The first, which provides a measure of composite strength under impact conditions, is a plot of composite flexural strength as a function of specimen span-to-depth ratio. This shows, as a function of test geometry and hence applied stress state, what stress must be applied before specimen failure will occur.

- The second, which provides a measure of composite energy dissipation capability, is a plot of dissipated impact energy as a function of specimen depth.

Together these two curves describe a material's ability to survive an impact event.

Experimental Procedure

Materials

A wide variety of fiber reinforced composites and one particulate reinforced metal matrix composite are considered. Also, all but one composite had aluminum alloys as their matrices. The one exception is a fiber reinforced titanium matrix composite. The composite compositions, along with their tensile properties, are presented in Table I. The 5.6, 5.7 and 8.0 shown before the boron and Borsic designations refer to the fiber diameters in mils. The FP alumina fiber is approximately 1 mil in diameter. The SiC particulate reinforced aluminum composite was obtained from Dolowy Webb Composite Specialties while the FP fiber reinforced aluminum composite was obtained from E. I. DuPont Co.

Table I. Fiber (f) and Particulate (p) Reinforced Metals

Matrix	Reinforcement		Tensile Strength (10^3 psi)		Elastic Modulus (10^6 psi)	
	Type	v/o	0°	90°	0°	90°
6061-0	None	-	18	18	10	10
6061-T6	None	-	45	45	10	10
Ti-6Al-4V	None	-	150	150	16.5	16.5
6061-0	5.6B(f)	50	215	2-	35	20
2024-0	5.6B(f)	52	218	35	37	20
1100-0	5.6B(f)	53	230	13	34	20
Al-Li	Al_2O_3(f)	60	85	32	33	22
1100-0	8.0B(f)	60	229	5	-	-
6061-0	SiC(p)	30	42	42	18	18
Ti-6Al-4V	5.7 Borsic(f)	40-50	140-180	40-70	32-38	24-32

Pendulum Impact Test Technique

While the data for the SiC(p) and FP fiber reinforced aluminum composites are reported herein for the first time, the pendulum impact data for the B and Borsic (BS) reinforced composites are assembled from some of the author's previous publications (1-4). In all cases, however, the same test procedure for pendulum impact testing was used. This consisted of placing rectangular cross sectioned specimens in a standard Charpy impact tester which has been instrumented to provide an impact load vs time trace through the placement of strain gages on the impact striker, or tup (5). All specimens whether notched, like standard Charpy specimens, unnotched, or thinner than the standard were tested in the same manner. Since the standard Charpy impact tester is designed for specimens having a thickness of 0.394", one necessary change in procedure for the testing of thin specimens is the necessity to provide spacers behind the specimen edges to place the specimen surface at the desired bottom of the pendulum swing on impact.

From this simple instrumented impact test procedure several important pieces of data can be obtained. First, from the applied load vs time trace, the maximum impact load on the specimen can be obtained. This can give an indication of the level of imposed load required to inflict damage to the specimen and, as will be shown below, can be used to measure composite flexural strength. Second, the total energy dissipated by the specimen during fracture can be obtained, or if desired, an energy vs time trace can be generated and separated into various stages of specimen damage. For the purposes of this paper only the total specimen energy dissipated to cause total failure will be used.

Finally, it should be noted that the Charpy type pendulum impact test is nothing more than a rapid three-point bend test with a fixed span of 1.62". As such, the loads applied can be used to calculate the levels of stress imposed on the specimens. Simple equations exist for the prediction of the nominal levels of maximum shear stress and flexural stress generated during the three-point bend testing of a beam of rectangular cross section. Given a bending span of length L, specimen depth h and width b, the maximum shear stress occurring at the neutral axis can be given by

$$\tau_{max} = \tfrac{3}{4}P/bh. \tag{1}$$

The maximum flexural stress occurring at the same time is

$$\sigma_{max} = \tfrac{3}{2}(PL/bh^2), \tag{2}$$

and occurs at mid-span on the side away from the loading nose.

The ratio of maximum applied flexural stress to maximum shear stress is given by

$$\sigma_{max}/\tau_{max} = 2L/h. \tag{3}$$

Thus, depending on the relative magnitudes of composite material flexural and shear strengths, and the value of span-to-depth ratio, a specimen can fail in either shear or flexural tension. For the test data presented in this paper, the value of L was always 1.62", the value of b was generally 0.394" (for the SiC_p/6061 it was 0.2") and the value of h varied from a maximum of 0.394 (for unnotched specimens) to less than 0.07". As will be shown in Figure 4, the value of h for notched specimens does not include the notch depth which has been subtracted from the overall specimen dimension.

The use of these equations presupposes that, as in the case of most of the specimens described here, the composite principal axes of orthotropy coincide with the axes of symmetry of the test specimen. This is not true for the off axis reinforced specimens (6,7). Similarly, these equations neglect completely the effects of stress concentrations produced by the V-notch in several of the specimens. However, it has already been shown (2,3,4,8,9) and will be discussed here that despite these deficiencies these equations are very useful in rationalizing composite behavior.

Results and Discussion

Composite performance will be compared using the proposed format involving both the maximum stress to cause failure and the energy dissipated during failure.

Failure Stress as a Function of Span-to-Depth Ratio

As shown in the previous section, the stress state imposed on the composite three-point bend impact specimen varies with span (L) to depth (h) ratio. Thus for 6061 aluminum matrix composites, the calculated maximum flexural stress can vary significantly with L/h, Figure 1. This is best

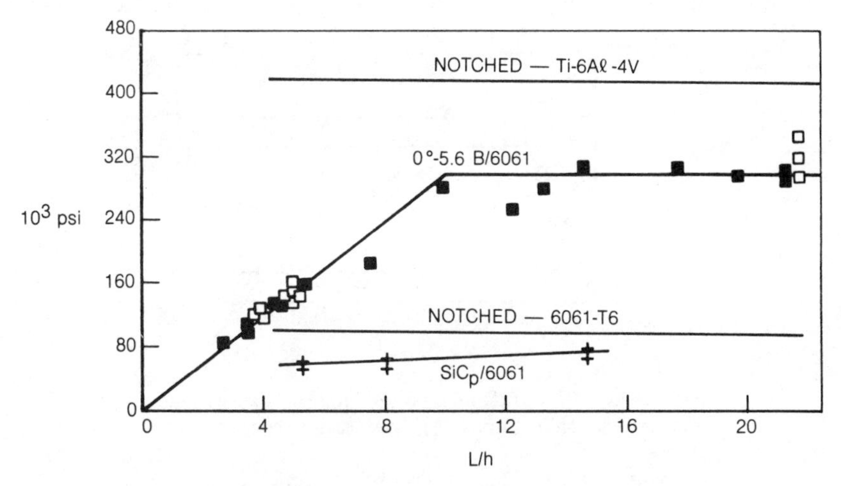

Figure 1 - Calculated maximum flexural stress vs span-to-depth ratio (L/h) for reinforced 6061 aluminum and unreinforced metals.
■ = slow bend of 0°-5.6B/6061 (unnotched)
□ = impact test of 0°-5.6B/6061 (notched & unnotched)
\+ = impact test of SiC_p/6061 (notched)

seen for 0°-5.6B reinforced 6061 aluminum composites where the specimens fail in shear in the low L/h region causing an apparent (but not real) drop in composite flexural strength (1-4). In fact, the sloped straight line portion of the composite curve can be obtained using equation (3) and the shear strength of the 6061 aluminum composite. At L/h values greater than approximately 10 the B reinforced composite flexural strength is independent of test condition due to the fact that it is controlled by true flexural failure. In contrast the other three materials represented in the figure all exhibited a nearly total independence of strength on L/h due to the fact that their shear strengths are quite high compared to their flexural strengths. Thus the notched SiC particulate reinforced 6061 aluminum specimens behaved in a manner very similar to the unreinforced titanium and aluminum alloys. For these non fiber reinforced materials only notched specimen data are presented in the figure and the details of the unreinforced metals testing can be found in Ref. 2. The metals and SiC_p reinforced 6061 were all found to be quite sensitive to the presence of these notches while the 0°-B reinforced aluminum composites were very insensitive to notch presence.

For the lower strength 1100 aluminum matrix composites reinforced with boron fibers the dependence of calculated flexural strength on L/h was similar to that of the B/6061 composite. In this case, however, the effects of fiber orientation were included, Figure 2. While there is a very strong dependence of flexural strength on fiber orientation (as would be expected) in the high L/h region, the shear controlled low L/h region specimen data all fall on the same line. This is also expected since all of these composites have the same interlaminar shear strength controlled by the 1100 aluminum.

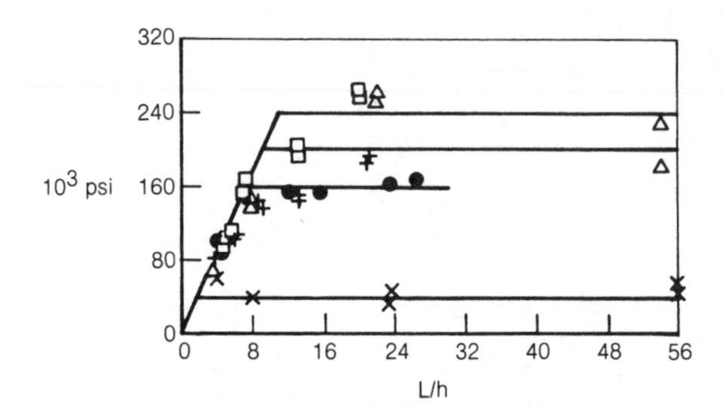

Figure 2 - Calculated maximum flexural stress vs span-to-depth ratio (L/h) for impact testing of 5.6B-1100.

□ = notched 0°, x = unnotched ± 45

\+ = notched and unnotched ± 22

Δ = unnotched ± 15, ● = notched 0°

Finally, Figure 3 presents the data for 5.7 mil diameter Borsic fiber reinforced Ti-6Al-4V composites tested in the 0° orientation. A behavior similar to that of the boron fiber reinforced composites is noted except for the fact that at low L/h values true specimen shear controlled failure was not achieved. This is due to the higher shear strength of the Ti-6Al-4V matrix.

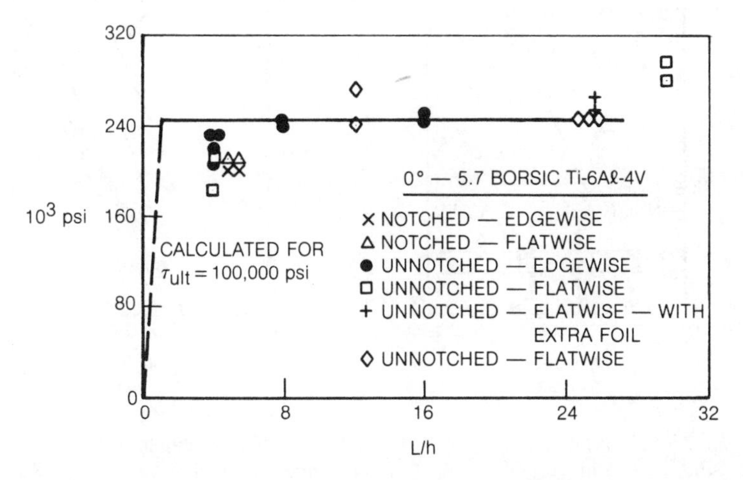

Figure 3 - Calculated maximum flexural stress vs span-to-depth ratio (L/h) for 0°-5.7 Borsic reinforced Ti-6Al-4V.

Impact Energy Dissipation

As shown in the previous section, composite failure mechanism can vary significantly with L/h due to the possibility of flexural or shear controlled failure. Since this failure mode can significantly alter the material energy dissipation capability, composite impact performance must also be described in terms of L/h. This has been shown in detail to be the case for 5.6B/6061 composites in the past (4), and is illustrated in Figure 4a for 0°-5.6B/1100 composites where the data are presented for both notched and unnotched specimens. As in the case of composite flexural strength, composite impact energy dissipation capability is independent of the notch presence as long as the value of specimen depth, h, is the net dimension, i.e. the notch depth has been subtracted. It should be noted that in this plot, specimen depth h, rather than L/h, is used. Since all specimens were tested at a constant value of L there is no ambiguity.

The impact energy data for unreinforced Ti-6Al-4V and 6061-T6 metal alloys are presented in Figure 4b along with a line representing the data for 0°-5.6B/6061. Although not indicated for the 5.6B/6061, this composite's performance is also independent of whether the specimens are notched or unnotched. On the other hand, the unreinforced metals are extremely notch sensitive. While the 0.4" thick specimens of Ti-6Al-4V and 6061-T6 could not even be broken during impact, the introduction of the standard notch caused a tenfold decrease in energy dissipation.

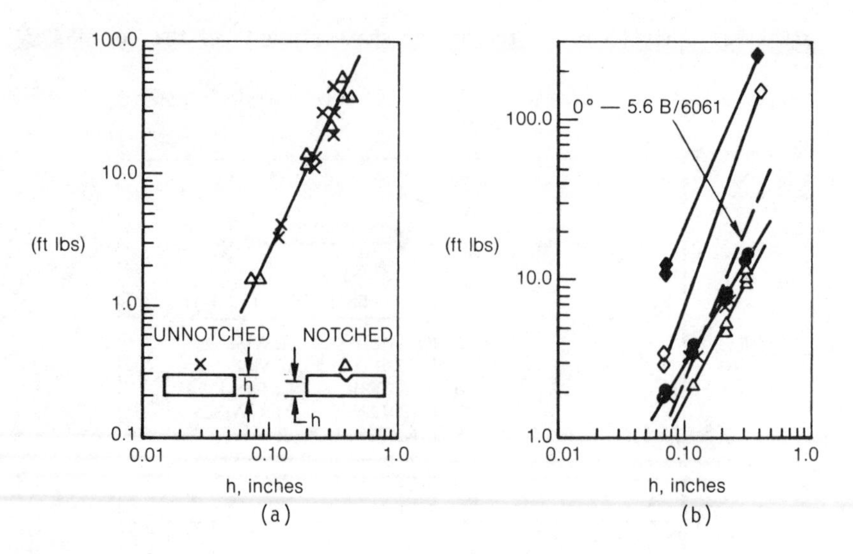

Figure 4 - Total impact energy dissipated vs specimen depth for (a) 0°-5.6B/1100 and (b) 0°-5.6B/6061 and unreinforced metals.
● = X=notched Ti-6Al-4V
△ = notched 6061-T6
♦ = unnotched Ti-6Al-4V
◊ = unnotched 6061-T6

The impact energy data for the potential low cost aluminum matrix composites and 5.6B/6061 are presented in Figure 5a. The first point to be noted is that the SiC particulate reinforced composite is extremely sensitive to the presence of the machined notch. Thus, there is nearly a tenfold decrease in impact energy level due to the introduction of the notch. This is very similar to the behavior of the unreinforced 6061-T6 aluminum described in Figure 4b, however, the levels of energy dissipated are much lower for the composite than the metal. Even when compared to the 0°-5.6B/6061 composite, over the entire range of specimen dimensions in both unnotched and notched conditions, the particulate reinforced material dissipates less energy. Also of interest is the observation that the energy vs h plot for the SiC_p/6061 is similar in shape to the other composite data trends. These data, along with the values for 0°-FP fiber reinforced aluminum are summarized in Table II. From the table, and also Figure 5, it can be seen that the FP fiber reinforced composite system is the least impact tolerant of all and in fact can be classified as a relatively brittle material. It will be shown later why this is the case.

The data in Table II also include calculated values of material fracture toughness for the two unreinforced metals and the SiC particulate reinforced aluminum. Because of the limited size of the specimens preventing full notch tip constraint, the values measured are somewhat larger than the handbook values and hence also, the value of 18 Ksi$\sqrt{in}$. for the SiC_p/6061 is likely an optimistic estimate.

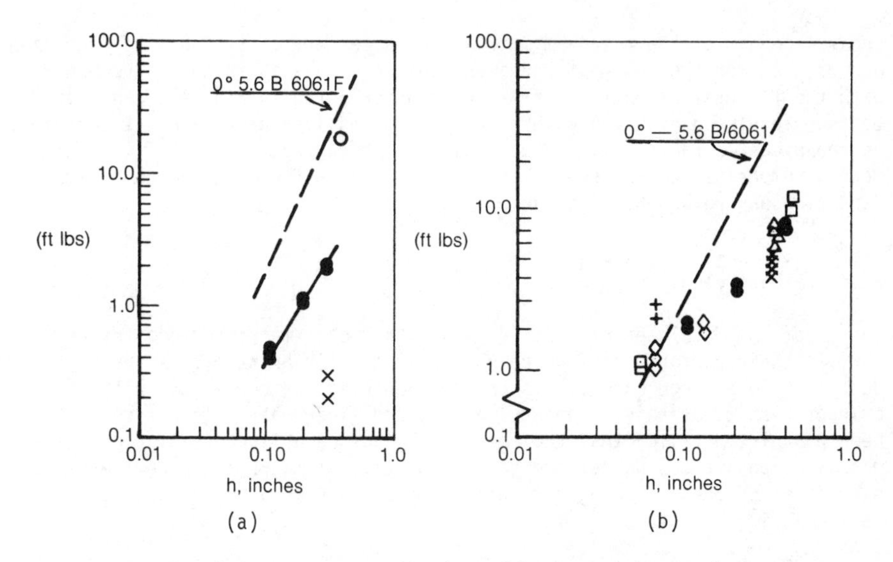

Figure 5 - Total impact energy dissipated vs. specimen depth for (a) aluminum matrix composites where ○ = unnotched SiC_p, X = notched FP/Al, ● = notched SiC_p/6061 and (b) 5.7 Borsic reinforced Ti-6Al-4V compared with 5.6B/6061. Symbols agree with those in Figure 3.

Table II. Full Sized Charpy Specimen Impact Test Data

Materials	K_{ID} ksi(in.)$^{1/2}$	Energy Dissipated Notched ft-lbs	Energy Dissipated Unnotched ft-lbs
6061-T6 aluminum	32	10	>150
	26*	-	-
Ti-6Al-4V	123	13	>260
	96*	-	-
30 v/o SiC_p-6061 Al (annealed)	18	2	18
50 v/o B-6061 Al (0°) (annealed)	-	20	40
60 v/o FP-Al (0°)	-	0.3	-

*from damage tolerant design handbook, MCIC-HD-01

Finally, in the case of titanium matrix composites, composite energy dissipation capability can also be readily described as a function of h using the same format, Figure 5b, with the possibly surprising result that over a wide range of conditions the 0°-5.6B/6061 is a superior system. The region of aluminum matrix composite superiority is that in which large specimen shear can take place due to the low value of L/h. Because of the higher shear strength of the titanium matrix composite very little shear deformation occurs even for the thickest specimens tested. This is in good agreement with the flexural strength controlled behavior noted in Figure 3 and points to the usefulness of using both types of data to interpret material behavior. At the high L/h region, i.e. small values of h in Figure 5b, the titanium matrix composites can be equivalent or superior to the aluminum composites depending on the amount of titanium foil present.

Fracture Morphology

From the data presented it is clear that metal matrix composites can vary in their performance from being nearly totally brittle to extremely ductile. To a large extent this can be predicted and correlated with the characteristics of reinforcement, matrix, and fiber-matrix interface. It has been postulated (1,10) that the energy to fracture a fiber reinforced metal matrix composite can be related to its constituent material properties by an expression of the following form which has often been used in the past for resin matrix composites.

$$\text{Energy per unit area of fracture} = \frac{\sigma_f^2\, d_f\, V_f}{24\tau} \qquad (4)$$

In this expression, which assumes a fiber pullout mechanism of energy dissipation, σ_f refers to the effective fiber strength, d_f the fiber diameter, V_f the fiber volume fraction, and τ the shear strength of the region which is pulling out. In a high fiber-matrix strength situation this would refer to the matrix shear strength since coated fibers would be pulling out. As shown in Figure 6 this is the case for the 5.6B/6061 aluminum matrix composite, while in the case of the 5.7 Borsic reinforced Ti-6Al-4V fracture morphology also shown in Figure 6, this more appropriately should relate to the fiber-matrix interfacial strength which is apparently below the matrix strength since the fibers pull out without any matrix adhering to them. This is also seen to be the case in transverse tension.

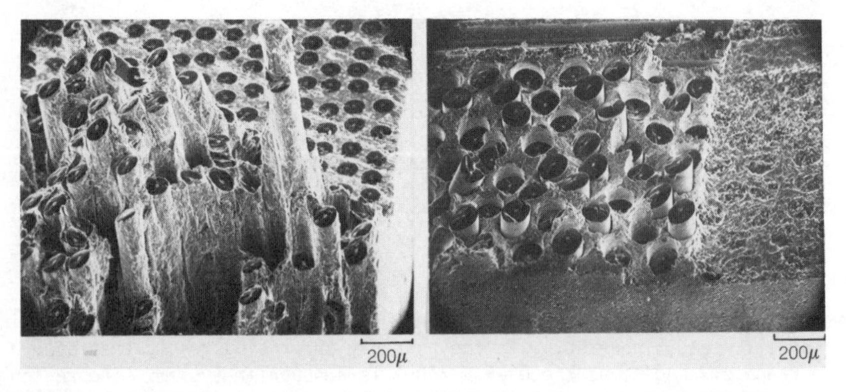

Figure 6 - Fracture surfaces of 5.6B/6061 (left) and 5.7/Ti-6Al-4V (right) showing region at edge of notch.

The almost complete absence of fiber pullout of any sort in the FP fiber reinforced aluminum specimens, Figure 7, is clearly due to the relatively low fiber strength and also the very small fiber diameter of approximately 1.0 mil. In contrast, with a much larger fiber diameter of 8.0 mils and a very low composite matrix strength it is possible to achieve extremely fibrous failure morphology, Figure 7.

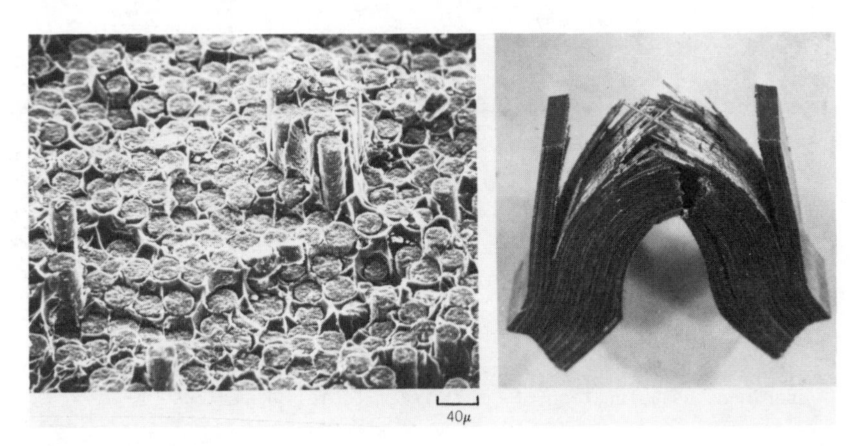

Figure 7 - Fracture surface of FP/Al (left) and overall view of 8B/1100 impact specimen (right) which disspated a total 78 ft-lbs of energy.

Equation (4) has been particularly useful if one wishes to assess the effects of a wide range of materials variables on specimen energy dissipation when impact is taking place under stresses where significant shear deformation can take place, i.e. in the region of low values of L/h. To illustrate this point the full sized notched impacted specimen energy data for the fiber reinforced composites are plotted in Figure 8. In this case the value of τ was replaced by one-half the value of composite transverse tensile strength since this will accurately represent the important failure mode, i.e. fiber-matrix interfacial or matrix shear. The value of fiber strength, σ_f, was simply taken from the composite axial tensile strength using $\sigma_f = \sigma_{0°}/V_f$. The data in the figure show exceptionally good agreement with the thus modified expression (4). This is somewhat surprising when one considers the large scale deformation of some specimens, Figure 7, that is not just fiber pullout.

The particulate reinforced aluminum composite cannot, of course, be strictly compared using this expression. However, because the SiC particulate size is very small and because the particulate to matrix bond strength is quite high it can be expected that very little large scale deformation takes place on the fracture surface, even for the low L/h specimens. This is clearly the case as shown in Figure 9 where the limited local matrix plasticity is not sufficient to overcome the embrittling effect of the 30 volume percent SiC ceramic.

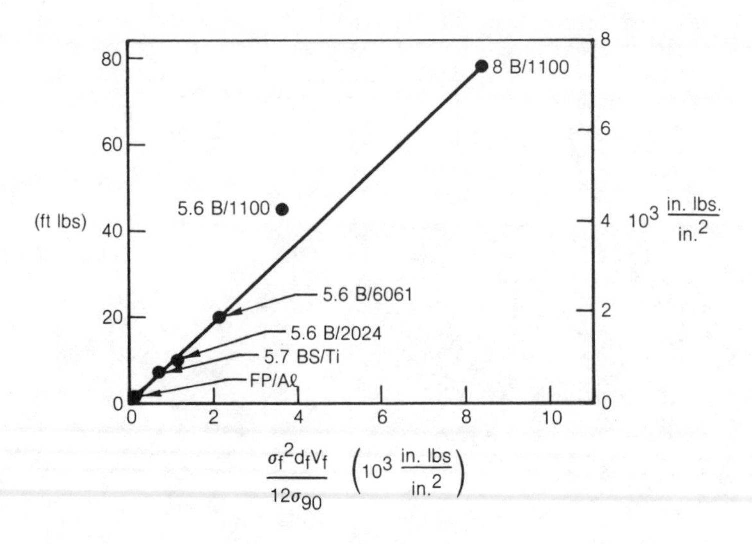

Figure 8 - Impact energy dissipated by full sized notched impact specimens as a function of composite material properties.

Figure 9 - Fracture surface of SiC particulate reinforced aluminum.

Conclusions

It is concluded that, although metal matrix composites can exhibit exceptional levels of impact energy dissipation capability, this is not always the case. By testing in a regime where significant shear stresses cannot be generated, i.e. high values of L/h in pendulum impact, specimen failure is controlled primarily by the brittle reinforcing fibers. In contrast, specimens subjected to high interlaminar shear stresses were shown to dissipate more energy than unreinforced metals. The following can be concluded.

1. Because of the importance of stress state it is necessary to provide impact specimen performance over a range of test conditions. Generalizations made using one test regime will not be valid in an intended application having a different stress state.

2. A format for comparing composite material performance is proposed based on both composite failure strength and impact energy dissipation.

3. Just as machining a V-notch into a metal drastically reduces its ability to plastically deform, the introduction of brittle fibers and particulates can cause the same deleterious effect; gross plasticity is prevented by the stabilization of local crack growth at the tip of the machined notch.

4. In contrast to the above, by using strong large diameter fibers in a low shear strength matrix it is possible to distribute matrix shear over large distances removed from the notch tip. By this process the remarkable result is achieved that a notched aluminum matrix composite containing 50% by volume brittle boron fibers can appear ductile in its fracture and dissipate more energy than notched monolithic Ti-6Al-4V.

5. Small diameter, relatively low strength FP alumina fibers severely embrittled their parent aluminum matrix while large diameter boron fibers achieved the opposite effect.

6. Caution should be exercised in dealing with SiC particulate (and probably whisker) reinforced aluminum since large scale matrix plasticity is severely limited by the reinforcement.

References

1. K. M. Prewo, Journal Composite Materials, 6 (1972) p. 442.

2. K. M. Prewo, AFML-TR-75-216, March 1976.

3. K. M. Prewo, Proceedings Failure Modes in Composites III, edited by T. Chiao and D. Schuster, AIME, 1976.

4. K. M. Prewo, Phil. Trans. R. Soc. Lond, A, 294, 551, 1980.

5. R. A. Wullaert, ASTM-STP-466, 1970, p. 148.

6. S. A. Sattar and D. H. Kellog, ASTM-STP-460, 1969, p. 62.

7. K. T. Kedward, Fiber Science Tech, 5 (1972) p. 85.

8. A. W. Christiansen, Fiber Science Tech, 7 (1974) p. 1.

9. J. V. Mullin and A. C. Knoell, Mat. Res. Std., 10 (1970) p. 16.

10. K. G. Kreider, L. Dardi and K. M. Prewo, AFML-TR-71-204.

THERMAL-MECHANICAL BEHAVIOR OF GRAPHITE/MAGNESIUM COMPOSITES

B.J. Maclean and M.S. Misra
Martin Marietta Aerospace
Denver, Colorado 80201

Continuous-filament, graphite-reinforced magnesium composites exhibit exceptional unidirectional mechanical properties for their weight. For aerospace applications requiring high modulus, light-weight, thermal deformation-resistant materials, Gr/Mg promises substantial payoffs and is compared to other more conventional materials. Single-ply and 3-ply VSB32/AZ91C/AZ31B panels and single ply VS0054/AZ91C/AZ31B panels were evaluated. NDI techniques of x-radiography, ultrasonic C-scan, and liquid penetrant were used to assess filament collimation, face sheet disbonds, and sheet tears or pitting, respectively. Ambient temperature tensile testing yielded longitudinal elastic modulus values predicted by the rule-of-mixtures (32.4 Msi for $28.4^{v}/o$ Pitch 100 Gr/Mg) and ultimate tensile strengths of better than 83 ksi. Failure analysis and fractography were then conducted to determine failure modes.

Introduction

High performance space structures and antenna systems must meet increasingly stringent requirements for weight-savings, dimensional stability, extended service life, and survivability. These requirements are most pronounced in the development of large space structures (LSS) where diameters may range from 20 meters to 200 meters or more. One such generic design prototype of a 1/15th scale 12-bay deployable box-truss is shown in Figure 1 in its final deployed configuration. The structural technology for LSS has progressed to the fabrication of full scale single cubes and is demonstrated in the 15 foot (4.6 meter) cube deployed in Figure 2. This cube was produced from state-of-the-art graphite/epoxy (Gr/E) composites. It is dimensionally stable, exhibits a high degree of stiffness and weighs less than sixty nine pounds (31 kg).

Organic-matrix composites such as graphite/epoxy offer high specific stiffness (elastic modulus divided by density) and near-zero coefficient of thermal expansion (CTE). However, they suffer from limited temperature capability, outgassing in space vacuum, low resistance to radiation damage, and problems with dimensional stability due to moisture absorption, creep, and micro-cracking and -yielding. Other problems stem from graphite/epoxy's low thermal and electrical conductivity including overheating and space charging.

Figure 1 -- Scale model of a 12-bay deployable box truss antenna system measuring 20 meters in diameter.

Figure 2 -- This full-scale single cube, made of graphite/epoxy, is dimensionally stable and weighs less than 69 lbs. (31 kg.).

It has been established under DARPA funded programs(1,2,3) that the Graphite/Magnesium (Gr/Mg) composite is most ideally suited for these high-performance applications. Gr/Mg offers higher specific strength and stiffness over Gr/E, with the inherent environmental stability found in metal systems. With the metal matrix's higher thermal and electrical conductivity, and temperature capability, Gr/Mg promises minimized material property degradation and extended mission lives in space. If zero or near-zero CTE can be obtained in this composite, Gr/Mg has tremendous potential as a primary structural element material for large spacecraft.

The objective of this ongoing study is to characterize the mechanical and thermal behavior of Gr/Mg composites and evaluate their application as a structural material for large space structures.

Technical Background

Material Description

Graphite/Magnesium composite panels are produced by first infiltrating graphite fiber tows. To facilitate wetting and prevent chemical attack by the molten matrix, an activation layer of TiB_2 is applied using chemical vapor deposition (CVD). The fiber tow is then run through a molten bath of AZ91C magnesium casting alloy to form a Gr/Mg composite filament. Infiltration levels for the various Pitch fibers are about 50 volume percent (v/o). To produce the Gr/Mg composite panels, the filaments are inspected, collimated, and then vacuum hot-pressed between magnesium face sheets. The face sheets provide the transverse and compressive strengths to the panels after this sandwich construction has been consolidated by diffusion bonding under pressure at elevated temperature. Panels can be produced with single, or multiple ply unidirectional orientations.

Obtaining Zero CTE

In order to utilize Gr/Mg as a structural material for dimensionally stable large spacecraft it must have zero or near zero thermal expansion. The graphite fiber volume fraction necessary to achieve zero CTE in Gr/Mg can be determined from the following rule-of-mixtures (ROM) relation[3]:

$$\frac{1}{v_{f_{Gr}}} = 1 - \frac{E_{Gr} CTE_{Gr}}{E_{Mg} CTE_{Mg}}$$

where

CTE = Coefficient of Thermal Expansion

E = Elastic Modulus

$v_{f_{Gr}}$ = Volume Fraction of Graphite

Table I presents the measured properties of magnesium alloys and several graphite fiber types with a tabulation of the fiber volume percents in magnesium necessary to obtain zero CTE. According to the ROM relation, 48 v/o of the Pitch 100 fibers will be needed to achieve zero CTE in Gr/Mg. Since Gr/Mg filament infiltration levels are presently limited to about 48 - 50 v/o graphite, the magnesium alloy face sheets would have to be all but completely eliminated to produce a dimensionally stable panel.

To overcome this tradeoff between dimensional stability and panel integrity the following approaches can be considered:

1. Increase Gr/Mg filament infiltration levels to offset the presence of bonded face sheets. Some filaments have been infiltrated to 60 v/o graphite but consistent quality is difficult to maintain as individual fibers begin to contact each other and the graphite/magnesium interface is lost.

2. Increase the negativity of CTE in the Pitch fibers. This is not likely to be possible since the CTE of single crystal graphite has already been reached in the Pitch 100 graphite fibers.

3. Increase the modulus of the graphite fibers. Pitch fibers with an elastic modulus of 120 x 10^6 psi (827 GPa), for example, are being developed and have been produced on a limited basis.(4,5)

4. Increase face sheet alloy strength or decrease face sheet alloy CTE. Magnesium face sheets can be strengthened to some extent by heat treatment but other possibilities might include hybridization.

Table I - Volume Percent Graphite In Magnesium Necessary to Achieve Zero CTE

Constituent	Modulus, E x 10^6 psi (GPa)	CTE x 10^{-6} $^oF^{-1}$ (x10^{-6} $^oC^{-1}$)	v/o Gr for zero CTE percent
Magnesium, AZ31B or AZ91C	6.5 (45)	+13 (+23)	NA
Graphite, Pitch 55 (VSB32)	55 (379)	-0.5 (-0.9)	77%
Graphite, Pitch 75S (VSC-32-S)	75 (517)	-0.7 (-1.3)	62%
Graphite, Pitch 100 (VS0054)	100 (690)	-0.9 (-1.6)	48%

Experimental Procedure

Material Selection

The materials evaluated in this study are listed in Table II. The Pitch 55 filaments contained only 33 v/o graphite and had a range of tensile strengths from 92 to 152 ksi (634 to 1048 MPa). These were used to produce a single-ply and a triple-ply unidirectional panel, each with face sheets of 0.011 inch (0.28 mm) thickness. The Pitch 100 filaments were infiltrated to an average of 47 v/o graphite with a range of tensile strengths from 110 to 160 ksi (758 to 1103 MPa). Two single-ply undirectional panels were produced from these filaments with a face sheet thickness of .008 inch (0.20 mm) each. The wrought magnesium alloy AZ31B was used for the face sheets of all the panels and each panel measured approximately 1 foot by 2 feet (30 cm by 60 cm) in size.

Table II -- Graphite/Magnesium Panels Selected for Evaluation

No. of Panels	Lay-Up	Fiber	Face Sheet Thickness	Panel Thickness
1	1-ply	P55	.011 inch (.28 mm)	.038 inch (.97 mm)
1	3-ply	P55	.011 inch (.28 mm)	.078 inch (1.91 mm)
2	1-ply	P100	.008 inch (.20 mm)	.032 inch (.81 mm)

Test Procedures

Testing of the panels included microstructural investigations, nondestructive evaluation (NDE), fiber volume fraction analysis, tensile testing, and fractography on the scanning electron microscope (SEM).

In order to assess the quality and integrity of the Gr/Mg panels, various NDE techniques were employed. Each of the four panels was x-rayed to inspect for filament spacing, filament breakage or other damage during consolidation, and inclusions. However, X-radiography was unable to detect delaminations or panel disbonds. Ultrasonic C-scanning was used on each panel to search for these disbonds. As a final assessment of face sheet quality, penetrant inspection was used to detect fissures, cracks, or tears on the surface of the panels.

Fiber volume fraction, a necessary parameter in describing/predicting mechanical and material properties, was determined by chemical dissolution of the magnesium matrix. Specimens were sectioned from each of the four panels and their specific gravity determined by Archimedes principle. The specimens were then dissolved in hydrochloric acid, dried, and weighed. The graphite fiber volume fractions, assuming no voids in the panels, were found to be 28.4% for the Pitch 100 panels, and 12.7% and 23.2% for the single-ply and three-ply Pitch 55 panels, respectively.

Tensile tests were performed on each of the four Gr/Mg panels in both transverse and longitudinal directions. To prevent the grip section of the test specimens from being crushed (a common occurrence in composites testing) aluminum end tabs were bonded with epoxy as shown in Figure 3. The gage sections were 0.50 inch (1.27 cm) wide by 1.00 inch (2.54 cm) long and instrumented with both strain gages and extensometer. Tests were conducted at a constant crosshead velocity of 0.05 inches/minute (1.27 mm/minute) while recording load vs. strain gage and extensometer outputs. Agreement between these two plots was excellent.

Fractography was conducted by scanning electron microscopy on the sections excised from tensile specimens.

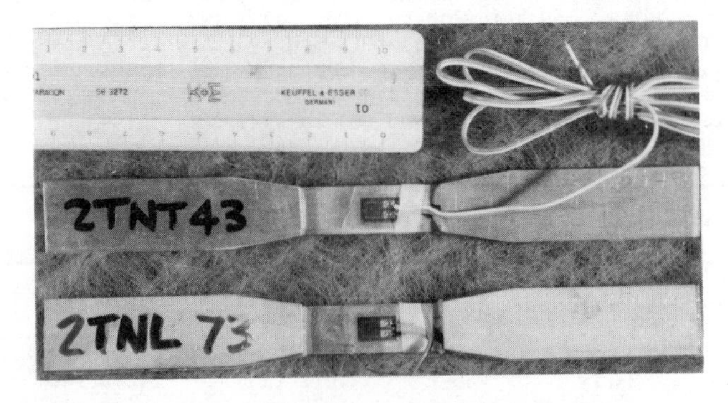

Figure 3 -- Gr/Mg tensile specimens with epoxied aluminum end tabs and bonded strain gages.

Results and Discussion

Microstructure

The microstructure of the Pitch 55 single-ply panel is presented in Figure 4 showing the face sheets bonded to the Gr/Mg filaments. A higher magnification view shows the interface between fiber tows and face sheet. The diffusion bond line is clearly visible running through the thin region between the fiber tows, which is rich in magnesium and devoid of fibers, and between the tows and face sheet. The bond appears to be less than optimum. This raises questions concerning the ability of the composite to sustain interlaminar shear forces induced during flexure and compression.

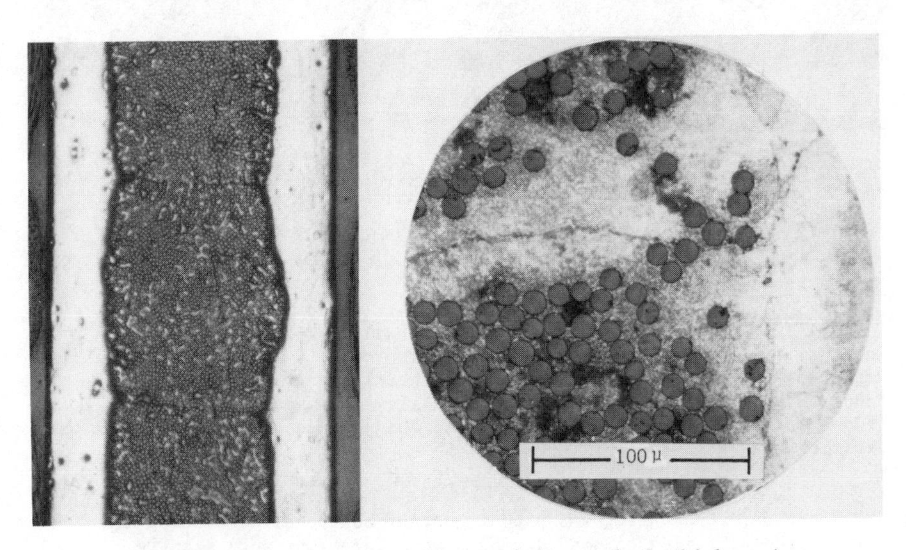

Figure 4 -- Single-ply Pitch 55 Gr/Mg panel (0.038 inch thickness) showing face sheets and filament tows. Note the diffusion bond lines in the enlargement.

The three-ply Pitch 55 panel is shown in Figure 5 where the Gr/Mg filaments are stacked together between the face sheets. The thin region between the fiber tows is also visible in this panel, illustrating the need for minimizing residual magnesium coatings on the filaments before consolidation and/or providing some degree of hot working to disperse this demarcation. In addition, a higher magnification view of apparent cracks through the center of the specimen is of obvious concern. Whether this crack was created during the processing of the composite, or was induced during the preparation of these microstructure specimens, is of question. It should be noted that no cracks were found in any of the single ply microstructure specimens. Finally, the infiltration of the graphite fiber tows with magnesium appears fairly uniform with no apparent voids and good fiber distribution.

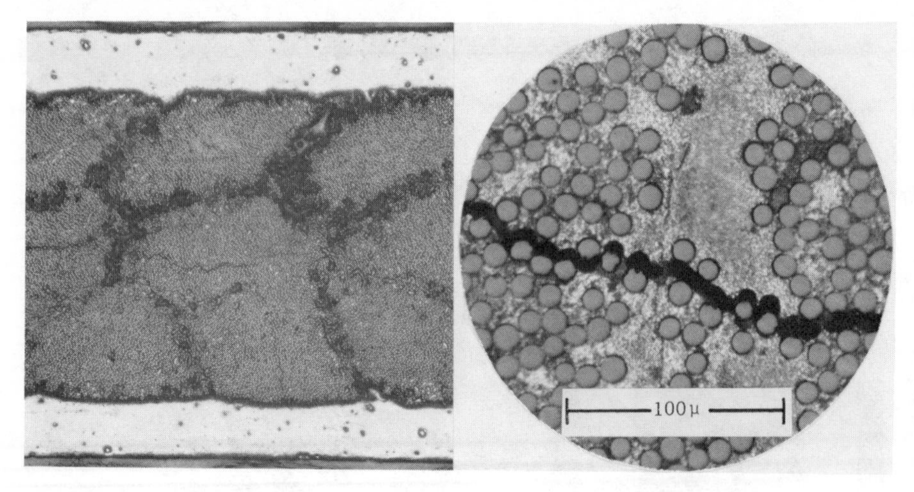

Figure 5 -- Three-ply Pitch 55 Gr/Mg panel (0.078 inch thickness) showing face sheets and fiber tows. The small crack in the center tow is enlarged at the right.

Non-Destructive Evaluation

All four Gr/Mg panels were radiographically inspected to reveal internal filament spacing and breakage. The results indicated a very high degree of quality in the collimation of the filaments with only rare occurrences of breakage (average of two filaments breaks per panel). No filament cross-overs were discovered and the absense of any inclusions was verified.

Ultrasonic C-scanning was used to detect any panel areas that had delaminated or failed to bond during processing. One such scan is presented in Figure 6. Lead arrows and circles were applied to the back surface of the panel before scanning to provide a reference in adjusting the resolution of the scan. The disbonds were detected almost exclusively around the edges of the panels where platten pressures during fabrication may have been non-uniform. The balance of the interior of the panels showed no disbonds or other anomolies.

The results of the penetrant inspection showed the absense of any cracks, tears, or fissures in the face sheets of the panels. The only surface irregularities found included minimal pitting (the Pitch 100 panels were protected with an acrylic laquer coating to minimize corrosion during handling) and imprints from flaws in the manufacturer's plattens.

Tensile Tests

The tensile data for the Gr/Mg composite panels are presented in Table III which includes numbers of specimens tested, panel orientations, and respective panel/face sheet thicknesses and fiber volume fractions.

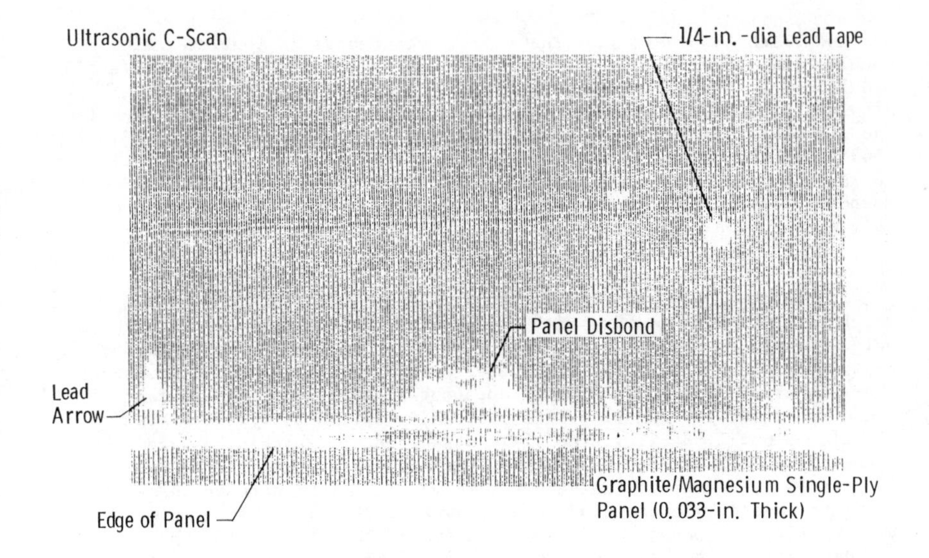

Figure 6 -- Ultrasonic C-Scan of a Pitch 100 Gr/Mg panel showing disbond along panel edge and images of lead tape used in adjusting C-Scan resolution.

Table III -- Gr/Mg Panel Tensile Data

	no. of tensile specimens	Total Panel Thickness (Face Sheet) (inches)	Graphite Fiber Vol. (%)	Elastic Modulus ($x10^6$psi)	Ultimate Tensile Strength ($x10^3$psi)	Yield Strength ($x10^3$psi)	Elong (%)
Pitch 55 single-ply							
Long. --	5	.040	12.7	12.4	53.2	**	.29
Trans. -	5	(.011)		4.6	18.2	14.6	1.28
Pitch 55 3-ply							
Long. --	4	.078	23.3	19.6	78.8	**	.48
Trans. -	4	(.011)		4.1	9.6	7.4	1.43
Pitch 100 single-ply							
Long. --	12	.033	28.4	32.4	83.2	**	.42
Trans. -	12	(.008)		4.8	14.6	9.9	.89

** Could not be determined

The Pitch 55 panels show the expected increases in longitudinal modulus and strength with increase in fiber volume fraction. The elastic modulus in this direction increased from 12.4 Msi (86 MPa) for the single-ply panel to 19.6 Msi (135 GPa) for the three-ply panel. Likewise, the ultimate tensile strength improved from 53.2 ksi (367 MPa) to 78.8 ksi (543 MPa), respectively. All longitudinal specimens were linearly elastic to failure, exhibiting the nominal 0.3 to 0.5% strain to failure of most graphite fibers. Therefore, a yield strength could not be determined for this orientation.

In the transverse direction, the Pitch 55 panels show a slight decrease in elastic modulus with increase in fiber volume fraction. The actual values of 4.6 (32 GPa) and 4.1 Msi (28 GPa) appear to be rather low considering the much higher transverse modulus of graphite itself. However, it can be assumed that this apparent elastic region is in fact a yielding of the fiber/matrix interface and only the face sheets are fully elastic up until the indicated yield strengths are reached. Since the three-ply and single-ply panels have the same face sheet thicknesses, it is reasonable to expect the decrease in modulus with increase in overall panel thickness. In addition, the transverse tensile failure loads for the two Pitch 55 panels were identical (about 300 lbs. or 1300 N) and only when normalizing to specimen cross sectional area does the decrease in tensile stress with increase in fiber volume fraction appear. This is a strong indication that the Gr/Mg filament plies contribute little to transverse strength. The ultimate tensile stress is seen to decrease from 18.2 ksi (125 MPa) to 9.6 ksi (66 MPa) and yield stress from 14.6 ksi (101 MPa) to 7.4 ksi (51 MPa) for the single-ply and three-ply panels, respectively.

For the Pitch 100 panels, the increase in filament fiber volume fraction (from 33 to 48 v/o), the increase in fiber modulus (from 55 to 100 Msi), and decrease in face sheet thickness (from .011 inch to .008 inch), compared to the Pitch 55 panels, has resulted in marked improvements in longitudinal properties. An elastic modulus of 32.4 Msi (196 GPa) was obtained and the ultimate tensile strength was 83 ksi (574 MPa). The transverse modulus was similar to those of the Pitch 55 panels (4.8 Msi or 33 GPa) and the material yielded in this direction at 9.9 ksi (68 MPa) with an ultimate tensile stress of 14.6 ksi (101 MPa). The transverse tensile failure load was only 175 lbs. (778 N) because of the thinner .008 inch (.20 mm) face sheets.

Fractography

The fractograph of a longitudinal Pitch 55 specimen is presented in Figure 7 where the Gr/Mg filaments and the fracture surfaces of the face sheets are clearly visible. The higher magnification views show the fiber breakage occuring either individually (Figure 7a), creating a random surface with fibers of different lengths and holes from fiber pullouts, or in groups (Figure 7b), leaving colonies of fibers in steps. A closeup of fibers on one of these steps is presented in Figure 8. However, even in group failure, individual fiber weak-points have caused sub-surface fiber fracture and pullout indicating the weak bond between fiber and matrix, as shown in Figure 8. The other fibers in this area have also disbonded with the matrix. This is probably due to the released elastic strain energy during fracture which propogates the disbonds downward along the fiber/matrix interface. It is to be expected, also, that residual stresses due to the thermal mismatch between the fiber and magnesium matrix after consolidation at elevated temperature are released during these room temperature fractures.

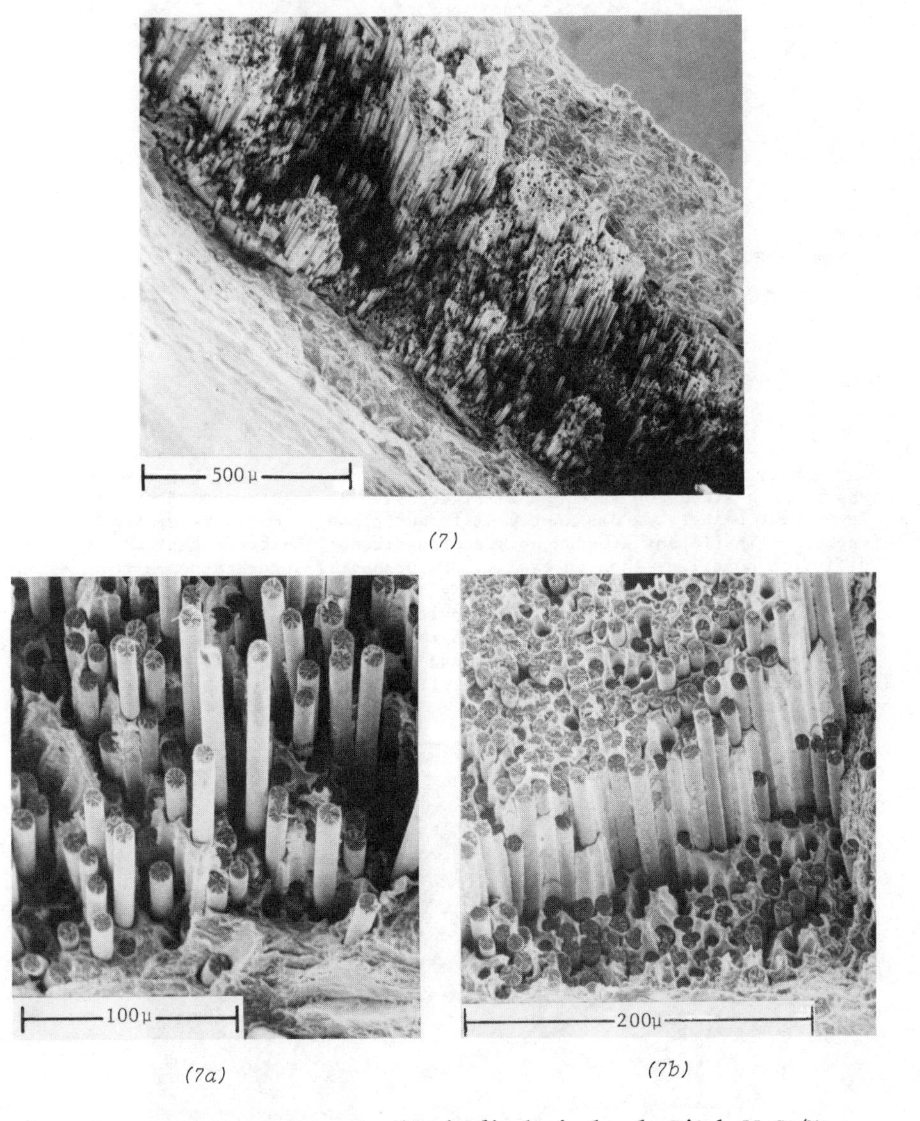

Figure 7 -- Fracture surface of a longitudinal single-ply Pitch 55 Gr/Mg specimen. The graphite fibers failed either individually (7a) or as stepped colonies (7b).

Figure 8 -- Fiber/matrix disbonds within a fracture step.

A view of the fracture transition from face sheet to fiber tow is shown for a Pitch 100 specimen in Figure 9. The fracture surface of the face sheet being at a distinctly different plane from the remaining magnesium infiltrant around the standing fibers, indicates that the diffusion bond line sheared easily. Face sheet fracture appears to have behaved fully independent of the Gr/Mg filament ply. Considering the 0.3 to 0.5% strain to failure of the graphite filaments, it is clear that the face sheets are still elastically deforming long after fiber fracture has occurred. After final tensile overload of the face sheets, delaminations are found along the entire face-sheet/fiber-tow bond line as shown in Figure 10.

Figure 9 -- Independent fracture behavior of face sheet (in the foreground) and graphite filament. Note the sheared diffusion bond interface.

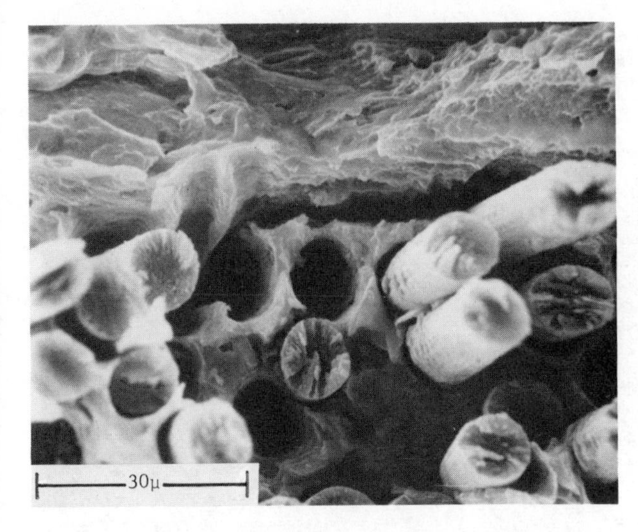

Figure 10 -- Delamination along diffusion bond line between face sheet and fiber tow.

The fracture of the face sheets appears to initiate from the inner ply surface and propogate outward. Figure 11 shows the difference in the fracture surface with a high degree of shearing (Figure 11a) transforming to a more dimpled, tensile overload surface as the outside of the face sheet is reached. In most cases, the face sheets were found to have fractured with a surface inclined 45° to the tensile axis.

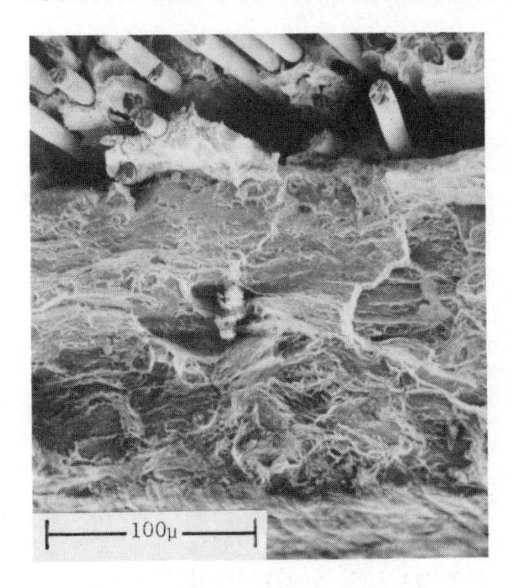

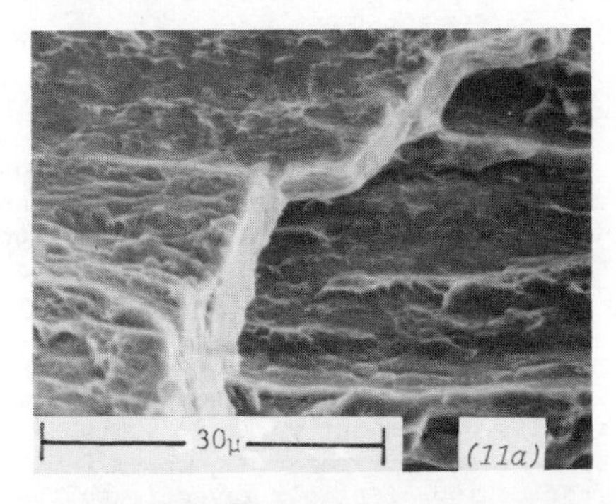

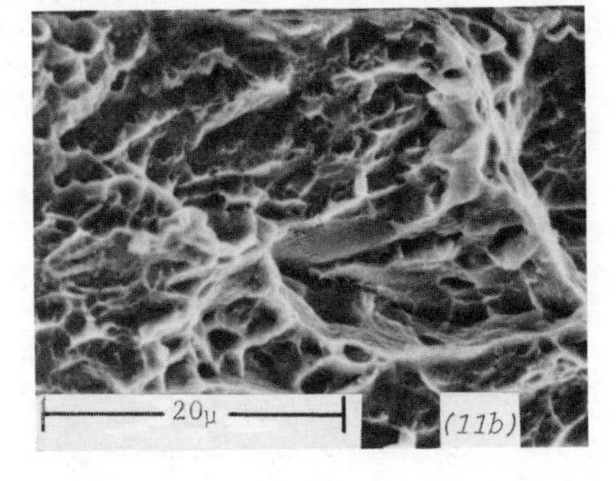

Figure 11 -- Face sheet fracture transition from shear, near the graphite (11a), to a dimpled, tensile overload structure, near the outside edge (11b).

The fracture surfaces of transverse oriented tensile specimens also show the 45° shear failure of the face sheets. A Pitch 100 transverse specimen fracture surface is presented in Figure 12. A higher magnification view of the Gr/Mg filament shows the individual fibers pulled apart.

Figure 12 -- A transverse specimen fracture surface showing 45° shear of the face sheets and weakness of the Gr/Mg filaments in this direction.

Discussion

An important use of the tensile data obtained during this study is to correlate experimental values with those predicted by the rule-of-mixtures (ROM). In Figure 13 such a comparison is made for the elastic moduli of Gr/Mg composites of two different graphite fiber types (Pitch 55 and Pitch 100) and several graphite fiber volume fractions. Using a modulus value of 6.5 Msi (45 GPa) for magnesium alloys at 0% graphite, lines are extrapolated to the moduli values for the respective graphite fiber types at 100% graphite. The experimental data provide a good fit and included are results of another work.(6) This particular material was made from the same size and type of magnesium-infiltrated Pitch 100 fibers used in the fabrication of panels investigated in this study. Since these panels had thinner face sheets (panel thicknesses averaged 0.025 inch compared to 0.032 inch for the Pitch 100 panels in this study) the graphite fiber volume fraction was much higher and averaged 34%. A corresponding increase in elastic modulus and strength resulted: 43 Msi (296 GPa) and 105 ksi (724 MPa), respectively.

Of particular interest is the prediction of the longitudinal elastic modulus of 48 v/o Pitch 100 Gr/Mg necessary for zero CTE. The ROM prediction is 51 Msi (352 GPa).

Another concern is the decrease in transverse tensile strength with increase in graphite fiber fraction. Experimental results are correlated with a simple ROM prediction of transverse ultimate tensile strength vs. volume fraction graphite in Figure 14. Here, the tensile strength of the AZ31B alloy face sheets is noted at 0% graphite. Assuming that the Gr/Mg

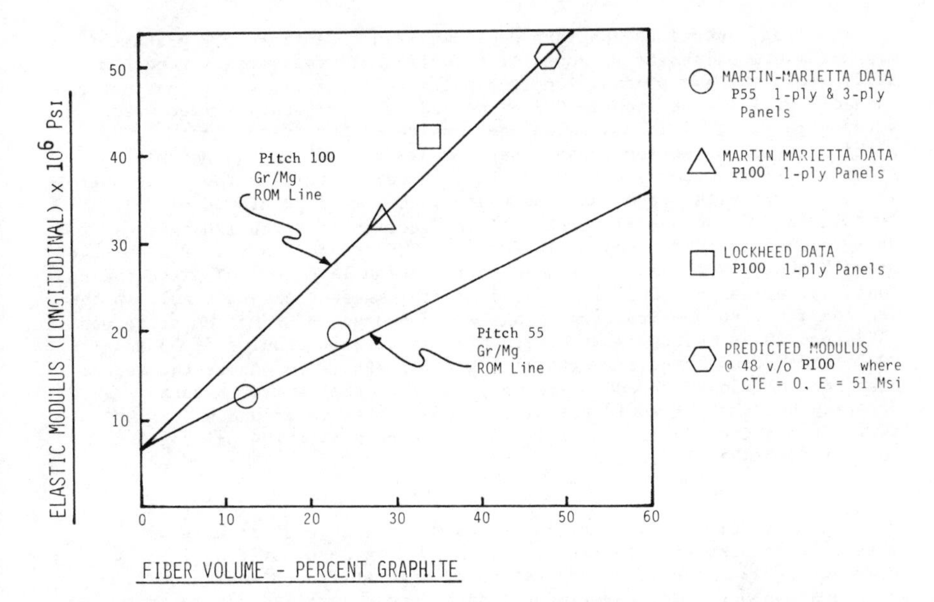

Figure 13 -- Rule-of-mixtures correlation for elastic modulus data.

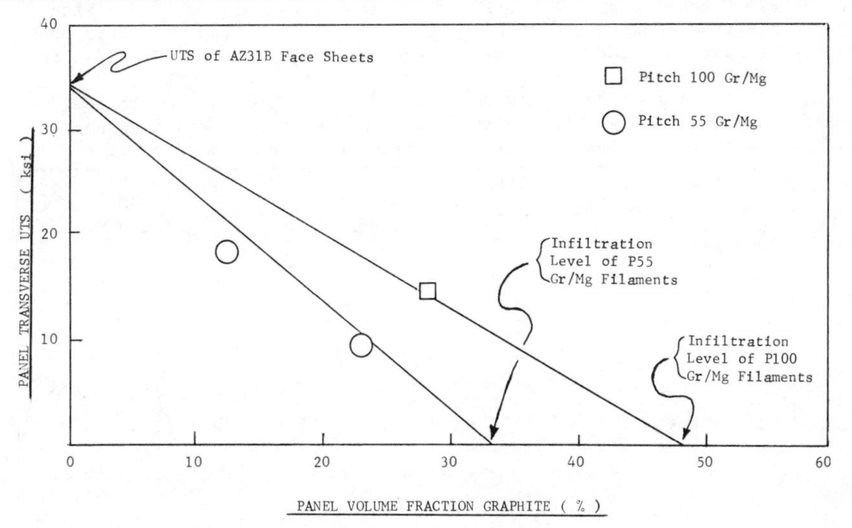

Figure 14 -- Rule-of-mixtures correlation for transverse tensile strength data.

filament plies offer no strength, lines are drawn down to the respective infiltration levels of the Pitch 55 and Pitch 100 graphite/magnesium filaments. The experimental values for transverse tensile strength of the panels tested in this study correlate very well with these predictions. It can be concluded from these results, as well as those from the fractography, that zero CTE Gr/Mg utilizing Pitch 100 fibers at present infiltration levels will have little or no transverse strength.

The final questions can then be formulated: Can zero CTE graphite/magnesium composites be produced with sufficient transverse strength and still yield the exceptional longitudinal specific stiffness values now obtained? Will a sacrifice in CTE be needed to obtain the minimum transverse strengths? A look back at the considerations listed in the Technical Background section of this paper provides some hope. Assuming a Pitch 120 fiber becomes available, that it can be infiltrated to an average of 50 v/o with magnesium, and that its CTE will be around -1.0×10^{-6} in/in ^{o}F, the overall fiber volume fraction of Pitch 120 in a Gr/Mg composite need be only 41% for zero CTE. Based on a .019 inch (.48 mm) filament ply thickness with an infiltration level of 50%, face sheets could be applied at .0021 inch (.053 mm) thickness each and result in the desired fiber volume fraction. This would produce a Gr/Mg composite panel .023 inch (0.58 mm) thick with zero CTE, an elastic modulus of 53 Msi (365 GPa), and a transverse strength of 6 ksi (41 MPa). To answer the second question, 41 v/o Pitch 100 Gr/Mg would provide the same transverse strength but the CTE would rise from zero to a value around 0.3×10^{-6} in/in ^{o}F and the elastic modulus would drop from 53×10^{6} psi to 45×10^{6} psi (365 GPa to 310 GPa).

A summary of specific stiffness and CTE for several structural materials is presented in Figure 15. As can be seen, Gr/E and Gr/Metal composites offer the best combination of high stiffness and low CTE. Points are plotted for experimental data from several volume fractions of Pitch 100 fibers in magnesium and aluminum. In addition, the ROM prediction of specific modulus and fiber volume fraction for zero CTE is shown for 41 v/o Pitch 120 and 48 v/o Pitch 100 in magnesium, and 59 v/o Pitch 100 in aluminum (However, with graphite filament infiltration levels presently limited to about 50 volume percent, the 59 v/o Pitch 100 Gr/Al is clearly unobtainable). The specific modulus of such zero CTE graphite/metal composites would be about 780×10^{6} inches and provide nearly a 60% improvement in specific stiffness over the graphite/epoxy laminate used in the LSS prototype box-truss cube introduced earlier.

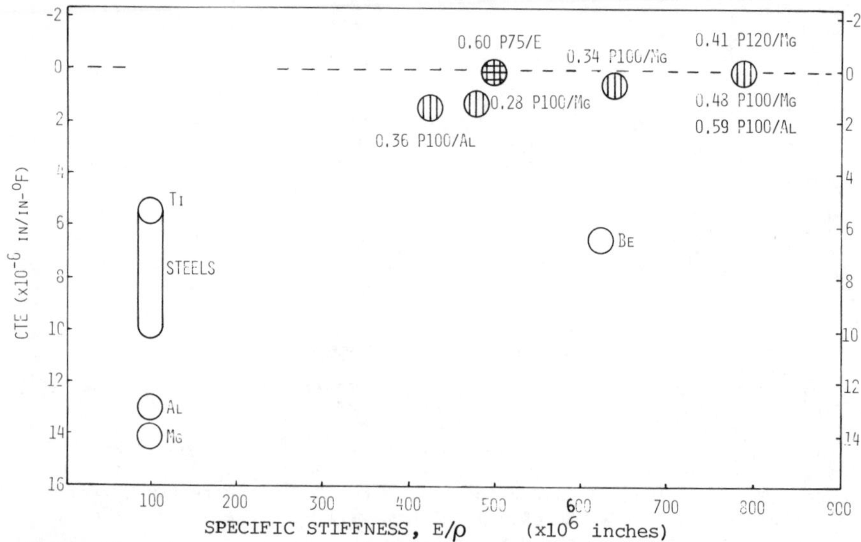

Figure 15
Thermal stability/specific stiffness comparison of structural materials.

Summary/Conclusions

Single-ply and 3-ply Pitch 55, and single-ply P100, undirectional graphite-reinforced magnesium panels were evaluated in this study. Non-destructive evaluation techniques were successfully employed to verify that the panels were free from serious voids or delaminations, and filament breakage. Microstructural examination revealed, however, that attention should be directed during panel fabriction to improve the diffusion bond lines between filaments and face sheets. Fractography studies confirmed the weakness of this interface. Extensive fiber pullout and fiber/matrix disbonds were also found, indicating the inability of fractured fibers to redistribute tensile stresses through the matrix to adjacent fibers. Furthermore, transverse specimen fractographs, as well as mechanical property data, showed the lack of filament contribution to transverse strength with face sheets bearing the load to failure.

Tensile data for both specimen orientations correlated well with rule-of-mixtures predictions of longitudinal elastic modulus and transverse strength. While the Pitch 100 panels had .008 inch face sheets with Gr/Mg filaments infiltrated to 47% graphite, the elastic modulus was greater than 32×10^6 psi at this overall fiber volume fraction of 28%. Longitudinal and transverse ultimate tensile strengths were 83 and 15 ksi, respectively. The lower modulus graphite fibers and fiber volume fractions of the Pitch 55 single and 3-ply panels yielded correspondingly lower longitudinal mechanical properties.

Rule-of-mixtures calculations using fiber and matrix properties can be made to predict the necessary fiber volume fraction of graphite to obtain zero CTE, and the resulting elastic modulus. The Pitch 100 graphite fibers (100×10^6 psi fiber modulus) require the least fiber volume fraction in magnesium (48%) to obtain zero thermal expansion. However, infiltration levels of Gr/Mg filaments are presently about 49%. The face sheets that are diffusion bonded to the collimated filaments during fabrication of Gr/Mg panels would have to be virtually eliminated to produce a composite with zero CTE. Fibers with greater elastic modulus and/or higher filament infiltration levels will be needed to obtain a zero CTE Gr/Mg composite with sufficiently thick face sheets to meet transverse strength and panel integrity requirements. Zero CTE Gr/Mg composites are predicted to have a modulus of greater than 50×10^6 psi (350 GPa).

Continuous filament graphite-reinforced magnesium composites demonstrate exceptional promise as a structural material for spacecraft requiring high specific stiffness and strength with zero or near-zero thermal expansion. With the inherent environmental stability of a metal matrix, Gr/Mg does not possess the problems associated with Gr/Epoxy composites such as outgassing, moisture absorption, low temperature capability, low resistance to radiation damage, and low thermal and electrical conductivity.

References

1. "Satellite Application of Metal-Matrix Composites" Technical Report AFML-TR-78-38, Space Division, Rockwell International Corp., Downey, California, May 1978.

2. "Satellite Applications of Metal-Matrix Composites", Technical Report AFML-TR-78-9, Hughes Aircraft Company, Culver City, California, March 1978.

3. "Satellite Applications of Metal Matrix Composites", Technical Report AFML-TR-79-4007, Space Technology, Space Systems Division, Lockheed Missiles & Space Co., Sunnyvale, California, May 1979.

4. Private Communication; J.A. Tallon, Division Technical Lab, Union Carbide, Parma, Ohio, January 1982.

5. Private Communication; S.J. Paprocki, Material Concepts, Inc., Columbus, Ohio, February 1982.

6. "Development of Graphite/Metal Advanced Composites for Spacecraft Applications", Quarterly Interim Technical Report LMSC-D667848, Space Technology, Space Systems Division, Lockheed Missiles & Space Co., Sunnyvale, California, 15 January 1980.

TRIBOLOGICAL BEHAVIOR OF GRAPHITE-FIBER COMPOSITES

M. F. Amateau

International Harvester Company
Components Engineering and Development
Hinsdale, Illinois

The tribological behavior of graphite-fiber metal matrix composites is significantly dependent on fiber type. The dependence, however, is unlike that of polymer matrix composites in that metal matrix composites exhibit greater wear resistance for the high strength (HS) fibers compared to that for the high modulus (HM) fibers. The wear mechanisms for graphite-aluminum alloy and graphite-copper alloy composites are not the same. Wear and friction of graphite-aluminum is dominated by a debris layer consisting of oxide and ground-up graphite fibers which mask the effect of fiber orientation. Graphite-copper composites have a very thin debris layer, if at all, and wear by a delamination mechanism. A significant fiber orientation effect is present for this composite with greatest wear resistance for fibers oriented normal to the sliding surface. The delamination behavior is controlled by both the concentration of graphite fibers at the surface and the properties of the fiber. High fiber concentrations and strengths tend to limit the area and depth of wear particles, thus reducing the wear rate.

Introduction

One year after the Royal Aircraft Establishment announced the development of the high-strength and high-modulus graphite fibers[1] results of a study on friction and wear of polymer composites reinforced with these fibers appeared in the literature[2]. This study confirmed that graphite fibers improved both the wear and friction behavior of a variety of polymer matrix materials. A year later work was published on the tribological behavior of chopped graphite fiber-metal matrix composites. Soon after the successful processing of graphite-metal matrix composites by liquid metal infiltration, work was carried out at The Aerospace Corporation[4] and the University of Texas[5] on the tribological behavior of continuous graphite fiber reinforced metals. Although the introduction of graphite fibers significantly improves the wear and friction behavior of polymers, metals and glass, the mechanisms of wear are far from understood and a number of results were found which run counter to expected behavior.

Tribology of Polymer Matrix Composites

A considerable mass of data has been gathered on the wear and friction behavior of graphite fiber reinforced polymers, however, no clear understanding of the mechanisms of wear have evolved. Properties of the counterface material are generally the dominating factors in the tribological behavior of polymers in the early stages of sliding. Wear rate under such circumstances is found to be proportional to the roughness of the counterface raised to the power of 1.5 to 3. As sliding continues, wear rate decreases to a steady state value caused by either abrasion of the asperities or filling the regions between asperities. These processes are quite dependent on the properties of the polymer and any filler materials which they may contain. The steady state wear rate is very dependent upon the nature of the transfer film. A series of studies by Giltrow and Lancaster [6,7,8,9,10] on graphite fiber reinforced polymers, consistently confirmed the importance of the transfer film properties on wear. Graphite fiber reinforcements generally reduce both the wear and friction of polymers but the degree of this reduction depends upon the properties of the fibers. Most of what is known about the tribology of graphite fibers-polymer composites has been determined on polyacylonitrile (PAN) precursor fibers, and may not apply to rayon or mesophase pitch precursor fibers.

A comparison of the wear and friction behavior between high modulus (HM) and high strength (HS) graphite fiber-polymer composites is given in Table 1. High modulus PAN fibers impart low friction to composite independent of the matrix polymer and counterface material. This has been attributed to role of high modulus fibers in the formation of the transfer film. On the other hand, high strength PAN fibers reduce the wear and friction of polymers to a much lesser degree. The higher wear rate is attributed to the higher abrasiveness of the HS fiber on the counterface.

The friction behavior is independent of the type of polymer matrix for either type of fiber reinforcement. The wear rates, however, are generally lower for thermoset than for thermoplastic polymer matrices.

Table 1

Summary of Wear and Friction of PAN Graphite Fiber/Polymer Composites

Fiber Type	Friction		Wear	
	Value	Counterface Dependence	Range	Surface Roughness Dependence
HM	low (.15 to .3)	no	low (10^{-8} mm^3/Nm)	yes
HS	high (.4 - .7)	yes (.13 - .2)	high (10^{-6} mm^3/Nm)	no

Neither the hardness nor the chemical content of the counterface appears to influence the friction of the HM fiber-containing composites. For HS fiber composites, high chromium content or high hardness counterface materials result in low friction. The wear rates for both fiber types are counterface dependent. Another significant difference between each type of fiber composite is the wear dependence on surface roughness. Initial surface roughness is generally smoothed much more rapidly and effectively by the more abrasive HS fibers. Below 40 volume percent fibers, the wear rate depends upon orientation. Sliding perpendicular to the fiber axis results in the lowest wear rate. The effect appears to apply to both HS and HM fiber composites.

Tribology of Metal Matrix Composites

Early studies on the wear and friction of HM graphite fiber-metal matrix composites was also performed by Giltrow and Lancaster[3]. They found that a significant decrease in wear occurred with increased chopped fiber volume fraction in lead, copper, nickel, cobalt and silver matrices. The friction behavior was not consistently influenced by the fiber content for each of the matrices. These materials were hot pressed and had poor mechanical properties resulting from less than optimum processing.

Fiber Property Effects

Continuous graphite-aluminum matrix composites produced by liquid infiltration which had rule-of-mixture mechanical properties were examined in dry sliding and found to have significantly different behavior from polymer composites[4]. Both rayon precursor and HS PAN fibers in the range of 20 to 30 volume percent in various aluminum alloys were examined. Wear and friction of these materials depend upon fiber type as was the case for polymer composites, however contrary to polymer composites, the HS fibers resulted in lower wear compared to the HM fibers. The effect on friction was reversed. That is, HM fibers resulted in the low friction values similar to the behavior in polymer composites. Since the comparison of aluminum matrix and polymer matrix composites involved large variation in matrix properties, fiber fractions, and experimental sliding conditions, it is difficult to make valid broad generalizations regarding basic composite behavior. There are even significant differences in the nature and properties of fibers which fall under the same broad classification. For instance, observations made on the behavior of the HM fibers in the polymer matrix composites involved only PAN precursor fibers while the HM fiber studies in metal matrix composites included both PAN and rayon precursor fibers.

For the case of metal composites, the higher wear rate in the HM fiber composites may be due to the larger wear debris particle. The larger particle may be the consequence of the larger crystallite or grain size which is characteristic of these fibers. Once the crystallite has been separated from its fiber by a grain boundary fracture it can act as a lubricating particle since it consists of very highly oriented graphite layers. The HS fibers fracture as smaller particles during grain boundary fracture and consist of substantial portions of non-lubricating graphiteacious material in addition to oriented graphite layers. The difference in wear rate between HS and HM fiber composites is consistent with the assumption that the elastic strain energy in the fibers near the sliding surface contributes to the work of fracture to create the wear debris particle. The HS fibers have 30 percent higher tensile strength and 50 percent higher fracture strain than the HM fibers. Therefore, the work of fracture to generate a wear particle is considerably greater for the composites containing the HS fibers.

The sliding behavior of graphite-copper matrix composites is somewhat different than graphite-aluminum. In this case both the wear and friction of continuous fiber-copper alloy matrix composites exhibits parallel behavior with respect to fiber type [11]. The wear rates are particularly sensitive to the degree of graphitization or the elastic modulus. The high wear and friction for composites containing HM fibers such as Type P, T50, HM 3000 and Modmor I(U), and low wear rates and friction for the HS composites such as T300 and Celion 6000 have been found in a variety of graphite-copper composites (Table 2).

Table 2
Effect of Fiber Type on
Friction and Wear of Copper Alloy Composites

Fiber Class	Fiber Type	Relative Wear Rate	Friction
High Modulus	T50	8-20	.4
High Modulus	HM3000	5-10	.2-.5
High Modulus	Modmor I	5	.3
High Strength	T300	1	.1
High Strength	Celion 6000	1	.1

Fiber Orientation Effects

The three orthogonal fiber orientations are generally considered including (1) fibers normal to the sliding direction (referred to as normal), (2) fibers parallel to the sliding surface but perpendicular to the sliding direction (referred to as antiparallel), (3) and fibers parallel to both surface and sliding direction (referred to as parallel).

For graphite-aluminum composites there are no significant fiber orientation effects for either friction or wear. The wear behavior of copper alloy matrix composites however are quite fiber-orientation dependent, even though friction behavior is not[4]. Typical behavior for high strength graphite-copper matrix composites containing 44 volume percent fibers is shown in Table 3. The lowest wear rate occurs for the fiber normal orientation. Antiparallel orientations generally result in lower wear rates than the parallel orientation.

Table 3

Dependence of Friction and Wear
On Fiber Orientation in Graphite/Copper Alloy Composite

Orientation	Relative Wear Rate	Coefficient Of Friction
Normal	4	.2
Anti-Parallel	8	.2
Parallel	12	.2

Wear Mechanisms

Graphite-aluminum composites behave considerably different than the aluminum matrix material under various sliding conditions[12]. The effect of sliding time is significantly different in each case. The unreinforced alloy sliding against pure iron has a constant but high coefficient of friction. The composites have an initially high friction and wear which decreases with time thus exhibiting a break-in period. This behavior suggests a surface modification, most likely the development of a transfer film. It is this transfer layer which is probably responsible for the enhanced friction and wear behavior of graphite-aluminum composite compared to aluminum alloys. No changes in the basic nature of the surface film, with changes in velocity, are reflected in the friction behavior, however wear rate is dependent on sliding velocity for sliding speeds in the range from .15 to .25 ms^{-1}. At the present time there is no satisfactory explanation for this velocity effect on wear rate. The cross section of the surface of worn graphite aluminum layer, as shown in Fig. 1, consists of a mixture of graphite, aluminum and aluminum oxide extending to a depth of 10 μm. This is confirmed by scanning Auger analysis of the surface composition of the graphite-aluminum composite. The transfer film is sufficiently established on the counterface to prevent iron from adhering the composite surface layer. This film must be highly abrasive since it consists of ground-up and embedded oxide and graphite debris in a highly worked matrix. Below this layer extends a region of 30 μm in which plastic deformation of the aluminum matrix is sufficient to cause measurable bending and fracture of the graphite fiber. Elements of many wear modes may be identified in this figure, such as subsurface shear, material transfer, and particle grooving by oxide and graphite debris.

Another interesting behavior of graphite-aluminum has been discovered in simulated brake tests against cast iron[13]. In these tests the brake material is pressed against a flywheel rotating at some initial velocity V_0 until the flywheel comes to rest. The sliding times during these tests are relatively short, in the order of 15 to 35 seconds. For pure aluminum the amount of material removed during the braking cycle is consistent with a parabolic relation between wear and initial angular velocity. This is expected since the sliding distance during the deceleration period is proportional to the square of time:

$$x = \frac{1}{2} r \omega t^2 \qquad [1]$$

where r is the radius of the friction track and ω is the angular deceleration. This is in accordance with all of the simple models of wear. Graphite-aluminum, however, shows a linear dependence of wear on initial velocity implying that wear, V, is proportional to the square root of sliding distance.

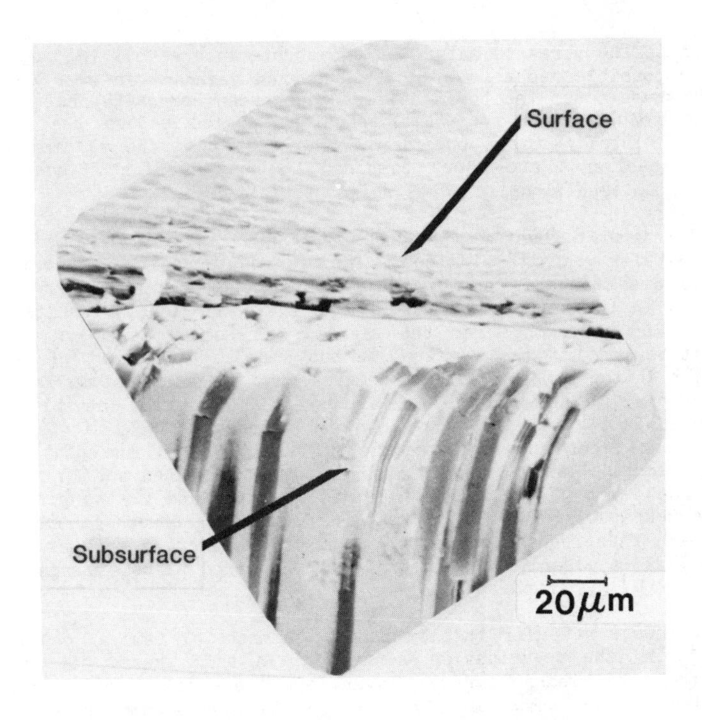

Fig. 1 Surface and subsurface views of graphite-aluminum composite after steady state sliding conditions established against an iron counterface.

Empirically the wear law to describe this behavior is:

$$V = k_c W x^{\frac{1}{2}} \qquad [2]$$

where k_c is the wear coefficient for composites. Since the sliding distance dependence in most wear laws is a result of geometrical accounting for volume the apparent square root, dependence may be a result of a distance dependence in the wear coefficient. Thus the empirical equation could just as easily be written as:

$$V = k'_c W x \qquad [3]$$

where k'_c now contains a sliding distance term (specifically, $x^{-\frac{1}{2}}$) unlike the wear coefficients for conventional materials. Two possible occurances taking place on the composite surface to account for this are (1) crack healing or (2) layer hardening so that wear rate is decreased with sliding distance.

Changes in the abrasive nature of the counterface as sliding proceeds are another possible mechanism for the observed decrease in wear rate. For any of these cases the basic difference between composite and metal would be the result of a prolonged transient period before the composites reached a steady-state condition. The fact that the sliding times in these experiments are short compared to the transient times to reach equilibrium conditions supports this speculation.

The wear mechanism for graphite-copper is considerably different than that for graphite-aluminum. The wear surfaces for the normal orientations are characterized by small fiber debris particles which remain embedded in the matrix as seen in Fig. 2. Fiber debris results from the fracture and plucking from the matrix. These particles of fibers can constitute part of the wear volume but only to the extent of its volume fraction in the composite. The remainder of the wear volume is obviously the matrix. The matrix debris leaves the surface as platelets as indicated in Fig. 3. The platelets can consist of pure matrix or matrix mixed with fiber and oxide particles. Outlines of the platelets on a sliding surface can be seen in Fig. 4. The fibers can act in two ways to influence wear: (1) to restrict platelet size and (2) to pin platelets on the wear surface. A small mean-free-distance between fibers resulting from high fiber concentration will reduce the average area of the platelet. This is clearly seen in Fig. 5 where large platelets are formed in regions with few graphite fibers.

The wear rate for this mechanism of material removal can be calculated using the delamination theory[14]. In this theory, it is assumed that the metal wears layer by layer, and there are S wear platelets in each layer. The number of platelets, S, is also assumed proportional to the number of asperity contacts. Finally, there is a critical sliding distance x_c required to remove a complete layer. The wear volume is then given as:

$$V = S\overline{V}_s L_s \quad [4]$$

where $\overline{V}_s = h\overline{A}$ is the average volume of wear platelets, and $L_s = x/x_c$ is the number of wear layers removed. x is the sliding distance, h the sheet thickness and $\overline{A}$ the average wear platelet area. The wear rate is given as:

$$\frac{dV}{dx} = \frac{\overline{A}h}{x_c} \quad [5]$$

hence, the decrease in wear rate with decreased area is clear. The presence of fibers can also maximize the critical sliding distance x_c by pinning wear and/or debris between them even after a considerable amount of subsurface damage has taken place. In Fig. 3 two types of areas are apparent: those in which delamination has occurred and those that are still intact. The areas of high fiber concentration are the ones that have not delaminated. The wear sheet thickness, h, must now be a function of the distance beneath the surface where fracture of the graphite fiber can occur. The more brittle, lower strength HM fibers are more likely to undergo fracture at larger bend radii than the HS fibers.

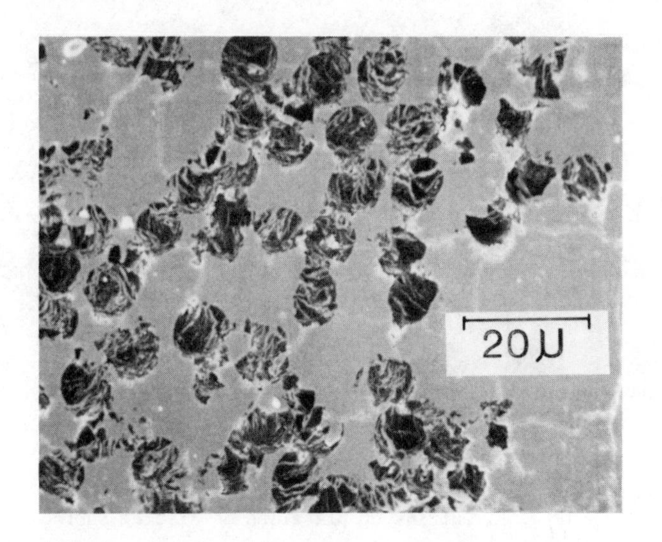

Fig. 2. Sliding surface of graphite-copper for normal sliding orientation

Fig 3. Wear platelets removed from the surface of graphite-copper after sliding

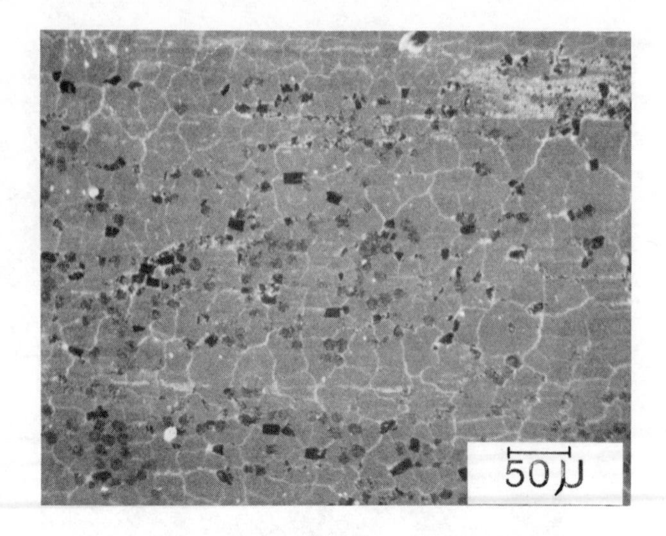

Fig. 4. Outline of platelets on sliding surface

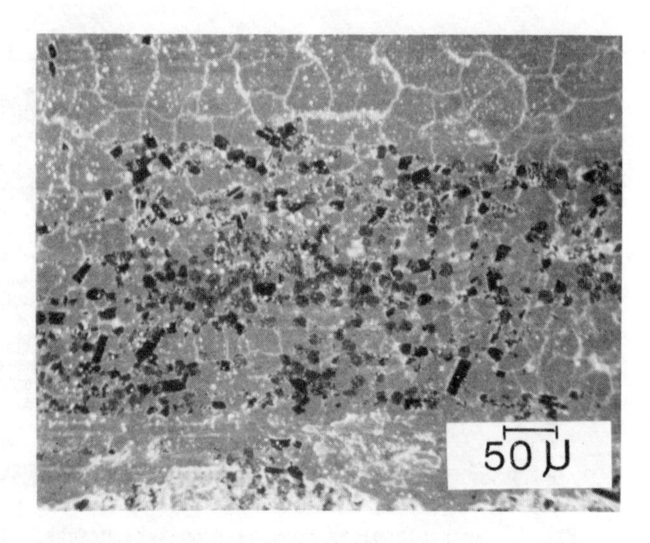

Fig. 5. Dependence of wear platelet size on fiber fraction

Thus, the average depth from the surface for fracture could be greater in the HM fibers. This is consistent with the greater wear rate for the HM composites. For pure bending the critical bend radius ρ_F is given as:

$$\rho_F = \frac{Ed}{2\sigma_F} \qquad [6]$$

where d is the fiber diameter and σ_F is the fracture strength of the fiber. Simple beam theory can be used to find the depth beneath the surface where the critical bend radius of the fiber (beam) extends if the bending moments can be accurately determined.

Subsurface cross section of a graphite-copper alloy composite after sliding reveals no thick debris layer as is found in graphite-aluminum composites (Fig. 6). The wear surface of graphite-copper alloy composites should then be sensitive to fiber orientation unlike the case for graphite-aluminum.

For the case of antiparallel fiber orientation sliding a completely different mechanism of wear takes place but the delamination model can still be applied. The platelets in this case are long and narrow corresponding to fiber segments and areas of matrix between these fibers. Fig. 7 shows a typical wear surface for this orientation. The platelet depth, h, is obviously the fiber diameter, i.e., 6 μm but the wear platelet area may be very large. The fiber debris size is likewise very long as the sliding action can pluck out long segments of fiber. Large fiber debris particles can be generated in this sliding orientation.

Conclusions

Polymer matrix composites have high wear rates for HS fibers while the metal matrix composites have low wear rates. In graphite-copper alloy composites the role of the fiber can be rationalized in terms of the delamination mode of wear where the thickness of the wear platelets depends upon the fracture depth of the fiber and the area of the platelets depend on fiber density. Orientation effects for wear and friction are different for graphite-aluminum and graphite-copper alloy composites. A thick debris layer develops on graphite-aluminum which dominates tribological behavior and masks fiber orientation effects. Graphite-copper alloy composites result in very little adhered debris layer and exhibits a strong fiber orientation effect. The lowest wear occurs when the fibers are oriented normal to the sliding surface.

Maximizing desirable tribological behavior of graphite fiber composites by material design has been shown to be quite feasible by the existing data. However, at the present time there is still an insufficient understanding of the wear mechanisms in these materials to achieve this goal effectively.

Fig. 6. Subsurface bending of graphite fibers showing different radii of bending

Fig. 7. Sliding surface of graphite-copper composite for anti-parallel orientation

References

1. W. Watt, L.N. Phillips and W. Johnson, The Engineer (May 26, 1966).

2. J. P. Giltrow and J. K. Lancaster, Nature, 214 (June 10, 1967), 1107.

3. J. P. Giltrow and J. K. Lancaster, Wear, 12 (1968), 91.

4. M. F. Amateau, W. W. French and D. M. Goddard, "Friction and Wear Behavior of Metal Matrix-Graphite Fiber Composites," Proc. ICCM, Vol. 2, AIME, N.Y. (1976), 623.

5. J. M. Casstevens, "Measurement of the Friction and Wear Characteristics of Copper Graphite Sliding Electrical Contact Materials at Very High Sliding Speeds and Current Densities", Masters Thesis, University of Texas, 1976.

6. J. P. Giltrow, Tribology, (February 1971), 21.

7. J. P. Giltrow and J. K. Lancaster, Wear, 16 (1974), 359.

8. J. P. Giltrow and J. K. Lancaster, "Graphite Fibre Reinforced Polymers As Self-Lubricating Materials," Institute of Mech. Engineers 6th Tribology Convention, Pitlochry (May 1968), Paper No. 18.

9. J. P. Giltrow and J. K. Lancaster, "Graphite Fibers in Tribology," Soc. Chem. Ind. 3rd Conf. on Graphite and Graphite, London (April 1970).

10. J. P. Giltrow and J. K. Lancaster, "Friction and Wear on Graphite Fiber-Reinforced PTFE," Proc. AFML and MRI Conf. on Solid Lubrication, Kansas City (1969), 305.

11. M. F. Amateau, R. H. Flowers and Z. Eliezer, Wear, 54 (1979), 175.

12. K. J. Pearsall, A. Eliezer and M. F. Amateau, Wear, 63 (1950).

13. Z. Eliezer, C. J. Schulz and H. E. Mecredy, Wear, 52 (1979), 133.

14. N. P. Suh, Wear, 25 (1973), 111.

FRICTION AND WEAR OF GLASS-MATRIX/GRAPHITE-FIBER COMPOSITES

V.D. Khanna and Z. Eliezer
Department of Mechanical Engineering and
Materials Science and Engineering
The University of Texas
Austin, TX 78712

and

Joanna McKittrick and M.F. Amateau
International Harvester Company
Hinsdale, IL 60521

Summary

Friction and wear experiments of glass matrix-graphite fiber composites against a tool steel counterface have been conducted. The friction machine was of the pin-on-disc type. It has been found that the unidirectional graphite composites have superior tribological characteristics compared to the bare glass matrix. Chopped fibers have virtually no effect. The relative humidity of the environment is a powerful factor in determining the friction and wear behavior of all the specimens tested. No influence of the elastic modulus of the graphite fiber could be detected.

Introduction

Composite materials based on a glass matrix reinforced with unidirectional graphite fibers have been recently developed (1). Although their attractive high-temperature properties make them potential candidates for structural applications, the presence of graphite (which imparts the inherent lubricating properties) makes them also potential candidates for tribological applications.

This paper is an attempt to describe the wear mechanism in this type of composite under different operational and environmental conditions. For comparison purposes, the friction and wear properties of the bare glass matrix, as well as those of polymer and metal-matrix composites will also be discussed.

Experimental

Apparatus

The friction and wear machine used in this investigation was of the pin-on-disc type. It has the capability of varying the sliding speed, normal load and environment.

Materials

Four types of specimens (two unidirectional composites, one chopped composite, and the bare glass matrix) were used in these experiments. These specimens were machined in the form of cylindrical pins with a diameter of approximately 4 mm. Their principal characteristics are shown in Table I. In all cases the counterface was a disc made of O_1 tool steel.

Experimental Procedure

After mounting the pin in the collet and the disc on the turntable, they were both polished down to 600 grit and thoroughly cleaned with freon and dry methanol. When the friction and wear tests were completed, the pin, the disc and the wear debris were stored in a dessicator for further optical, x-ray, SEM and AES analysis.

Results

The principal results of this investigation are presented in Figs. 1 through 5. Figure 1 shows the friction and wear values exhibited by the unidirectional and the chopped fiber composites in a typical laboratory environment. It can be seen that the unidirectional composites exhibit much lower friction and wear values than the chopped fiber composite.

Figure 2 depicts the degree of damage to the counterface (tool steel disc). It is evident that the wear rate of the counterface is high when sliding occurs against the chopped fiber composite. The effect of humidity under conditions of constant load, velocity, and temperature is shown in Fig. 3. The wear rates of both the composite and the steel counterface are very low at relative humidities higher than 40% and increase with decreasing humidity. The friction coefficient however remains virtually unchanged. The corresponding behavior of the bare glass matrix is

Table 1 - Characteristic Properties of Samples Used in This Investigation

Composite Pin No.	Fibers								Matrix*
	Type	Commercial Designation	Diameter (μm)	Density (g cm^{-3})	Strength (GPa)	Young's Modulus (GPa)	Volume Fraction (%)	Orientation	
1	Pitch	VSB-0054	11	2.16	2.24	690	67	Unidirectional	Pyrex
2	PAN	Thornel 50	6.5	1.80	2.42	393	71.8	Unidirectional	Pyrex
3	PAN	Celanese 6000	7	1.77	2.76	235	31.5	Chopped	Pyrex

*Typical properties of the matrix (Pyrex 7740) were: Density ≃ 2.2 g cm^{-3}; Young's modulus ≃ 63 GPa; Strength ≃ 7-70 MPa.

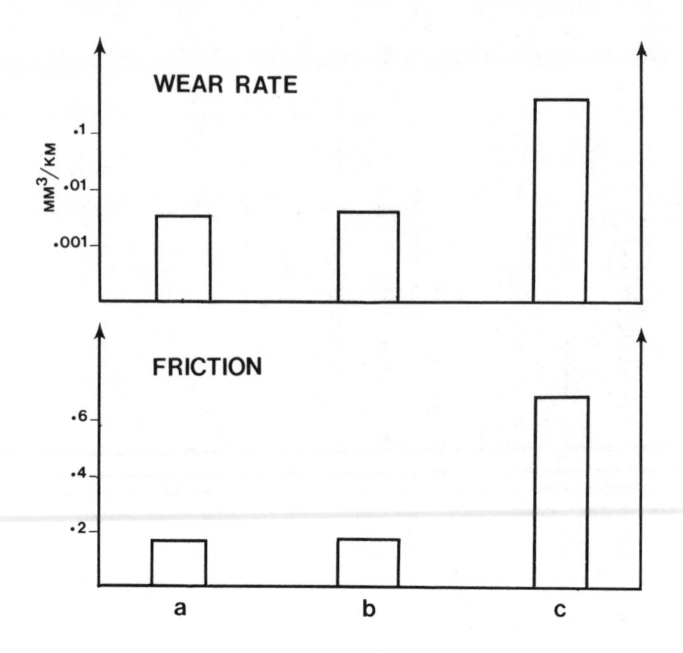

Figure 1 - Wear rate and friction coefficient of (a) pin no. 1 (unidirectional composite); (b) pin no. 2 (unidirectional composite); and (c) pin no. 3 (chopped fiber composite). Sliding velocity: 2.5 m/sec. Load: 13.36 N; Relative humidity: 65%.

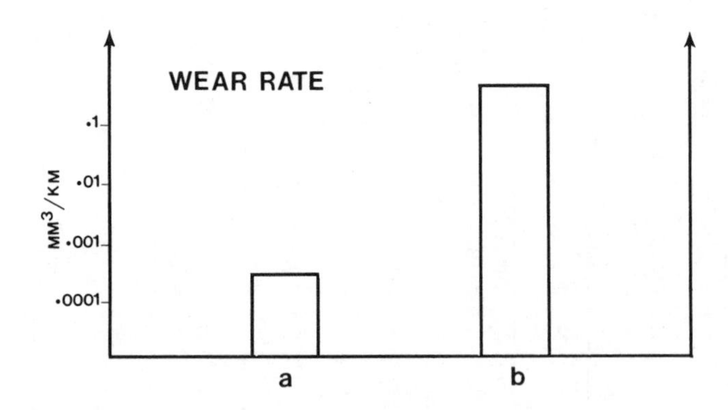

Figure 2 - Wear rate of the steel counterface; (a) against unidirectional composite; (b) against chopped fiber composite.

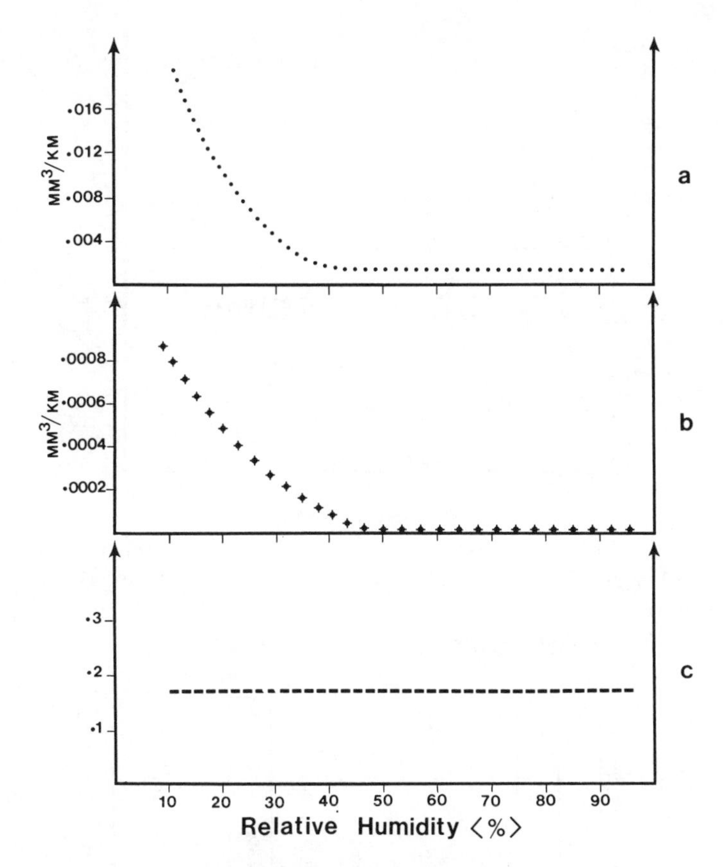

Figure 3 - The dependence on relative humidity of:
(a) Wear rate of unidirectional composite
(b) Wear rate of steel counterface
(c) Friction coefficient
Sliding velocity: 2.2 m/sec; Load: 4.45 N; Temperature: 24°C.

depicted in Fig. 4. It can be seen that the wear rate is high at low humidity, low for intermediate humidity values, and increases again at high humidity. The coefficient of friction decreases with increasing relative humidity and stays virtually constant at intermediate and high humidity values. It should be noted that the coefficient of friction is about 0.15-0.2 for the composite-steel system compared to 0.8-1.2 for the glass-steel system. Similar differences can be seen between the corresponding wear rates (Fig. 5).

Discussion

As stated in the introduction, the presence of graphite in the glass matrix composites is expected to be beneficial to their tribological behavior. This is definitely demonstrated by the very low friction and wear values displayed by the unidirectional composites compared

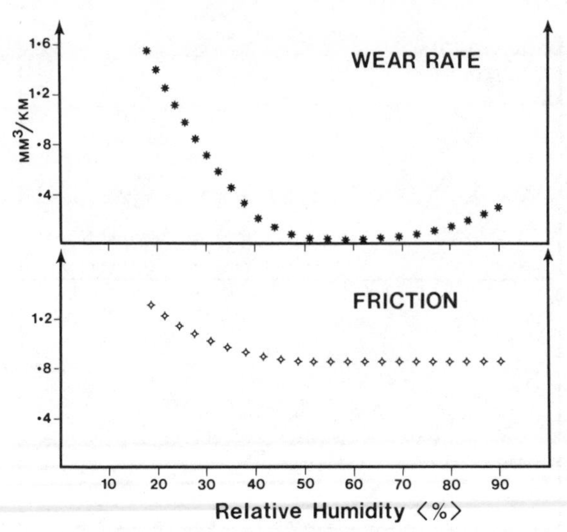

Figure 4 - Friction and wear values of pyrex pin as a function of relative humidity. Sliding velocity: 0.054 m/sec; Load: 4.45 N; Temperature: 24°C.

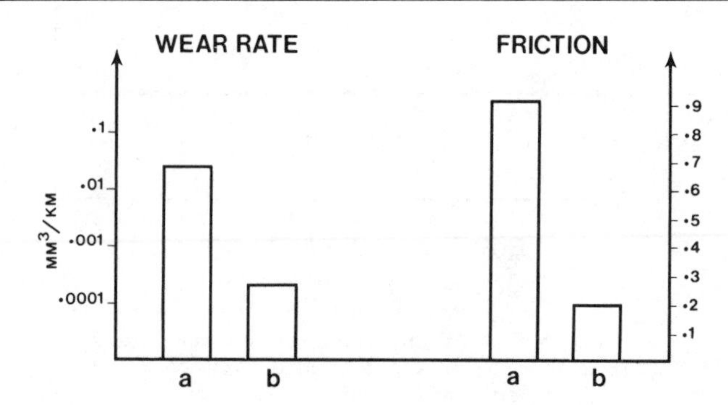

Figure 5 - A comparison between the friction and wear values exhibited by bare glass matrix (a) and unidirectional graphite composite (b).

to the bare glass matrix (Fig. 5). Optical and SEM studies of the interface revealed the existence of a graphite-containing film which explains the good frictional characteristics of the unidirectional composite. A close examination of Fig. 1 however, shows that the mere presence of graphite is not enough for imparting good tribological behavior. Indeed, the chopped fiber composite displays friction and wear values comparable to those of the pure glass matrix. The conclusion can be drawn that the mechanical properties of the composites (lower in the case of chopped fibers than in that of unidirectional fibers) influence the tribological characteristics. It is suggested that large wear glass particles are embedded in the interface film, decreasing the contribution of the graphitic component. Such large glass particles might have a pronounced abrasive effect, leading to a drastic increase in the friction and wear of both the composite pin and the steel counterface (Fig. 1c and 2b). The effect of relative humidity on the frictional behavior of glass and of graphite has been well documented (2,3). Strong adhesion bonds are known to develop between clean glass and metal surfaces. When glass is slid against a metal surface in vacuum, glass particles transfer to the metal. This transfer indicates that the cohesive fracture strength in the amorphous glass is less than the interfacial glass-metal bond. If similar experiments are conducted in a humid environment, metal is usually transferred to glass. The humid environment, therefore, seems to favor easier shear in the metal.

The frictional properties of graphite are also heavily dependent on the environmental conditions. Low friction and wear is developed only in the presence of moisture or of some volatile organic material. The function of these substances is, apparently, to help in splitting off the layer-lattice platelets.

In our experiments (pyrex-on-tool steel), the friction coefficient is virtually independent of relative humidity, except for low humidity values. On the other hand, the wear rate is significantly dependent on relative humidity (Fig. 4). This complex behavior may result from the formation of different types of iron oxide at the friction interface, as revealed by optical, SEM, and EDAX observations.

Glass matrix graphite fiber composites would be expected to exhibit tribological characteristics that would reflect the properties of their glass and graphite components. Fig. 3 shows that this is indeed the case. At intermediate and high humidity values, the graphite film acts as an effective solid lubricant and leads to very low friction and wear. At low humidity values, however, although the friction coefficient remains low, the wear rate of both the pin and the disc increase. The increase in the wear rate of glass is, undoubtedly, a result of a weakening effect on glass in the absence of water vapors. As a consequence, glass particles are removed at a higher rate, which in turn may increase the abrasion on the counterface.

It is interesting to note that in polymer matrix, and metal matrix-graphite fiber composites the elastic modulus of the fiber determines to a large extent their tribological properties (4,5). In the case of metal-matrix composites, for instance, it was found that the lower the elastic modulus of the fiber the lower the wear rate. Since in such composites the matrix is heavily plastically deformed during friction and the fibers bend and eventually break in the direction of motion (6), a dependence of the wear rate on the elastic properties of the fibers can be expected. In the case of glass matrix composites, at least under the conditions used in our experiments, no dependence of the wear rate on fiber modulus could be found (Fig. 1). A transverse section of the pin specimen revealed that no bending or breaking of the fibers took place, thus no influence of their elastic properties would be expected.

Conclusions

1. The friction and wear characteristics of unidirectional glass matrix-graphite fiber composites are superior to those of the bare glass matrix.
2. Incorporation of chopped graphite fibers in a glass matrix has no beneficial effect on the tribological properties.

3. The relative humidity of the environment strongly influences the friction and wear behavior of both the glass matrix and the glass-graphite composites.
4. In the conditions of our experiments, no effect of the elastic modulus of the fibers could be detected.

Acknowledgment

The experimental work was performed at the University of Texas at Austin under a contract with the International Harvester Company.

References

1. Karl M. Prewo, James F. Bacon and Dennis L. Dicus, "Graphite Fiber Reinforced Glass Matrix Composites," SAMPE Quarterly (July 1979) pp. 42-47.

2. Donald H. Buckley, "Metal-Dielectric Interactions," NASA Technical Memorandum 79151 (1979).

3. R. Savage, "Graphite Lubrication," J. of Applied Physics 19 (1948) pp. 1-10.

4. T. Tsukizoe and N. Ohmae, "Wear Mechanism of Unidirectionally Oriented Fiber-Reinforced Plastics," Journal of Lubrication Technology (Oct. 1977) pp. 401-407.

5. M.F. Amateau, R.H. Flowers and Z. Eliezer, "Tribological Behavior of Metal Matrix Composites," Wear, 54 (1979) pp. 175-185.

6. K.J. Pearsall, Z. Eliezer and M.F. Amateau, "The Effect of Sliding Time and Speed on the Wear of Composite Materials," Wear, 63 (1980) pp. 121-130.

MICRO-MECHANICAL ANALYSIS OF FRACTURE IN COMPOSITE SYSTEMS CONTAINING DISPERSIVE STRENGTH FILAMENTS

Ming S. Chang
International Harvester Company
Hinsdale, Illinois 60521

The fracture process in composite systems containing dispersive strength filaments involves three steps: 1) the sudden fracture of a filament, 2) the filament crack propagation into the matrix and 3) possible debonding at the interface of adjacent filament and the matrix. The purpose of this study is to investigate the influence of the fracture process on the stress and deformation fields in the composite mentioned above. A simplified two-dimensional composite model consisting of strong laminae embedded in a soft matrix is presented for this study. By using finite-element analysis, stress and deformation fields of the model under uniaxial tensile loading were computed for the three fracture steps. It is found that the sudden fracture of the weakest laminate in the composite model causes an axial tensile stress concentration in the adjacent laminate and matrix around the crack. This stress concentration is responsible for the crack propagation into the matrix and adjacent laminate. Possible debonding is investigated by examining the induced transverse stress field around the crack in the fracture process. Comparison among the stress fields of the fracture models shows that stress concentrations are localized in between the crack tip and the adjacent laminate. There is no significant change of stress fields beyond the adjacent laminate of the crack as the crack propagates through the matrix and the subsequent debonding occurs.

Introduction

Composite materials, containing high strength filaments in soft matrix, are of interest for structural applications because of their high strength to density and stiffness to density ratios (1). Unfortunately, most high strength filaments are brittle and have dispersive strength (2). Therefore, the fracture of a composite containing dispersive strength filaments may involve the following steps (3): 1) the sudden fracture of a filament in the composite, 2) the filament crack propagates into the matrix and 3) possible debonding between adjacent filament and the matrix. Everyone of these fracture steps can affect the stress and deformation fields in a given composite (4). It is the purpose of this study to investigate the influence of each fracture step on the stress and deformation fields in the composite mentioned above.

A review of the literature indicates that it is difficult to use the method of classical stress analysis (5) to deal with the fracture behavior in the composites which are considered as heterogeneous systems. This paper utilizes the techniques of finite-element analysis (6) to study the influence of the fracture process on the stress and deformation fields in the composite subjected to uniaxial load.

A simplified two-dimensional model consisting of hard laminae embedded in a soft matrix is presented for this study (Fig. 1). The analysis accounts for the influence of nonhomogeneity in the composite, breakage of the laminate, interfacial debonding and fracture in the matrix. This investigation therefore becomes a model study of fracture behavior in composite which is considered as a heterogeneous system.

Analysis of the Models

The finite-element analysis was employed to overcome many of the well-known limitations of the solutions based on the classical theories of elasticity.

A simplified two-dimensional model consisting of strong laminae embedded in soft matrix is used for this study and is shown in Fig. 1. Models for the three fracture steps of filament breakage, fracture in the matrix and interfacial debonding are shown in Fig. 2(a) to Fig. 2(c).

Because the X and Y axes in Fig. 1 are lines of symetry, only one quarter of the model was analyzed. The resulting finite-element representation is shown in Fig. 3.

Loading conditions were modeled by imposing displacement boundary conditions in the analysis. The displacements on the boundary are

$$u=0 \text{ at } X=0,$$
$$v=0 \text{ at } Y=0 \text{ and}$$
$$v=\Delta \text{ at } Y=\pm a,$$

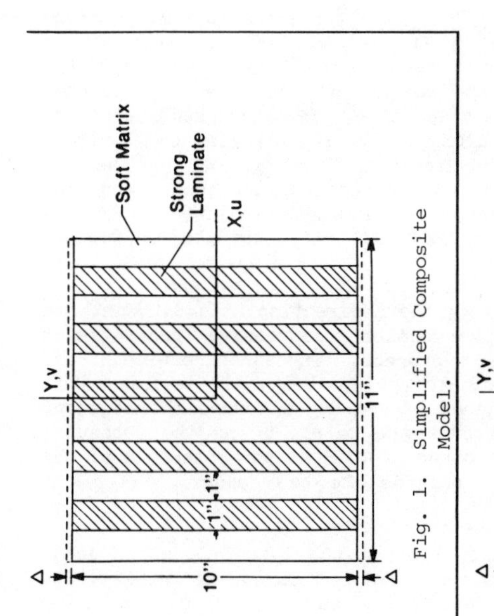

Fig. 1. Simplified Composite Model.

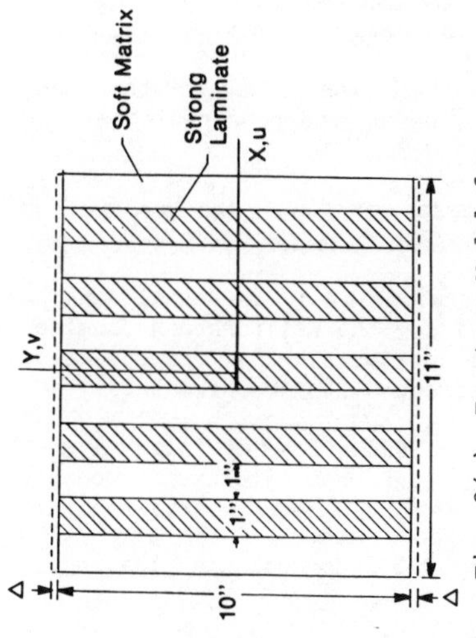

Fig. 2(a). Fracture Model for Laminate Crack.

Fig. 2(b). Fracture Model for Crack Propagation through the Matrix.

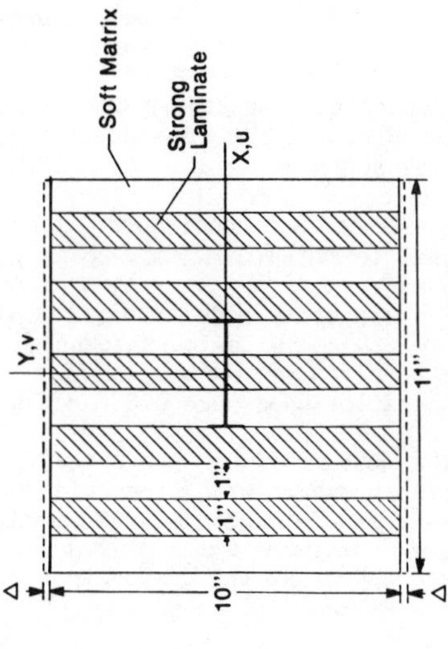

Fig. 2(c). Fracture Model for the Subsequent Debonding.

in which u=displacement in the X direction,
v=displacement in the Y direction,
Δ=displacement on the boundary,
a=half length of the composite model.

special care was taken to model the cracks in the models for the three fracture steps. They were determined by releasing the supports normal to the crack surfaces.

The boundary conditions along the crack surfaces are shown from Fig. 4(a) to Fig. 4(c).

The composite model was given the dimensions of 10 in. axial length, 11 in. width and 0.25 in. thickness. Moduli for the reinforcement laminae and soft matrix were taken as 60×10^6 psi and 10×10^6 psi respectively. Poisson ratios used were 0.2 for the laminae and 0.34 for the matrix.

The analysis was performed by using STRUDL-II (7) finite-element analysis program with the aid of the electronic digital computer. Linear-Strain-Triangle (LST) elements formulation (6) was chosen for this analysis. Stress and deformation fields were computed for the three fracture models at the composite strain of 0.04% under uniaxial loading.

Analytical Results

For the given three fracture models, with the selected mechanical property constants, stress and deformation fields were obtained.

Tensile axial stress fields (σ_{yy}) for the three fracture steps along X-direction (Y=0.00 in.) are shown in Fig. 5. It is seen that stress concentration exists in the adjacent laminate near the crack. As the laminate crack propagates through the matrix, the tensile axial stress (σ_{yy}) raises about 17% in the adjacent laminate. The subsequent debonding shows no significant change of σ_{yy} along Y=0.00 in.. Beyond the laminate adjacent to the crack, there is no significant change of σ_{yy} as the laminate crack propagates through the matrix and the subsequent debonding occurs. Increase in σ_{yy} in the adjacent laminate indicates that the breakage of additional laminate is likely if the composite is further loaded.

Transverse stress fields (σ_{xx}) in the X-direction at Y=0.00 in. are shown in Fig. 6 for the three fracture models. Compressive stress fields exist behind the crack in the fracture process. These compressive stress fields are induced as the broken laminate springs back. It is found that the position of peak tensile transverse stress moves toward the adjacent laminate as the crack propagates through the matrix or as the debonding occurs. The peak tensile transverse stress (σ_{xx}) at the interface of the matrix and the adjacent laminate is regarded to be responsible for the debonding in the composite model.

Axial tensile stress fields (σ_{yy}) along the interface of adjacent laminate and the matrix (X=0.150 in.) are shown in Fig. 7. As the laminate crack propagates into the matrix, higher stress concentration exists in the

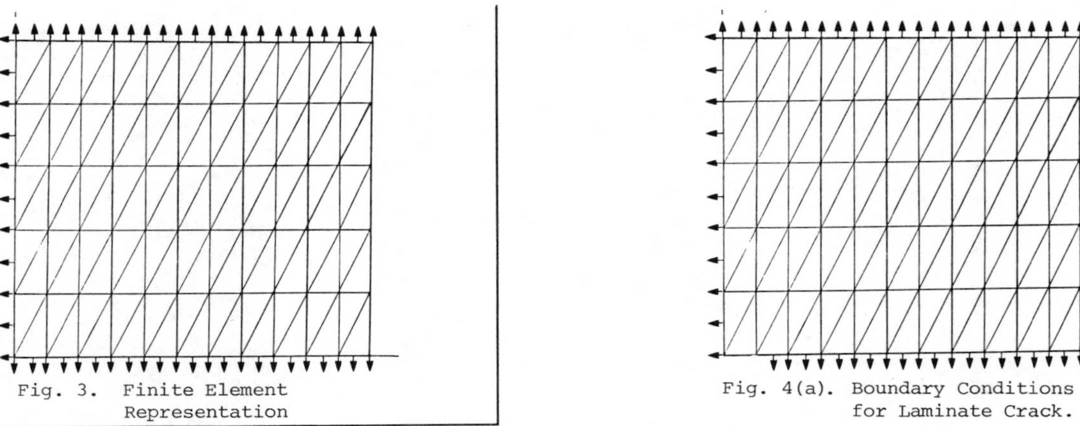

Fig. 3. Finite Element Representation

Fig. 4(a). Boundary Conditions for Laminate Crack.

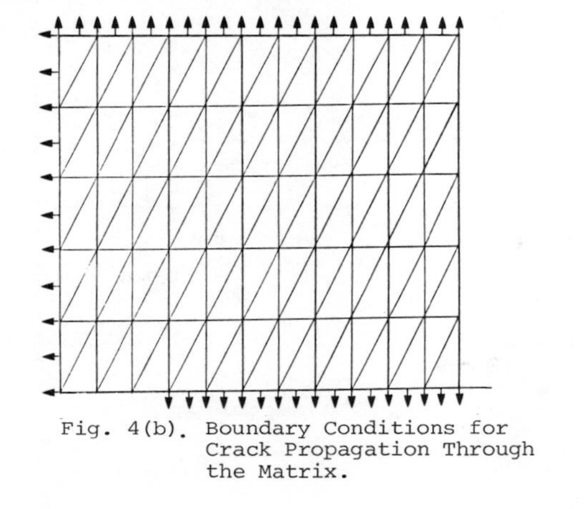

Fig. 4(b). Boundary Conditions for Crack Propagation Through the Matrix.

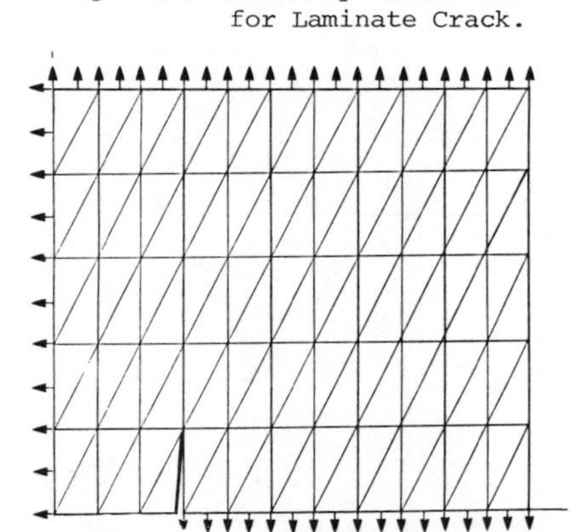

Fig. 4(c). Boundary Conditions for the Subsequent Debonding.

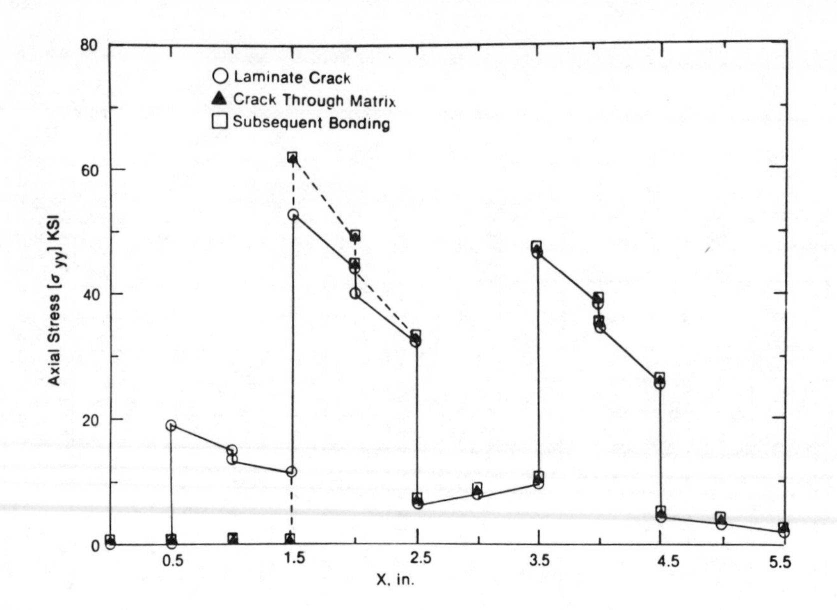

Fig. 5. Axial Stresses Along Y=0.00 in..

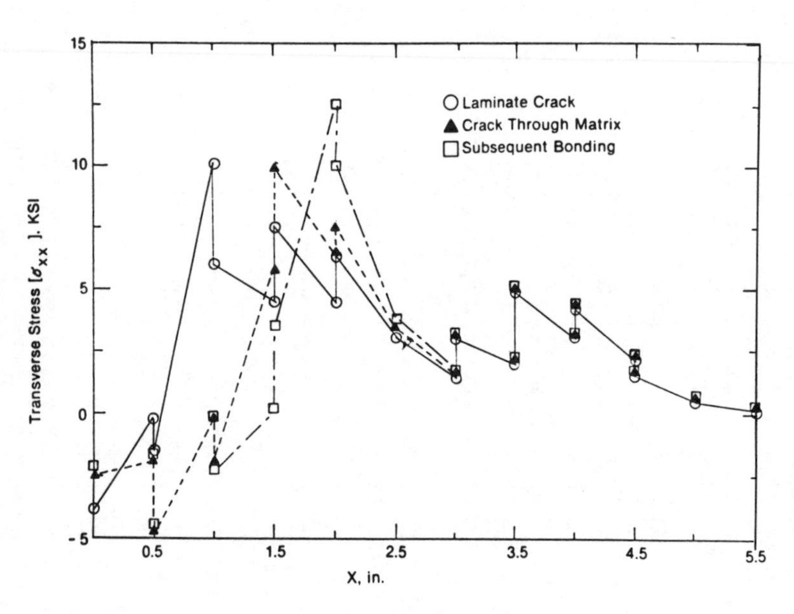

Fig. 6. Transverse Stresses Along Y=0.00 in..

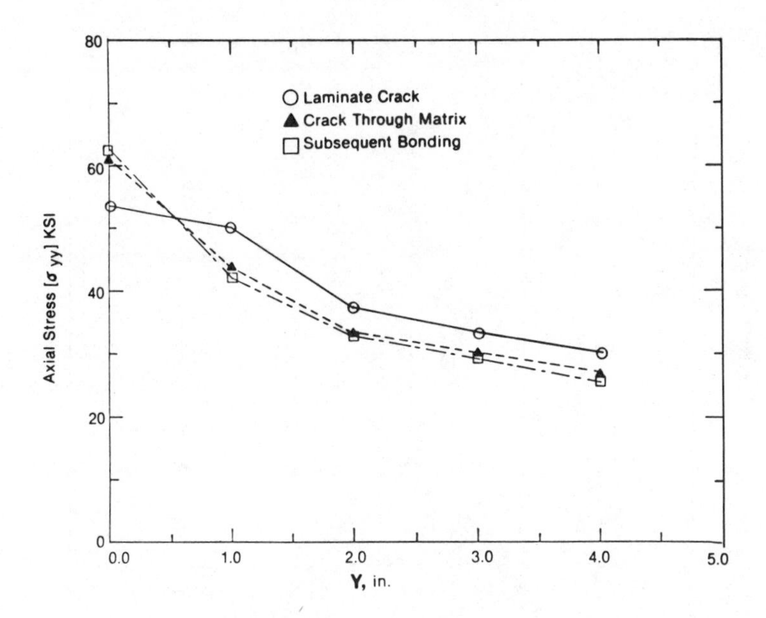

Fig. 7. Axial Stresses Along Adjacent Laminate(X=0.15 in.).

adjacent laminate near the crack tip. σ_{yy} relaxes beyond Y=0.50 in., this indicates that the breakage of additional laminate is less likely to occur beyond Y=0.50 in.

Stress fields of σ_{yy} and σ_{xx} along X-direction for Y=4.00 in. are shown in Fig. 8 and Fig. 9 respectively. No significant stress variation is found among the three fracture models.

Crack opening displacements (COD) for the three fracture steps in the composite model are shown in Fig. 10. It is observed that the broken laminate springs back after it fractures. COD continues to increase as the crack propagates through the matrix and the subsequent debonding occurs.

Conclusions

A simplified two-dimensional composite model, consisting of strong laminae embedded in a soft matrix, has been used to study the fracture behavior in a composite which contains dispersive strength filaments. This demonstrates that a qualitative micro-mechanical analysis in the composite mentioned above is possible by using finite-element analysis.

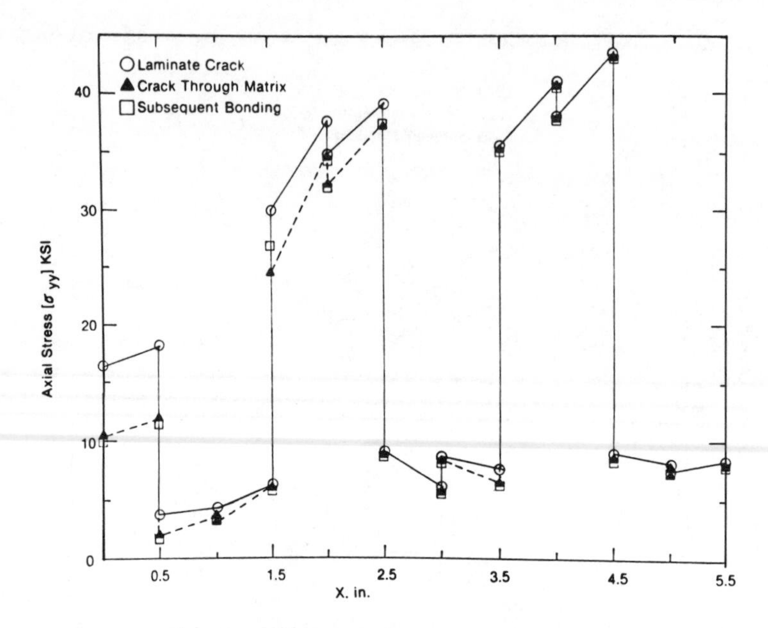

Fig. 8. Axial Stresses Along Y=4.00 in..

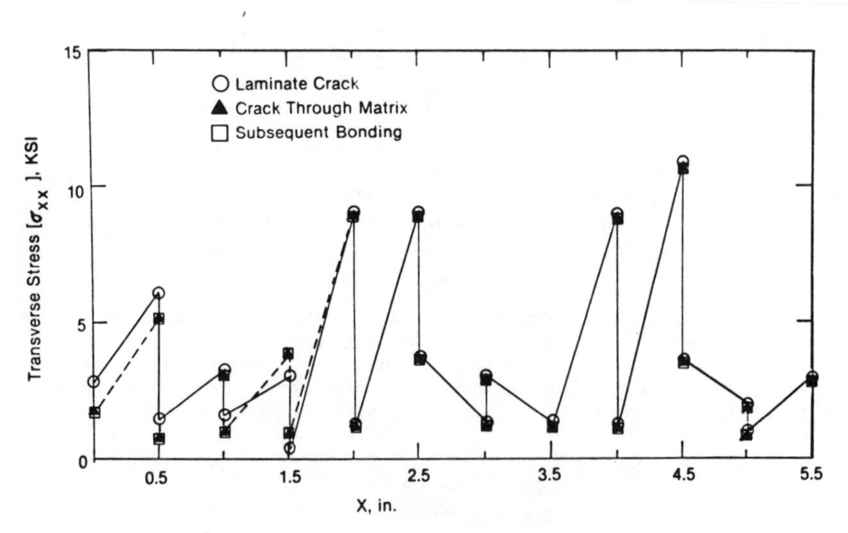

Fig. 9. Transverse Stresses Along Y=4.00 in..

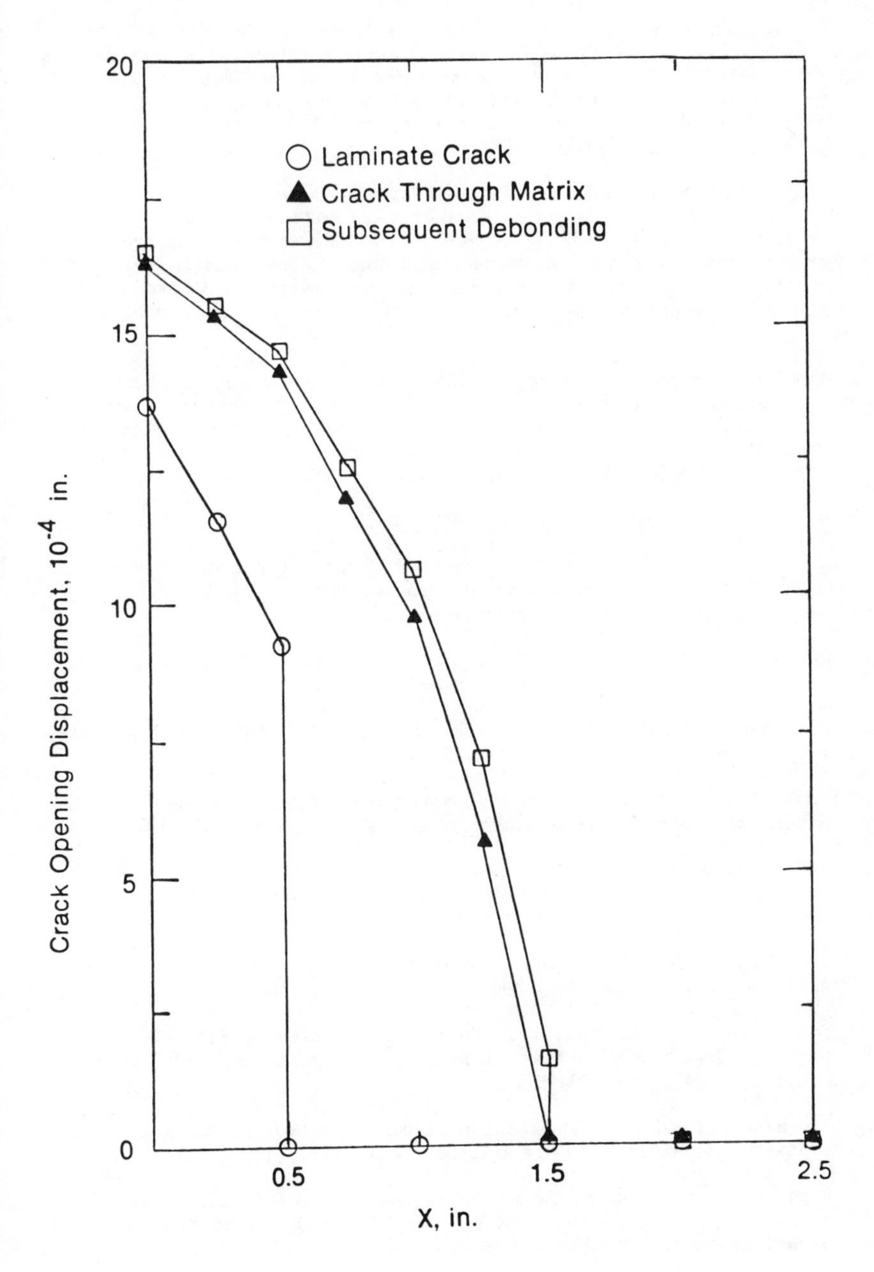

Fig. 10. Crack Opening Displacements Along the Crack

The sudden fracture of the weakest laminate in the composite model causes axial tensile stress concentration in the surrounding matrix and its adjacent laminate. This stress concentration is responsible for the crack propagation into the matrix and adjacent laminate. The subsequent debonding shows no significant change of axial stress fields in the composite model.

Behind the crack tip, compressive transverse stress is induced because of the spring-back of the broken laminate. Position of peak transverse tensile stress moves toward the adjacent laminate as the crack propagates through the matrix or as the debonding occurs. The tensile transverse stress concentratoin at the interface of the matrix and the adjacent laminate is regarded as responsible for the debonding in the composite model.

Comparison among the stress fields of the three fracture steps shows that stress concentrations are localized in between the crack tip and the adjacent laminate. There is no significant change of stress fields beyond the adjacent laminate of the crack as the crack propagates through the matrix or as the subsequent debonding occurs.

Acknowledgments

The author would like to express his appreciation to Professor R. Gallagher of Cornell University for his assistnace and Dr. E. Scala of Cortland Line Company for his encouragement.

References

1. Broutman, L. J., and Krock, R. H., Modern Composite Materials, Addison-Wesley, Reading, Mass., 1967.

2. Rosen, B. W., Mechanics of Composite Strengthening Fiber Composite Materials, American Society for Metals, Metals Park, Ohio 1965.

3. Tetelman, A. S., Fracture Processes in Fiber Composite Materials, Composite Materials: Testing and Design, ASTM STP 460, American Society for Testing and Materials, 1969, pp. 473-502.

4. Zweben, C., Tensile Failure Analysis of Fibrous Composite, Journal, AIAA, Vol. 6, No. 12, Dec. 1968.

5. Cooper, G. A., and Kelly, A., Tensile Properties of Fiber-Reinforced Metals: Fracture Mechanics, Journal of the Mechanics and Physics of Solids, Vo. 15, pp. 279-297.

6. Zienkiewicz, O. C., The Finite Element Method in Structural and Continuum Mechanics, McGraw Hill Book Co., N.Y., 1967.

7. Logcher, R. D., Connor, J. and Ferrante, A. J., ICES SSTRUDL-II. The structural Design Language, MIT Dept. of Civil Engineering Research Report R-68-92, June 1969.

FAILURE MECHANISMS IN VINYL-ESTER/GLASS COMPOSITES

N.S. Sridharan

Manager, Composites and Advanced Materials
International Harvester Company
Hinsdale, Illinois 60521

ABSTRACT

In a composite, failure can be governed by the fiber, the matrix, or the interface between the two. It is shown that each of these modes can be observed depending on the load and specimen geometry in vinyl ester/glass composites. Five failure modes are examined in a high strength sheet molding compound containing 60% by weight of continuous glass fiber and 5% chopped random glass fiber in a matrix of filled vinyl ester. The fracture surface morphology is examined using scanning electron microscopy, and morphology-property correlations are suggested. Directions for failure analysis of parts molded from this and other high strength sheet molding compounds are suggested.

INTRODUCTION

One approach for enhanced fuel efficiency in trucks and automobiles is the use of composites for weight reduction. Of the spectrum of composites available to the designer, sheet molding compounds (SMC) which contain glass fibers in a matrix of calcium carbonate filled polyester or vinyl ester combine low cost with ease of processing, and have been examined in some detail for potential applications in land vehicles. Conventional SMC materials contain 25% to 35% choppped glass fiber distributed in a random manner in the matrix, and have been used for noncritical applications for over a decade. Recently, a new class of high strength sheet molding compounds (HSMC) have been developed, and these contain up to 70% by weight of glass reinforcement(1,2).

The promise of high strength sheet molding compounds for structural and semistructural applications have led to extensive studies of their mechanical behavior for the development of a data base for design. Denton(3), Heimbuch and Sanders(4), Riegner and Sanders(5), and Walrath et al(6) have studied one or more of these materials. Loos and Springer(7) analyzed the moisture absorption and the effect on mechanical behavior for three commercially available materials. Sridharan (8) showed that the elastic and strength properties in the principal directions can be predicted using a combination of micromechanics and laminated plate theory. In all of these studies, the emphasis was on the mechanical behavior. The failure mode or fracture surface morphology was not examined. Edwards and Sridharan(9) found from scanning electron microscopy of the fracture surface that the degradation in mechanical properties in hydrothermal environments is due to the debonding at the fiber matrix interface. This behavior was observed in several HSMC's. McKittrick et al(10) showed that matrix dominated interlaminar shear can lead to failure at 1/3 to 1/2 the failure stress calculated from classical beam theory.

Although there have been some studies of the fracture surface morphology of advanced composites(11), similar studies for high strength sheet molding compounds are not available. The characterization of the failure modes and analysis of fracture surface morphology associated with each of these modes is critical for the analysis of the performance of parts molded from these materials. Such studies are essential, before widespread applications of these materials are realized. In this report, typical failure modes are induced under controlled conditions, and the fracture surface morphology is examined using scanning electron microscopy. Key morphological features that can be utilized in failure analysis are identified.

MATERIALS AND METHODS

A high strength sheet molding compound containing 60% by weight continuous glass fiber and 5% chopped random glass fiber (SMC-C60R5) was chosen as the model material for the study. The material as obtained from Owens-Corning Fiberglas Corporation is asymmetric (with continuous fiber on one side and random chopped fiber on the other) and was molded in balanced lay-ups in a compression molding press using a flat plate mold. Four plies represent a thickness of approximately 2.5 mm, and this was the thickness used for all specimens except the interlaminar shear specimen, where higher thicknesses were used to obtain the needed span-to-depth ratio.

Five distinct failure modes were induced by careful selection of specimen geometry, loading condition, and environmental pre-conditioning. Each of these modes had been observed in an extensive characterization study of this material reported elsewhere[9,10]. All specimens were failed in monotonic loading at a loading rate of 0.5 mm/sec. The failure mode, the specimen type, and the load geometry are shown in Table I.

TABLE I

Specimen and Load Geometries for Each Failure Mode

Failure Mode	Specimen	Loading Geometry
Fiber Dominated	STRIP (L/D=20)	3 Point Bend
Interlaminar	STRIP (L/D=6)	3 Point Bend
Buckling	IITRI	Axial Compression
Shear	IITRI	Axial Compression
Interfacial	End Tabbed Strip*	Axial Tension

* hygrothermally conditioned

Figure 1 shows a photograph of the failed samples from each of these tests. The three-point bend specimens (A and B in Figure 1) were 25 mm wide strips supported over a span of 75 mm. The thickness was chosen to provide the needed span-to-depth ratio to produce fiber dominated or interlaminar shear failure. The compression specimens were 12.5 x 2.5 mm end tabbed strips for use with the IITRI compression fixture.

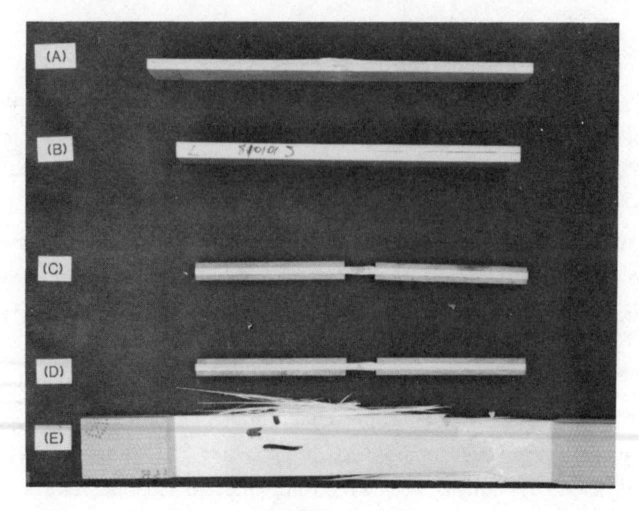

FIGURE 1
Failed Specimens

These were tested in both axial and transverse directions. The interfacial failure coupons were ASTM D3031 tensile specimens which were immersed to saturation in distilled water at 24°, 49°, and 82°C. All except the hygrothermally conditioned specimens were tested at room temperature and ambient humidity. The hygrothermally conditioned samples were tested at the temperature of exposure.

The failed specimens were visually examined and the fracture morphology was analyzed in an AMR scanning electron microscope. To render the surface conductive, a thin film of gold was evaporated onto the surface under high vacuum. A number of such fracture surfaces was analyzed to identify the key morphological features discussed here.

RESULTS

Figure 1 shows specimens failed in five different failure modes discussed. Table I lists the failure mode desired, the specimen type, and the load geometry chosen to induce the failure mode.

Figure 2
Fracture Surface in Fiber Dominated Failure

Figure 2 shows a fracture surface of the fiber dominated failure specimen. Substantial breakage of the fibers is seen and this results in the well known "brooming" effect observed at a macroscopic level. Figures 3 and 4 show low and high magnification views of an interlaminar shear failure surface. This was also induced in a three-point bend specimen by decreasing the span-to-depth ratio. Such interlaminar failures occur at loads one half that observed for fiber dominated failures. Although the failure occurs predominantly in the matrix, some failure of glass fiber interface is also seen in the higher magnification view. On a macroscopic level for the SMC C60R5 studies, the failure invariably took place on an interface between the continuous fiber layer and the chopped random fiber reinforced layer above the neutral axis of the specimen.

Fracture surfaces of a compressive buckling failure at low and high magnifications are seen in Figures 5 and 6. This mode of failure was seen routinely for this class of materials under axial compression. The amount of fiber breakage is much less than that seen in Figure 2, and the bending of the fibers is clearly seen. If the same material is tested in compression in a direction transverse to the continuous fiber axis, it fails in shear as seen in Figure 7. Again, the failure surface shows some evidence of interfacial failure in addition to the matrix failure.

Figure 8 is a photograph of the specimens that were hygrothermally conditioned and tested to failure at 25, 49, and 75°C. In an earlier study it has been shown that such conditioning can lead to rapid and significant losses in tensile properties (60-70% loss

Figure 3
Interlaminar Shear Failure - Fracture Surface

Figure 4
Higher Magnification View of Interlaminar Shear Fracture Surface Showing Matrix and Interface Failures

at 75°C). Figures 9, 10, and 11 show the micrographs of the failure surface. A trend from cohesive failure of the matrix to adhesive failure at the glass matrix interface is clearly seen with increasing temperatures of conditioning/testing.

Figure 5
Compressive Buckling Failure

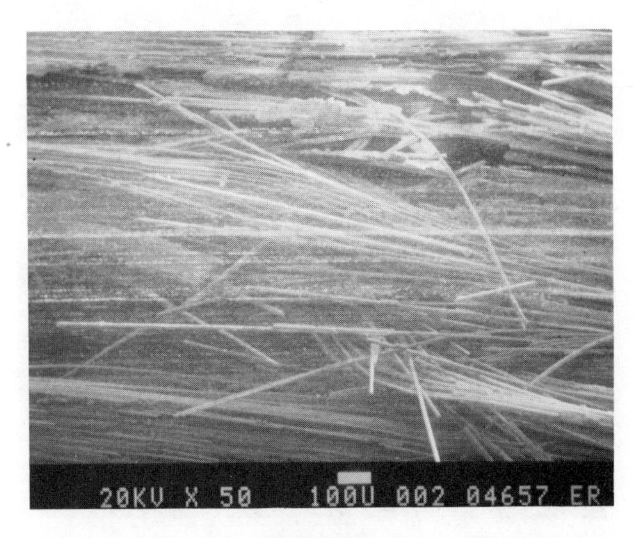

Figure 6
Higher Magnification of Compressive Buckling Failure Showing Kinked Fibers

Figure 7

Shear Failure Surface in Transverse Compression

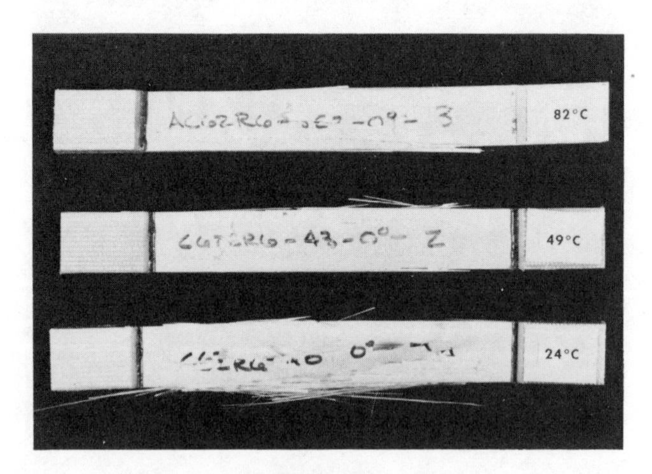

Figure 8

Failed Tensile Coupons Showing Diminished Brooming after Hygrothermal Conditioning

Figure 9

Fracture Surface of Specimen Conditioned at 25°C Indicating Cohesive Failure of Matrix

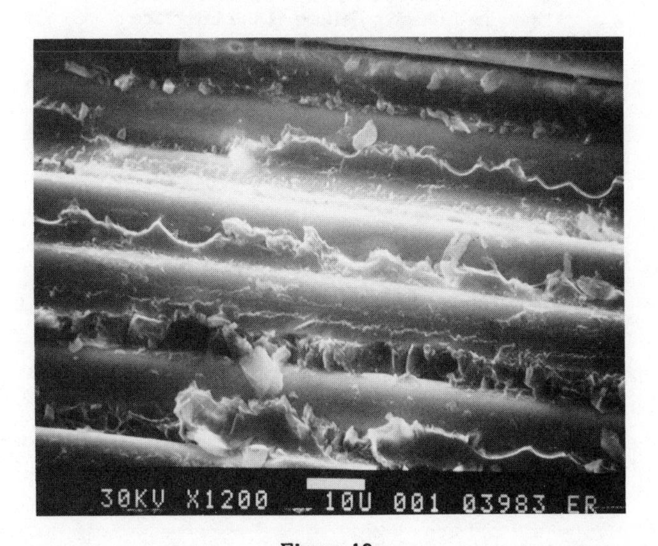

Figure 10

Tensile Failure at 49°C Showing Marginal Fiber Matrix Adhesion

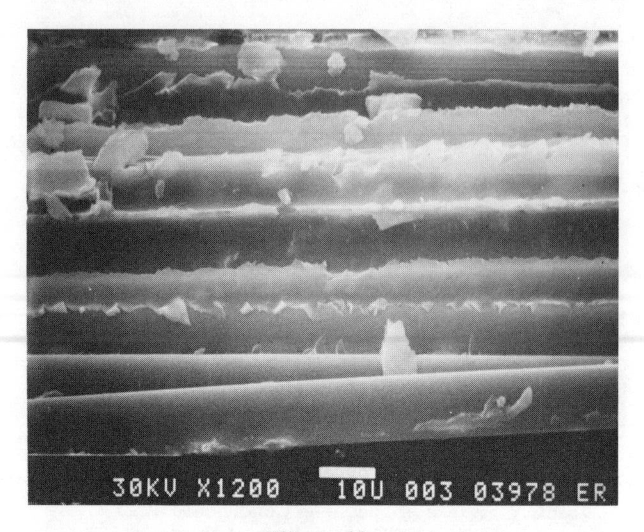

Figure 11

Tensile Failure at 82°C
Showing Adhesive Failure at the interface

DISCUSSION

The failure morphologies discussed thus far have been obtained under controlled conditions where the load geometry was chosen to produce the failure in question. In analysis of failures of structures, the situation is reversed. In such a case a stepwise analysis which identifies the primary mode of failure and indicates areas of concern should be followed. The approach is schematically indicated in Figure 12. This area of study for these materials is in its infancy and the approach suggested may be of limited utility in complex failures.

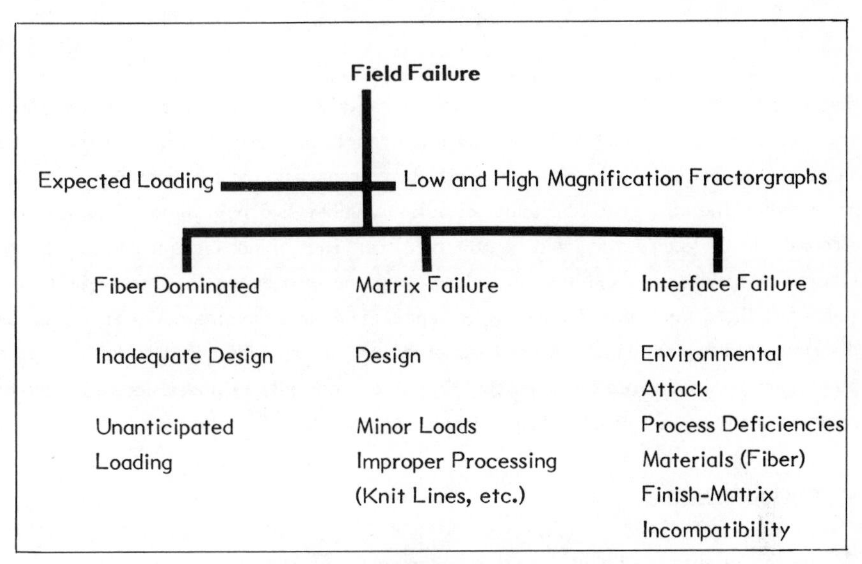

Figure 12

Analysis of Failure of Vinylester/Glass Composites

Fiber Dominated Failure

The key morphological features for a fiber dominated failure are: 1) Substantial fiber damage. 2) Powdery appearance of the matrix or total absence of matrix material in the vicinity of broken fibers (Figure 2). Since a composite is designed to enable the fibers to be substantially loaded, a fiber dominated failure is an indication of design inadequacy or major unanticipated loads.

Matrix Dominated Failure

The key indicators of a matrix dominated failure are essentially smooth, featureless fracture surface at lower magnifications (100X). Although some fibers and interfacial damage may be evident at higher magnifications (Figures 3, 4, 7). This type of fracture surface can arise from several conditions and design of the structure and should be critically evaluated. In some specific structures (leaf springs, cross members), the preferred mode of failure by design is interlaminar and will thus be matrix dominated. Matrix dominated failures can also arise from unanticipated minor transverse and shear loads in HSMC, as the strength of these materials in transverse shear or tension (which can lead to Mode I crack propagation in the matrix) can be 10% or less than the strength in principal directions, unlike isotropic materials where the strengths in these directions will be 50% or more. Thus, minor loads from assembly or vibrations, for example, can lead to failure. Improper processing can lead to failure at the knit lines, and the failure again appears matrix dominated. This can be discriminated by examining the propagation of the failure, whether it is confined to a plane parallel to the reinforcements. Knit lines generally meander across planes of reinforcement and are easily differentiated from interlaminar failures.

Interfacial Failures

Of the modes discussed thus far, this is the most difficult to clearly identify. Two clear indications of failure in the interface are: a) Presence of adhesive failure at the fiber matrix interface (on a plane parallel to the principal reinforcements - Figures 10 and 11) or substantial fiber pull-out on a failure surface. In both of these cases, the next question to be answered is whether poor fiber-matrix adhesion was inherent to the material or was induced by poor processing techniques, or because of severe service environments suich as high temperature/humidity conditions. It should also be kept in mind that the interfacial failures can be mistaken for matrix failures at low magnification and should, therefore, be examined at higher magnifications to discriminate between the two modes.

CONCLUSIONS

It is shown that a number of distinct failure modes can be observed in vinyester/glass composites depending on the loading and hygrothermal history of the material. Key morphological features that can be used to identify the different modes are discussed.

REFERENCES

1. Ackley, R. H. and Carey, E. P. in Proceedings, 34th SPI Conference 1979.

2. Jutte, R. B., Paper No. 780355, SAE 1978.

3. Denton, D. L. in Proceedings, 34th SPI Conference 1979.

4. Heimbuch, R. A. and Sanders, B. A., "Mechanical Properties of Automotive Chopped Fiber Reinforced Plastics," Report MID-78-032, General Motors Corp., Warren, Michigan (1978).

5. Riegner, D. A. and Sanders, B. A., Proceedings, 35th SPI Conference 1980.

6. Walrath, D. E., Adams, D. F., Riegner, D. A., and Sanders, B. A., ASTM STP 772, 113 (1982).

7. Loos, A. C. and Springer, L. S., J. Compos. Mater., 14, 142(1980).

8. Sridharan, N. S., ASTM STP 772, 167(1982).

9. Edwards, D. B., and Sridharan, N. S., Poly. Compos., Vol. 3, No. 1, 1(1982).

10. McKittrick, J. M., Sridharan, N.S., and Gujrati, B. D., Submitted for Publication.

11. Agaraval, B. D., and Broutman, L. J., in "Analysis of Performance of Fiber Composites," pp. 47, John-Wiley and Sons (1980).

NONDESTRUCTIVE EVALUATION OF BORON-CARBIDE-COATED, BORON-FIBER-REINFORCED TITANIUM

J. C. Duke, Jr., A. Govada, and A. Lemascon
Materials Response Group
Engineering Science and Mechanics Department
Virginia Polytechnic Institute and State University
Blacksburg, VA 24061 USA

ABSTRACT

Boron carbide coated boron fiber reinforced Ti 6Al-4V metal matrix material was characterized in the as-received condition using a number of nondestructive inspection (NDI) techniques: ultrasonic C-scanning, X-ray radiography, eddy current, and vibrothermography. Subsequent to this the material was machined into coupon type tensile specimens. During quasi-static tension of the specimens at constant rates of loading, acoustic emission (AE) and ultrasonic attenuation were monitored. Retrospective analysis of the initial NDI results in the vicinity of the final failure site did not support any contention regarding initially existing imperfections. Post test metallography, Auger electron spectroscopy (AES), and extensive analysis of the continuously monitored AE results provided insight into the microscopic deformation mechanisms active in this material; the ultrasonic attenuation results were found to be complementary.

Introduction

The continued need to improve the performance of flight vehicles places a premium on design of efficient, reliable airframe structural configurations. One key to better design is the effective application of advanced material systems. Advanced composite materials are being developed because they have the potential for enhanced structural characteristics to improve vehicle performance. A large problem inherent to these materials, however, is the difficulty of predicting the mechanical behavior deviations caused by internal defects. The result is a lack of validated design data which is especially true for metal matrix composites.

The nature of this problem is basically two fold. On the one hand an adequate description of the imperfections typically occurring in these materials which are inherent in the constituents or which result from their combination or during the subsequent manufacture of a part is not available. And on the other, and perhaps even a consequence, no mechanical theory exists which appropriately allows for predicting the strength, stiffness or life of these materials for actual service conditions. If such descriptions of the imperfections were available, and a suitable mechanical theory existed the situation regarding nondestructive evaluation would be that described pictorially in Fig. 1.

This however, is not the case, and can be viewed only as the goal. Table I lists areas where problems exist which must be overcome in order to attain this goal. A more realistic description of the situation is displayed in Fig. 2 where the emphasis is in the areas of technique development as well as the development of a mechanical model which would allow for adequate description of the materials response under actual service conditions.

Table I Identification of Nondestructive Evaluation Problem Areas*
(Assume all objects inspected are tested)

CASE	IMPERFECTION	NONCONFORMITY	DEFECT	EVALUATION
1	NO	NO	NO	RIGHT
2	YES	NO	NO	RIGHT
3	YES	YES	YES	RIGHT
4	YES	YES	NO	WRONG
5	YES	NO	YES	WRONG
6	NO	NO	YES	WRONG

CASE	PROBLEM AREA
2	NDI EFFICIENCY
3	SOURCE OF OBJECTS
4	MECHANICAL THEORY
5	MECHANICAL THEORY
6	NONDESTRUCTIVE INSPECTION

*To aid in understanding the table, consider as an example case 4. An object is inspected nondestructively, an imperfection is found. After considering the specifications for the object, the imperfection is found to be a nonconformity. It is tested to check the evaluation and it does not fail; the imperfection was not a defect.

The main objective of this effort was to apply advanced nondestructive evaluation (NDE) techniques to characterize the nature of internal imperfections in titanium matrix composites and correlate the presence of these imperfections to actual mechanical properties(1).

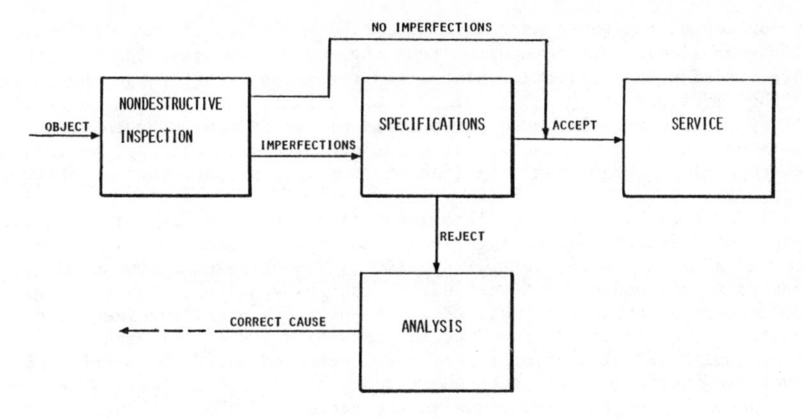

Figure 1 - The ideal nondestructive evaluation process.

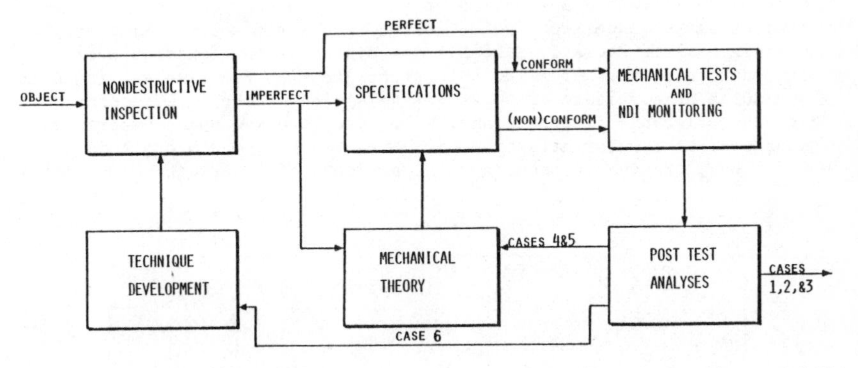

Figure 2 - The approach for developing a nondestructive evaluation capability.

Initial Evaluation and Specimen Preparation

Prior to any specimen preparation the test panel, a 30.5 cm x 30.5 cm x 1.5 mm plate of Ti 6Al-4V metal matrix material unidirectionally reinforced with boron carbide coated boron fibers, was inspected using a number of different NDI techniques.

Ultrasonic C-scanning

A conventional ultrasonic C-scan immersion system was used which consisted of a Sperry UM721 Reflectoscope and Automation US 450 Laboratory Scanner, along with a 10 MHz focused transduer. The panel was interrogated through the thickness. In order to obtain an integrated estimate of any internal inhomogeneities and because of limitations in spatial resolution caused by the relatively small thickness of the panel, the scanning was performed using the following procedure. First, the panel was positioned approximately 0.75 cm above a polished "reflecting" plate. The transducer position was then varied so that the ultrasonic pulse echo signal from the top surface of the panel was maximized. The amplitude of the ultrasonic pulse echo signal, which resulted from the sound wave traveling through the panel striking the reflecting plate, and returning back through the panel, was then monitored. By doing this, relative differences in the material's ability to transmit the sound waves could be determined. The individual specimens were also ultrasonically C-scanned prior to testing. No indication regarding damage resulting from the specimen preparation was observed.

To assume that a reproducible inspection procedure was being followed, one of the specimens was kept for use as a control specimen. Every C-scan included a C-scan of the control specimen. The ultrasonic system was adjusted to reproduce the C-scan of the control specimen. Now although each C-scanning procedure included a scan of the control specimen, the record is not included in the figures containing various C-scans. Variations of the amplitude below an arbitrary level were indicated as imperfections. Visual examination of the panel suggested that surface irregularities might be responsible for the variations in the results of the ultrasonic examination, as shown in Fig. 3.

X-ray Radiography

X-ray radiographic inspection of the as-received panel was conducted by subjecting the panel to a soft, low energy, X-ray source. The system used was a Hewlett Packard 43805 N X-ray system of the Faxitron Series. Using Kodak Industrex M-5 film, an appropriate exposure time was found to be 5 minutes at a voltage of 50 kV. Examination of the film records showed no noticeable imperfections, i.e. fiber breaks, voids, matrix rich regions etc. Particular attention was given to the areas brought into question by ultrasonic C-scanning. No evidence of imperfection was observed

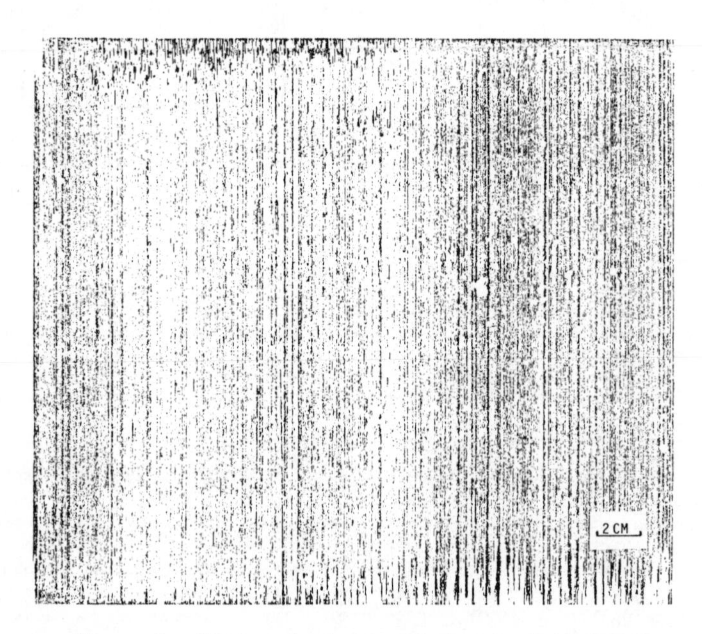

Figure 3 - Ultrasonic C-scan of the panel (reduced).

Figure 4 - Sample X-ray radiograph of a region of the panel.

as a result of this examination. Figure 4 shows a sample of the results of this examination.

Eddy Current

Preliminary eddy current inspection at an excitation frequency of 200 kHz showed the technique to be sensitive to surface irregularities observed on the panel visually. The equipment available was not capable of automatic scanning of the panel. (Due to the limited specimen size it was not possible with the eddy current probes of this device to adequately interrogate the specimens during the mechanical tests.)

Vibrothermography

Limited inspection of the specimens using the vibrothermography technique which involves the use of a thermal imaging camera to observe heating in materials subjected to low amplitude high frequency vibration proved inconclusive. This was a result of both the high quality of the material as well as the high thermal conductivity of the material.

Specimen Preparation

Specimens parallel and transverse to the fiber direction were cut from the panel of boron carbide coated boron/Ti 6Al-4V. Cutting was performed using a diamond impregnated cutting wheel and continuous coolant flow. Such a procedure provided a specimen of rectangular cross section with extremely smooth and burr free edges. Subsequent to cutting, each specimen was ultrasonically C-scanned and X-ray radiographed to determine if damage had occurred during preparation; no damage was observed.

Mechanical Tests and Nondestructive Inspection Monitoring

In order to study as precisely as possible the response of the metal matrix composite material to tensile deformation, continuous monitoring of acoustic emission activity and changes in ultrasonic attenuation were performed.

Quasi-Static Tension

Both specimen types, parallel to the fiber direction--"longitudinal" and perpendicular to the fiber direction--"transverse," were subjected to quasi-static tension with load increasing at a constant rate. Figs. 5 & 6 are typical examples of data obtained from such tests. Table II shows the mechanical results typical of both the specimen types examined.

Table II Average Mechanical Test Values

Specimen Type	E	σ_y	σ_{UTS}
Longitudinal	210.5 GPa	492.3 MPa	962.8 MPa
Transverse	195.0 GPa	107.2 MPa	408.7 MPa

The average initial stiffness value for longitudinal specimens shown, are especially low in light of a rule of mixtures prediction of 244 GPa for a material combining these constituents; the volume fraction of fibers in this panel was 0.5.

This panel was made by diffusion bonding, at elevated temperatures and pressures, sheets of matrix sandwiching layers of fibers. As a result, in some areas the fibers are poorly distributed and not completely surrounded by the matrix. The value obtained for the transverse specimens is less alarming, but it would be less sensitive to the problem just described; the strength for the transverse specimens would be sensitive, however.

Acoustic Emission Monitoring

Acoustic emission (AE) activity was monitored continuously during a number of tests. Figs. 7 & 8 are the typical results of longitudinal and

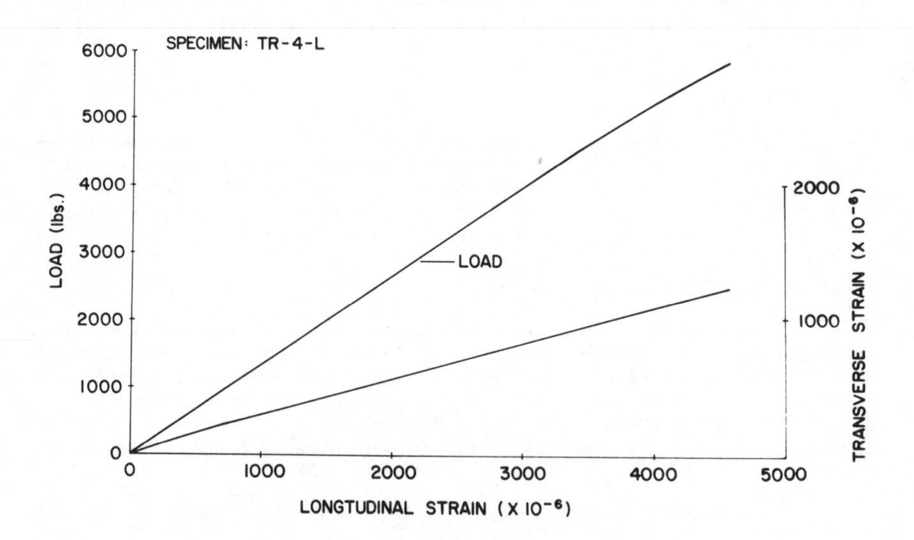

Figure 5 - Load and transverse strain vs. longitudinal strain typical of a "longitudinal" specimen.

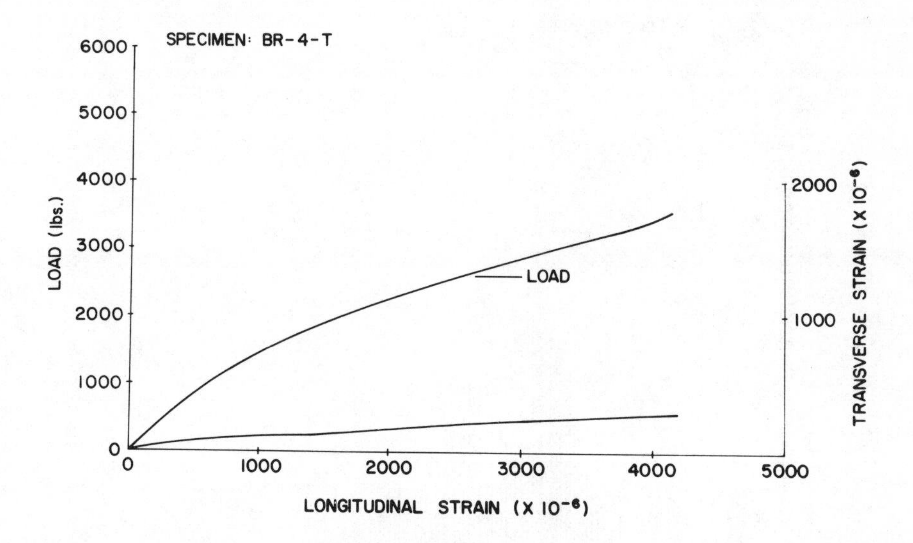

Figure 6 - Load and transverse strain vs. longitudinal strain for a "transverse" specimen.

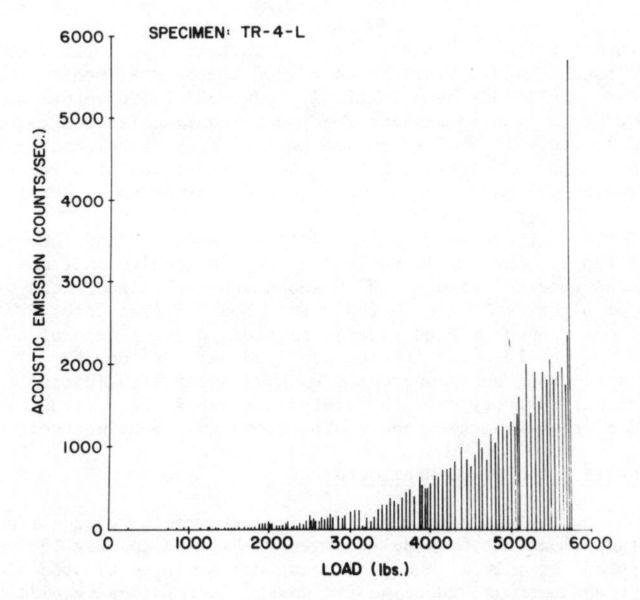

Figure 7 - AE count rate vs. load observed during a test on a longitudinal specimen.

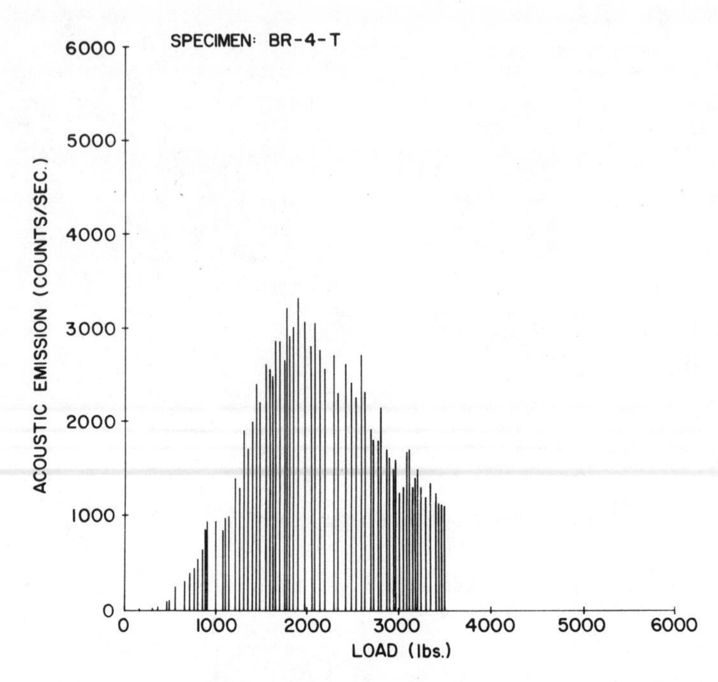

Figure 8 - AE count rate vs. load observed during a test on a longitudinal specimen.

transverse specimens, respectively. Monitoring was performed by attaching a wide band AE sensor, AET Model FC 500 to the specimen the output from which was amplified, by a Tektronix Model 502 Differential Amplifier and recorded by a high speed tape recorder, (maximum frequency sensitivity 300 kHz) Honeywell 5600 B; the count rate above a selectable voltage threshold (approximately 0.25V in these tests) was also recorded by means of a Hewlett-Packard 5326 B Timer Counter output to an XY plotter.

The tape recorded signals were later examined and three distinct signal types occurring in tests on both the longitudinal and transverse specimens were observed. Fig. 9 shows the real time record of the three types of signals along with their fast Fourier transform (FFT). The rate of the occurrence of each type is plotted on the record of load versus strain in Fig. 10 a,b,c for the longitudinal case and Fig. 11 a,b,c for the transverse case; superimposed on each is a curve indicating the total count rate of all three types. The distributions of the various types was obtained by reviewing the tape at low speed and counting their occurrence.

Ultrasonic Attenuation Monitoring

Changes in ultrasonic attenuation were monitored continuously for both longitudinal and transverse specimens: Fig. 12 and Fig.13, respectively are typical examples. The monitoring was achieved by attaching an ultrasonic transducer and fused quartz waveguide to opposite side of the specimen and monitoring the amplitude of successive echoes with an automatic attenuation recorder. The system used combined a Matec Model 6000 Pulse Generator and Receiver coupled with a Matec Model 2470A Attenuation Recorder.

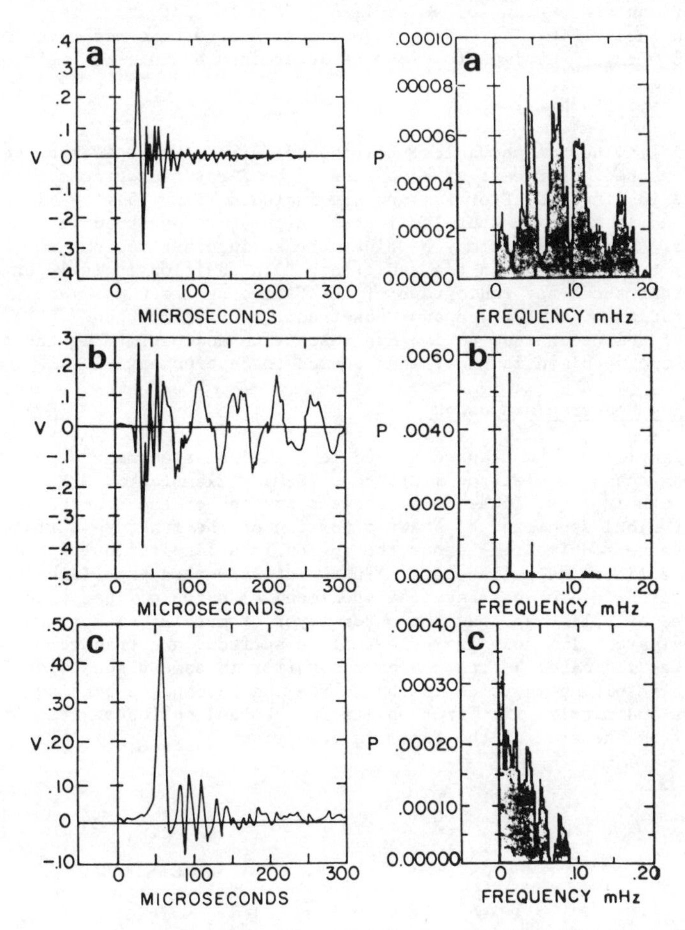

Figure 9 - AE signal typical of the three types observed in tests on both longitudinal and transverse specimens; the FFT's are also shown: a) Type 1, b) Type 2, c) Type 3.

Monitoring in this fashion has been shown to be extremely sensitive to cracking which occurs in epoxy matrix composites (2).

Because the echoes selected for monitoring travel an identical distance through the specimen but different distances through the undeforming waveguide, the sensitivity to damage in the specimen arises from diffraction effects.

Post Test Examination

Ultrasonic C-scan

Following the mechanical testing to failure the specimens were ultrasonically C-scanned following the procedure described earlier. For ease of comparison the C-scans of a specimen before loading and after failure are included (Fig. 14). Failure was observed to have occurred in a region without any imperfections as determined by this NDI method.

X-ray Radiography

Following the mechanical testing to failure the specimens were X-ray radiographed. For ease of comparison, the X-ray radiographs of a specimen before loading and after failure are included (Fig. 15). Failure has occurred in a region where no imperfections were observed in the radiographs. However, what may be fiber breaks in other regions of the specimen appear to be evident in the radiograph. In addition it is important to note that the X-ray radiograph of the failed specimen was made after the application of an X-ray opaque penetrant, tetrabromoethane (TBE). Development of this technique to improve resolution and performing the examination while the specimen is under load seemed to have potential.

Scanning Electron Microscopy

Samples of the transverse and longitudinal specimens were examined using a scanning electron microscope (SEM). Examples of SEM photographs are shown in Figs. 16 a-f: a) shows a portion of the fracture surface of a longitudinal specimen, b) shows a portion of the fracture surface of a transverse specimen, c) shows the region in a longitudinal specimen from which a fiber has pulled out, d) shows at increased magnification the failure surface of a transverse specimen, e) shows a fiber crack which was exposed by polishing away the outer layer of matrix, f) shows a region approximately 1mm away from the failure surface in a transverse specimen. Findings indicate: i) the specimens suffer in some places from lack of fusion resulting in voids in the matrix and incomplete bonding between fibers and matrix, ii) fiber splits, cracks and pullouts, iii) fiber damage away from the area of the final failure site.

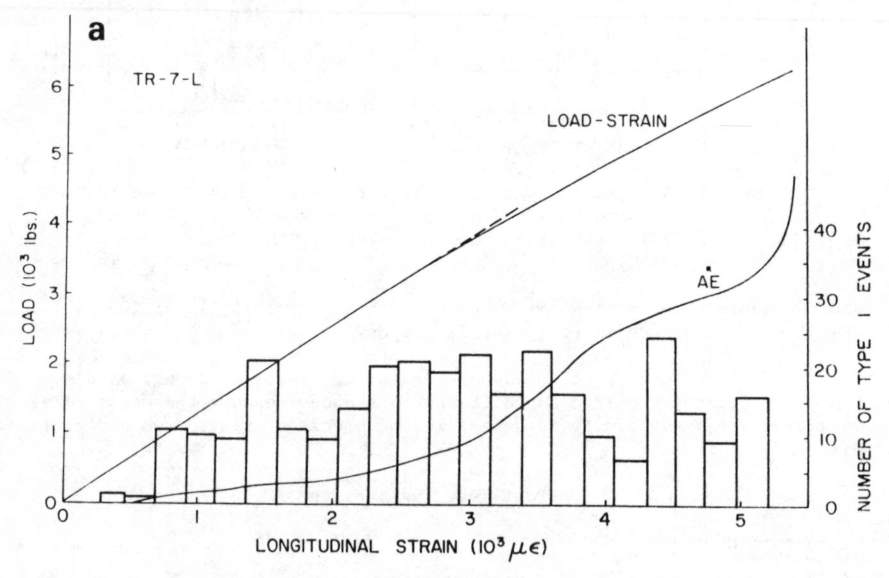

Figure 10 - Distrubution of various AE signal types as a function of longitudinal strain for a test on a "longitudinal" specimen. a) Type 1,

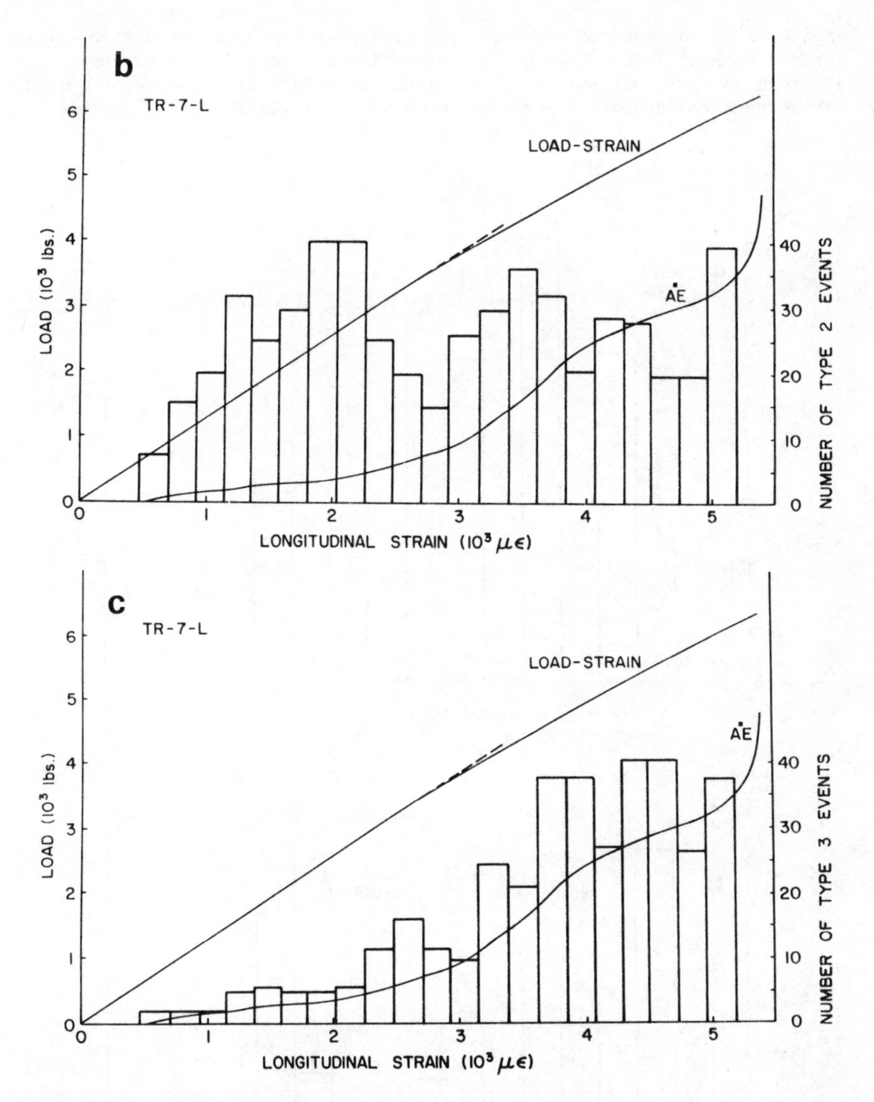

Figure 10 - Distribution of various AE signal types as a function of longi-
(contd.) tudinal strain for a test on a "longitudinal" specimen.
b) Type 2, c) Type 3.

Depth Profiling by Auger Electron Spectroscopy (AES)

The AES technique for chemical analysis of surfaces is based on the Auger relaxation process. When a core level of a surface atom is ionized by an impinging electron beam, relaxation to a lower energy state can occur through an electronic rearrangement which leaves the atom in a doubly ionized state. The energy difference between these two states is given to the ejected Auger electron which will have a kinetic energy characteristic of the parent atom. When the Auger transitions of atoms located within a few angstroms of the surface occur, Auger electrons may be ejected from the surface without loss of energy and give rise to peaks in the secondary electron energy distribution. The energy and shape of these Auger features can be used to identify the composition of the solid surface. Depth

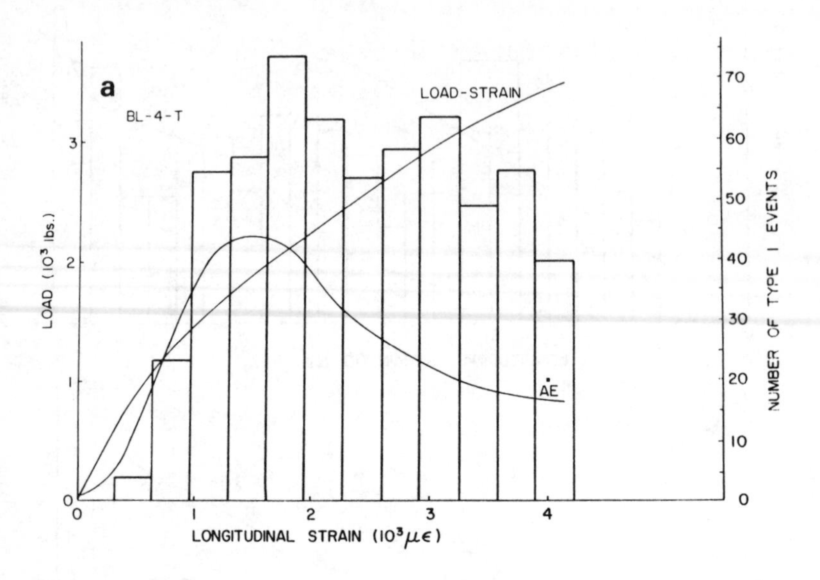

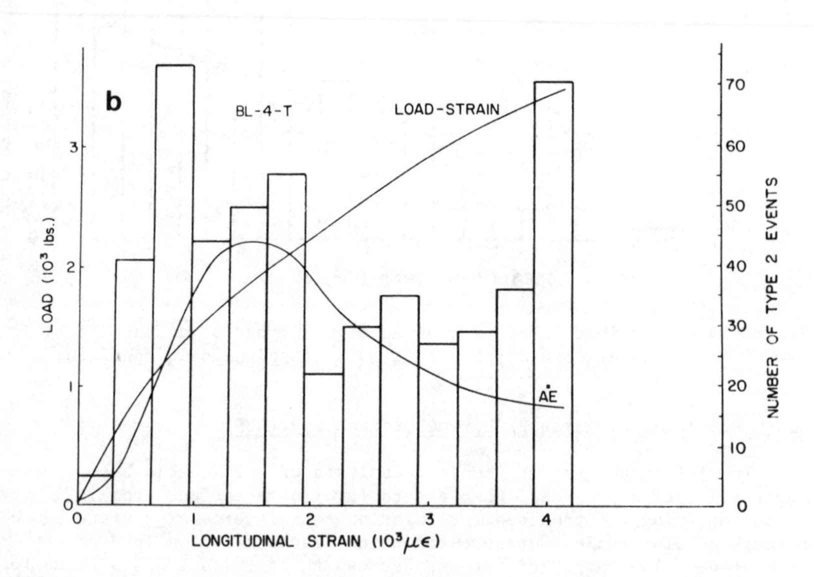

Figure 11 - Distribution of various AE signal types as a function of longitudinal strain for a test on a "transverse" specimen. a) Type 1, b) Type 2.

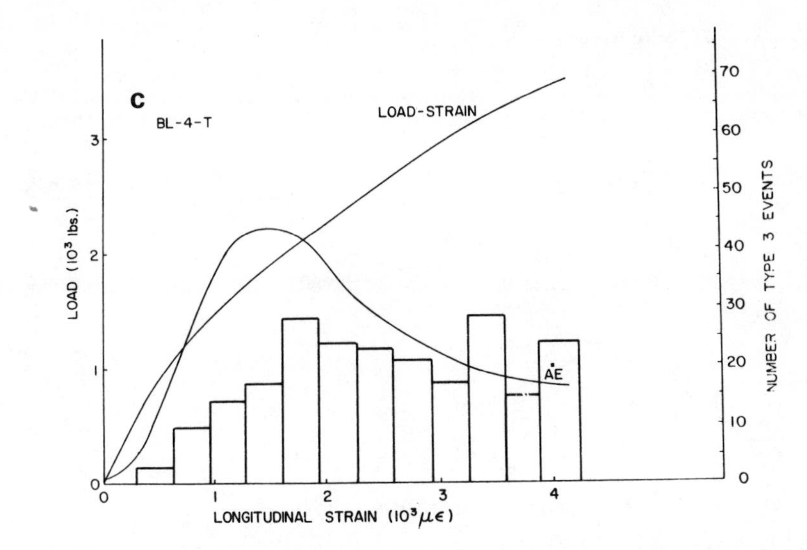

Figure 11 - Distribution of various AE signal types as a function of longitudinal strain for a test on a "transverse" specimen. c) Type 3.
(contd.)

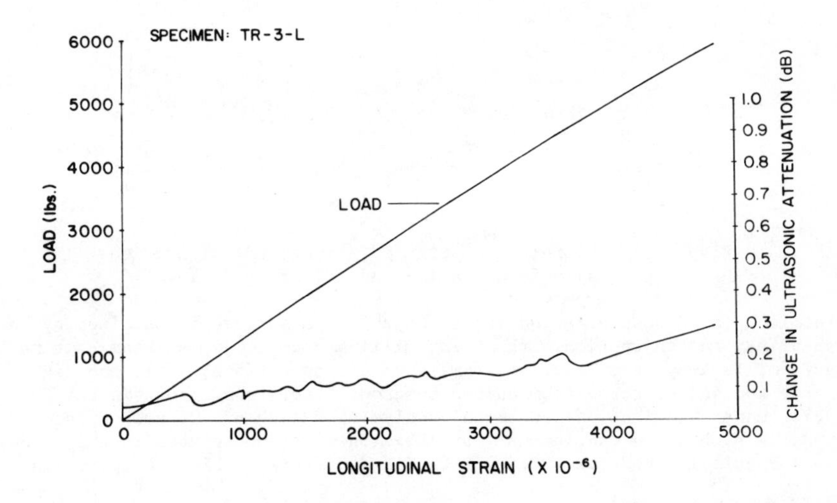

Figure 12 - Change in ultrasonic attenuation vs. longitudinal strain for a "longitudinal" specimen.

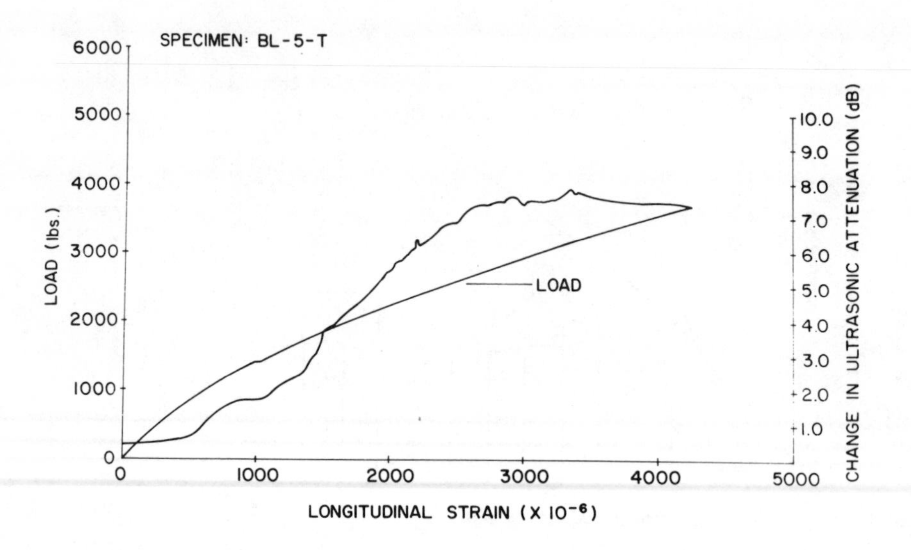

Figure 13 - Change in ultrasonic attenuation vs. longitudinal strain for a "transverse" specimen.

Figure 14 - Ultrasonic C-scan of a longitudinal specimen a) before loading and b) after failure.

profiling (atom concentration versus depth) of a surface is obtained by the use of AES in conjunction with an ion milling beam. The ion beam continuously mills away atom layer by atom layer of the surface, while the high energy focused electron beam causes numerous Auger transitions in the region under examination. Chemical analysis of each of the new surfaces exposed is obtained continuously by AES. The instrument used in this study was a Physical Electronics Model ESCA-SAM 550 (Perkins Elmer Corporation).

AES depth profiles were determined in selected regions of both the transverse and longitudinal specimens. It should be noted to facilitate examination by SEM, a gold-palladium coating was evaporated onto the surface. Fig. 17 shows an AES survey of a fiber which has pulled out; the AES depth profile of this region is shown in Fig. 18. Figure 19 shows an AES

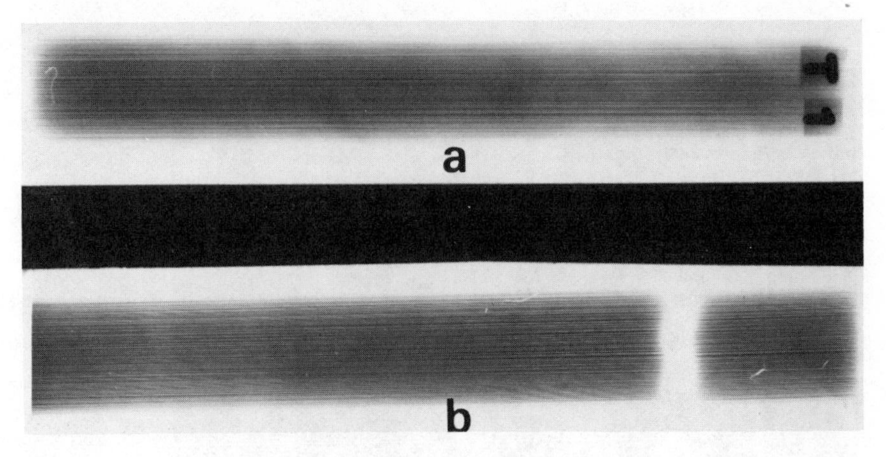

Figure 15 - X-ray radiographs of a longitudinal specimen a) before loading and b) after failure.

survey of the region from which a fiber has pulled out; this region is displayed in Fig. 16c. The associated AES depth profile is shown in Fig. 20.

Discussion of Results

It is necessary to note initially that the results of this study are for a single panel of material and do not reflect variations which might exist in other panels of the same type.

The basic objective of the study was to evaluate the panel regarding initial imperfections and to determine their influence on the mechanical response of this material. Several nondestructive evaluation techniques were used: ultrasonic C-scan, X-ray radiography, vibrothermography, eddy current. The results of the latter two techniques were very limited and they were not actively pursued beyond the preliminary panel examination. Ultrasonic C-scanning and X-ray radiography, however, were used extensively to evaluate both the as-received panel and the test specimens. Indications appearing in the C-scanning records coincided with regions of minor surface irregularities, and no correlation with final failure site was observed. Upon examination of the radiographs, the only imperfections apparent were those areas where fibers deviated from linearity over a small percentage of their length. In every case X-ray radiographs were made of the failed specimens and no clear indication of damage other than at the sight of failure was evident. In order to gain insight into the nature of the damage initiation and progression, nondestructive monitoring of continuous tests as well as examination of some specimens after interrupting the tests were performed.

Acoustic emission monitoring indicated that damage appeared to initiate early and continued to occur in a fashion which was not self-similar. The activity increased as the loading increased for the longitudinal specimens with what seems to be almost a bilinear pattern of AE count rate versus load occurring. The transition point appeared in every case to come before the point of the load strain curve, showing a distinct change in tangent modulus. This was not the case for the transverse specimens, although the AE was seen to vary as the loading increased. The peak in the count rate versus load did not appear to correspond to any distinctive feature of the

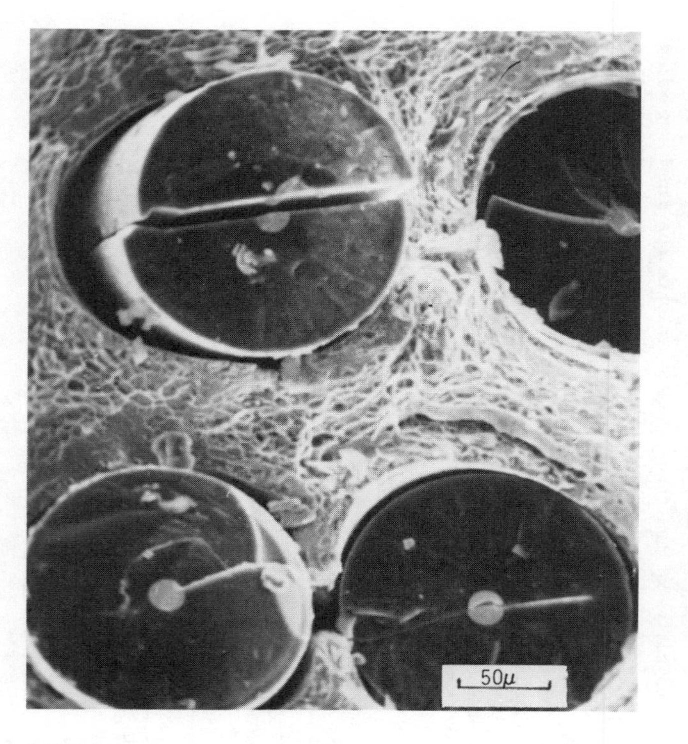

Figure 16a - Shows a portion of the fracture surface of a longitudinal specimen.

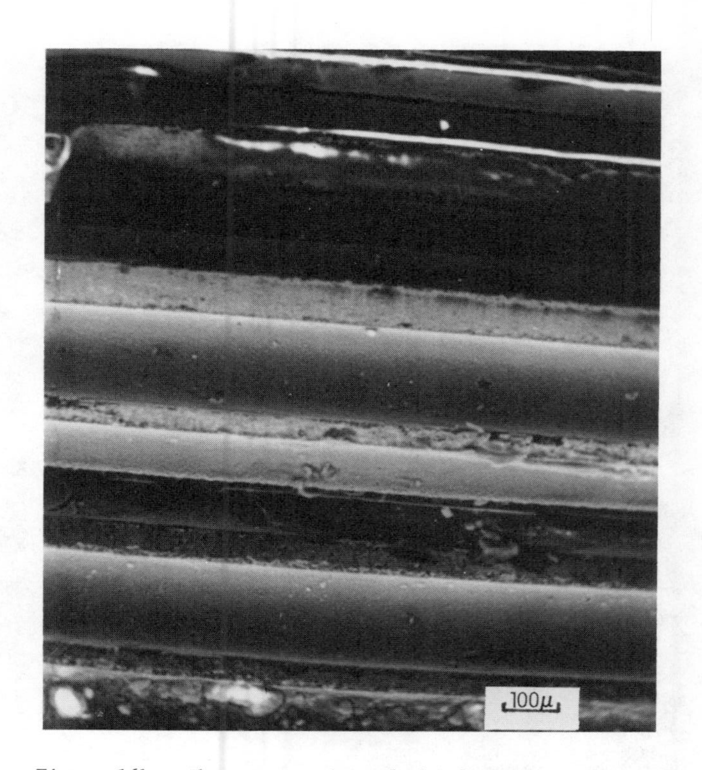

Figure 16b - Shows a portion of the fracture surface of a transverse specimen.

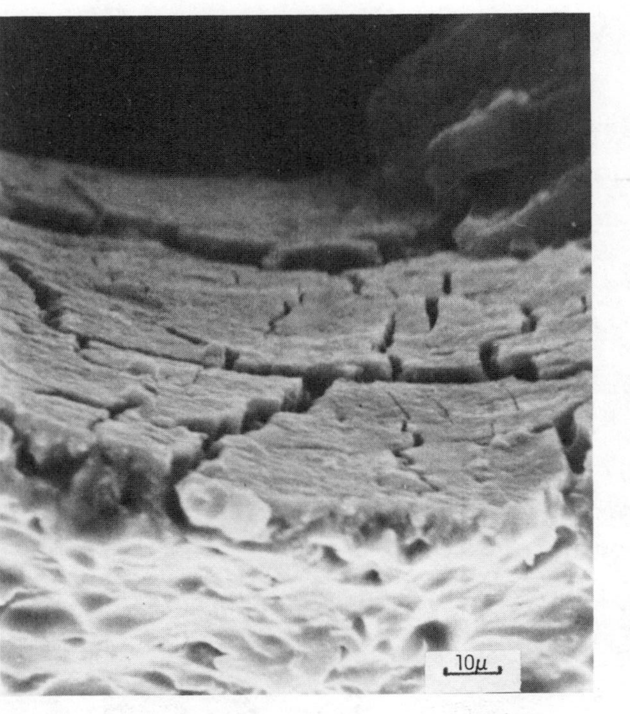

Figure 16c - Shows the region in a longitudinal specimen from which a fiber has pulled out.

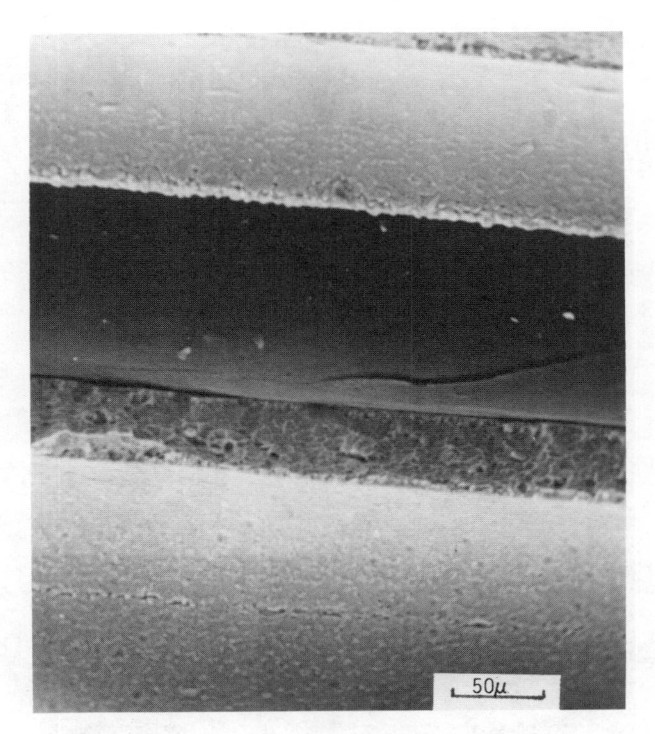

Figure 16d - Shows, at higher magnification than b) the failure surface of a transverse specimen.

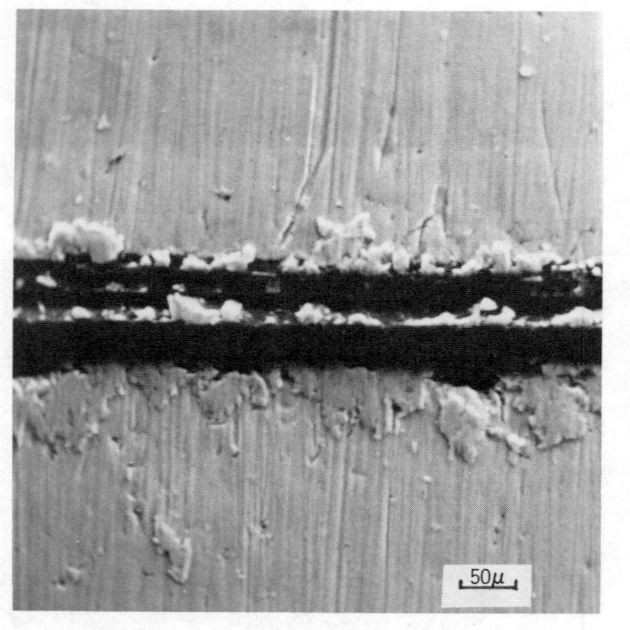

Figure 16e - Shows a fiber crack which was exposed by polishing away the outer layer of matrix.

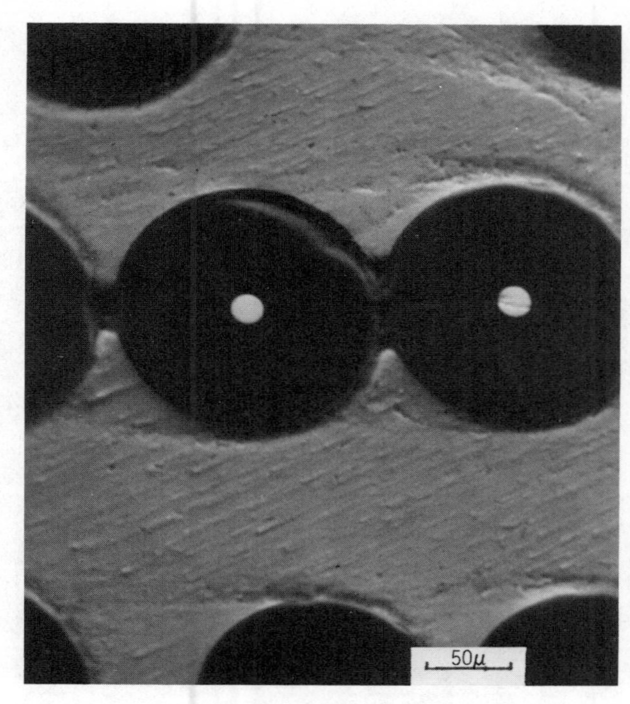

Figure 16f - Shows a region approximately 1mm away from the failure surface in a transverse specimen.

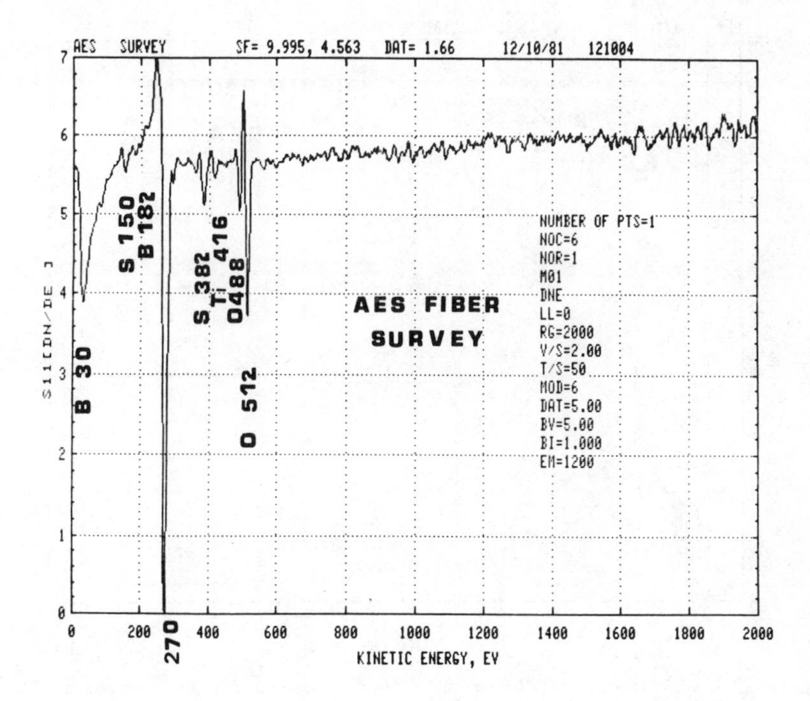

Figure 17 - AES survey of a fiber pulled out during the failure of a longitudinal specimen.

load strain curve. In an attempt to ascertain a closer correlation of the acoustic emission activity with material damage the acoustic emission recorded during the tests on both a longitudinal and transverse specimen were carefully examined. The results of this examination are provided in Figs. 10 and 11. The tape record was divided into segments and reviewed at a reduced playing speed. It was observed that of the discrete AE signals occurring during the tests, three distinct types occurred; the same three types in both the longitudinal and transverse cases were observed. The rate of occurrence of the different types have been plotted as histograms where the width indicates indirectly the length of the segment of the tape record that each value represents. Fig. 9 displays both the time and frequency domain records typical of these signals. Although a complete explanation of the frequency content was not the object of this study, it was determined from a simple beam vibration analysis that the peak frequency components appeared to be comparable to modes of specimen vibration, i.e. longitudinal, transverse (width, transverse vibration of the thickness dimension was beyond the tape recorder's 300 kHz upper frequency limit) and flexural modes. Such a correspondence seemed reasonable since the specimens were approximately the same size. The speculation regarding the source of such vibration is that various modes of deformation, fiber cracking, fiber-matrix separation, fiber breaking, whatever, would give rise to stress waves which would cause stress redistribution that would be different, i.e. in a longitudinal test a fiber break would affect stress redistribution predominantly along the length while fiber cracking would provide the greatest stress change perpendicular to the crack. From observing the scanning electron micrographs, it is clear that fiber cracking occurs in

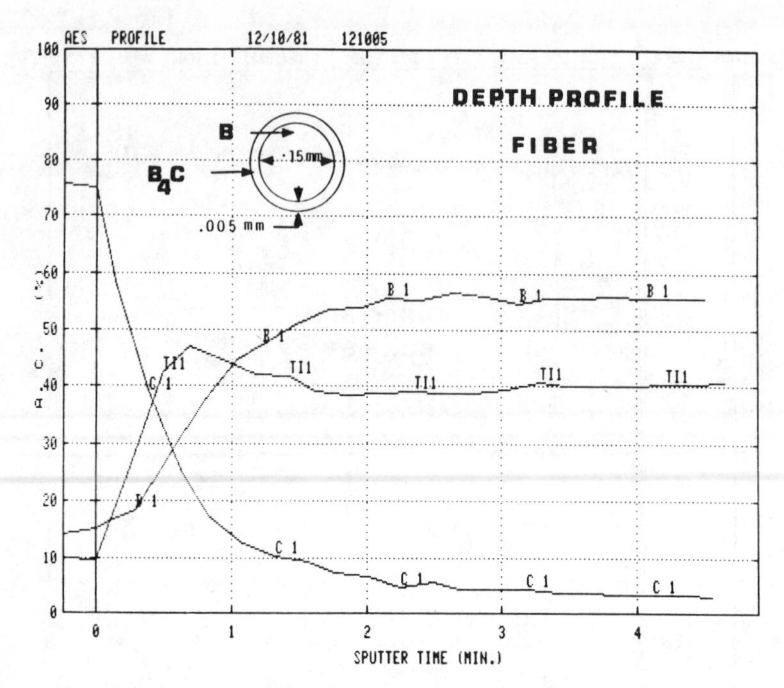

Figure 18 - AES depth profile of the fiber examined for the survey in Fig. 17.

both the longitudinal and transverse tests. In fact, such cracking is reported to occur during the fabrication of this material. Furthermore, examination of these micrographs allows one to see clear evidence of disbonding which may even initiate from regions where the metal layer failed to bond during processing. In an attempt to determine whether fiber fracture was occurring away from the failure sight in the transverse case and thereby ascertain whether a fiber crack was the failure initiator, a failed specimen was polished to remove the outer layer of titanium and reveal the fibers. Fig. 16e shows a crack in the fiber which runs across the width of the specimen. So although the micrograph of the transverse specimen failure surface shows cracked fibers, their existence is not limited to that region.

Further evidence in this regard appeared as a result of monitoring ultrasonic attenuation during loading. Previous work in graphite fiber reinforced epoxy at the Materials Response Laboratory has shown that a theory describing diffraction of the sound beam from matrix cracking explains "apparent" changes in ultrasonic attenuation under the monitoring conditions employed. "Apparent" is here used because the echoes being compared pass through the specimen the same distance but one is delayed in time by a quartz delay block. Figure 12 and Fig. 13 show changes occurring in tests on both longitudinal and transverse specimens. The rate of change of the attenuation increases near transitions in the load strain curves. These results seem to support the idea that fiber cracks and, or, matrix cracking are occurring early in the test. The increase could be due to increases in the number or the size of the opening of the cracks.

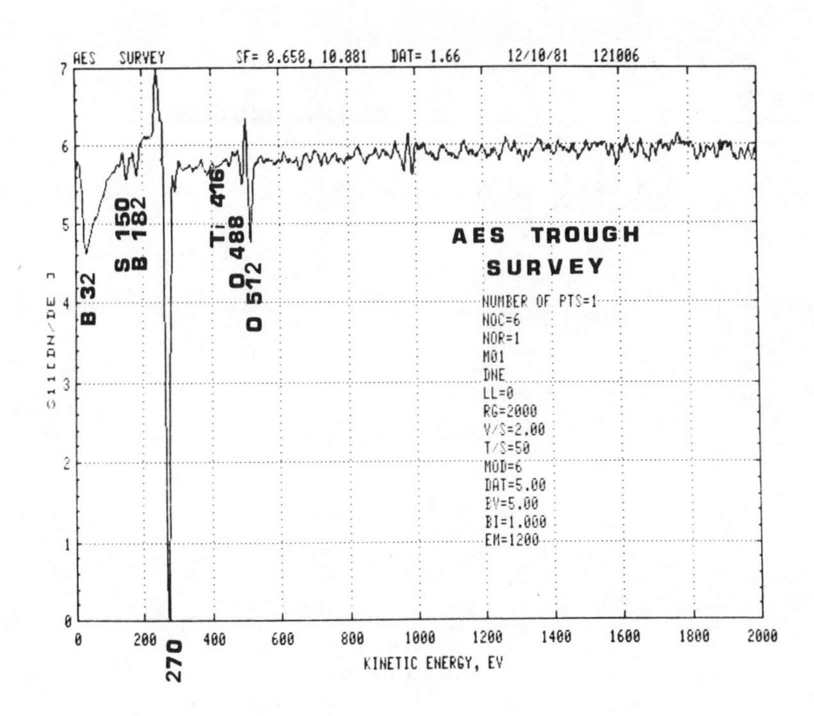

Figure 19 - AES survey of the region shown in Fig. 16c.

From the results of SEM, it is clear that failure of the matrix in both the longitudinal and transverse specimens has occurred in a ductile fashion. In the longitudinal case it is of course obvious that both matrix and fibers must fail ultimately. However, the sequence of events is the issue of importance. It would appear, though, that because of the greater ductility of the matrix that fiber failures are the precipitous event. But the appearance of the surface of the material left after a fiber has pulled out in longitudinal specimen, Fig. 16c, suggests that fiber/matrix separation has occurred prior to the fiber failure. To determine why such an event might have occurred, and to decide if poor fiber/matrix bonding was responsible the AES work was performed.

Depth profiles of a fiber and a trough, from which a fiber has pulled out (area shown in Fig. 16c) were obtained. Figure 17 and Fig. 19 show the AES surveys of the fiber and the trough respectively. The presence of carbon, oxygen, and sulfur is because of contamination and oxidation of the fracture surface. The carbon peak was used as a reference at 285 eV. Figure 18 is the AES depth profile of the trough. The ordinate and the abscissa indicate atom concentration in percent and the sputter time in minutes, respectively. Typically one minute of sputtering would erode 1000 Å of material. As is usually the case on the surface, the carbon content is relatively high as a result of the contamination mentioned above. However, as the profiling continues, the concentration of all atoms present becomes nearly constant. Those elements present include: boron (B1), carbon (C1), titanium (Ti 1), vanadium (V1), and aluminum (Al 1). If both the fiber and coating had pulled away, the composition of the material of the trough would be that of the titanium matrix and boron and carbon would

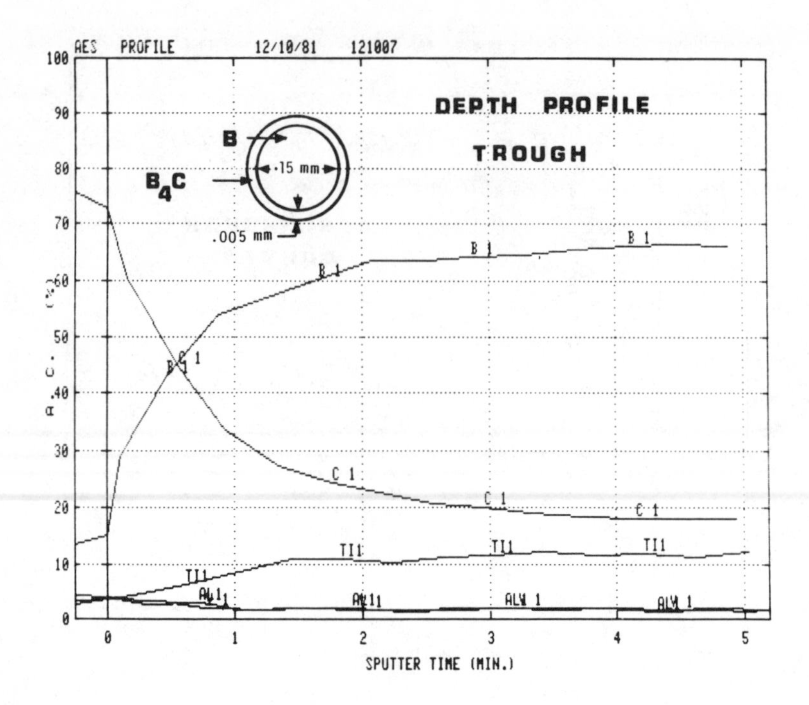

Figure 20 - AES depth profile of the region shown in Fig. 16c.

not be present. The corresponding profile of a fiber, Fig. 17, indicates no carbon (C 1) present aside from contamination; boron (B 1) and titanium (Ti 1) are the only elements present. The presence of titanium is a result of the back sputtering of matrix material. Combining the results from both the AES depth profiles, it is clear that the B_4C initially coating the boron fiber has separated and remained attached to the titanium matrix. This is indeed consistent with the brittle appearance of the surface containing both axial and circumferential cracks, Fig. 16c. A reasonable explanation for this appears to be that when the material is subjected to axial loading, both the fibers and matrix are stressed in proportion to their moduli, which are different. In addition, the Poisson effect creates stresses normal to the loading direction. These comments hold true as well for the coating which is clearly bonded to the matrix. As the strain levels increase, the fiber/coating interface fails, subsequently the deforming matrix to which the coating is attached causes it to crack; the circumferential cracks result from the applied load, the axial cracks from the Poisson effect on the cylindrical void. It would be expected that circumferential cracks running out into the matrix reduce its load bearing capacity and concentrate the stress on this region of the fiber. In many instances in this material, the close proximity of fibers to one another eliminates the necessity for these regions to link up laterally.

The failure of these interfaces is perhaps in part the reason that the ultrasonic attenuation changes for the longitudinal specimens, and in addition is very likely to be the source of one of the types of AE observed. The cracks in the coating as well as fiber cracks and breaks are other likely candidates for the longitudinal specimens.

For the transverse specimens, failure does not necessarily need to result from failure of fiber and matrix. Evidence, in this study, suggests that the failure only coincidently involves fibers and generally these appear to be fibers which were cracked and were conveniently located in the path of the failure. In most instances the failure proceeded by fiber/coating interface failure occurring, thereby eliminating the load carrying capacity in these regions and then the matrix fails; the ultimate strength value supports this conjecture. Again the poor fiber distribution creates an easy path for failure to take. Both the ultrasonic attenuation and AE results indicate that these events are occurring well before the specimen fails and in fact the peak of such activity is well removed from the failure load. The distribution of type 1 and 2 AE signals for the transverse specimen suggests that this mechanism may be responsible for their occurrence. Local matrix failure and the interaction of cracked fiber surfaces are speculated to be the source of AE as well in the transverse specimens.

Summary

Strong evidence indicates that damage occurs early in the loading of transverse and longitudinal specimens of this material. This evidence is provided by AE and ultrasonic attenuation monitoring, and is supported by SEM examination and AES depth profiling. The damage continues and culminates in final failure as a result of increasing load. However, an exact discernment of the sequence as well as complete correlation between deformation mechanisms and NDI observations demands that much more extensive metallography of tests interrupted at various stages of the deformation be performed. The actual significance of the early occurrence of damage is unclear but suggests that cyclic loading of the material may result in premature failure if its presence is ignored.

Conclusions

1. NDI was unable to detect imperfections responsible for any of the specimen failures.
2. AE and changes in ultrasonic attenuation indicated early and progressive microscopic damage.
3. Three distinct types of AE signals have been identified. The frequency spectrum suggests that the distinct nature of the signals results from the specimen geometry and is not in general representative of the source of the AE. The AE frequency is believed to be related to the nature of the stress redistribution caused by the damage mechanism which acts as the AE source.
4. Final failure in both longitudinal and transverse specimens is related in part to the failure of the fiber/coating interface.

Nomenclature

Specimen designations: For example, TR-4-L indicates the specimen came from the top (T) right (R) corner of the panel, and is the number 4 "longitudinal" (L) specimen.

E - initial stiffness in the direction of load application

σ_y - yield stress

σ_{UTS} - ultimate tensile strength

A.C. - atom concentration

S11 [DN/DE] - First derivative of Intensity vs. Kinetic Energy.

Acknowledgements

The authors wish to acknowledge the Martin Marietta Corporation for their partial support of this project. In addition, the efforts of Mrs. Barbara Wengert in the preparation of the manuscript have been sincerely appreciated.

References

1. J. C. Duke, Jr., A Govada and A. Lemascon, "Characterization and Evaluation of Advanced Composite Materials," Final Report to Martin Marietta Corporation 2FD/752321, VPI-E81-9, April 1981.

2. D. T. Hayford and E. G. Henneke, II, "A Model for Correlating Damage and Ultrasonic Attenuation in Composites," American Society for Testing and Materials STP 674, pp. 184-200, 1979.

TUNGSTEN-FIBER-REINFORCED SUPERALLOY COMPOSITE,

HIGH-TEMPERATURE COMPONENT DESIGN CONSIDERATIONS

Edward A. Winsa

National Aeronautics and Space Administration
Lewis Research Center
Cleveland, Ohio 44135

SUMMARY

Tungsten fiber reinforced superalloy composites (TFRS) are intended for use in high temperature turbine components. Current turbine component design methodology is based on applying the experience, sometimes semi-empirical, gained from over 30 years of superalloy component design. Current composite component design capability is generally limited to the methodology for low temperature resin matrix composites. Often the tendency is to treat TFRS as just another superalloy or low temperature composite. However, TFRS behavior is significantly different than that of superalloys, and the high temperature environment adds considerations not common in low temperature composite component design. This paper describes the methodology used for preliminary design of TFRS components. Considerations unique to TFRS are emphasized.

Introduction

Tungsten Fiber Reinforced Superalloy composites (TFRS) offer an alternative to monolithic superalloys when designing components for demanding high temperature applications. These members of the Fiber Reinforced Superalloy composite (FRS) family have exhibited use temperature potential up to 150 K greater than the best superalloys (Ref. 1). Moreover, cost effective fabrication feasibility has been demonstrated (Figs. 1 and 2, Refs. 2 and 3).

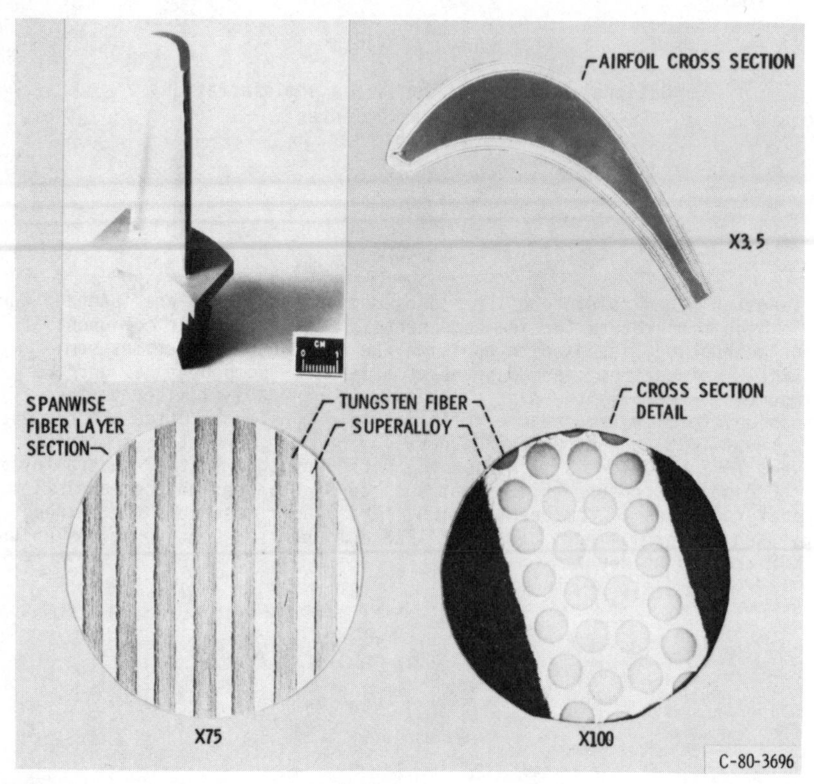

Figure 1 - TFRS blade frabrication was demonstrated using a convection cooled JT9D-like airfoil mated to a circular arc dovetail.

Despite their promise, TFRS will remain laboratory curiosities until rig or engine tests can confirm their utility in actual components. Recognising this fact, NASA contracted with the General Electric Co. - Aircraft Engine Group (Evandale, Ohio) to conduct a "Hardware Designers' Overview of Tungsten-Fiber Reinforced Superalloy Composites for Turbojet Engines". One objective of the "...Overview..." was to select three potentially practical TFRS engine components and develop preliminary designs using a first generation TFRS (W/FeCrAlY, Ref. 4). The most promising component could then be detail designed, fabricated, and rig tested in a possible following sequence of programs and contracts. Preliminary designs for a turbine blade, turbine vane, and an outlet guide vane (OGV) resulted from the contract.

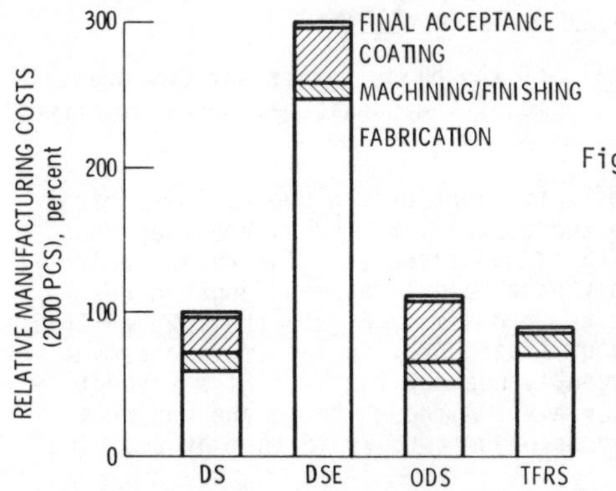

Figure 2 - TFRS fabrication costs are competitive with those of Directionally Solidified (DS) Superalloys, DS Eutectics (DSE), and Oxide Dispersion-Strengthened (ODS) Superalloys (Ref. 3).

Designing the "...Overview..." contract components necessitated a joint effort by General Electric (GE) and NASA (Fig. 3). GE and NASA collaborated to select engine components and establish the geometry and property requirements of the TFRS versions. Next, NASA determined the TFRS laminate configuration needed to meet the requirements and calculated the appropriate physical and mechanical properties. Then, GE used the calculated laminate properties in performing a structural and heat transfer analysis of the components. Because these were preliminary "screening" designs, no optimization of the components was attempted.

This report briefly reviews the major considerations and methodology used by NASA to design the TFRS laminated airfoils used in the three components. Emphasis is placed on those considerations which make TFRS components somewhat unique relative to superalloy or low temperature composite components.

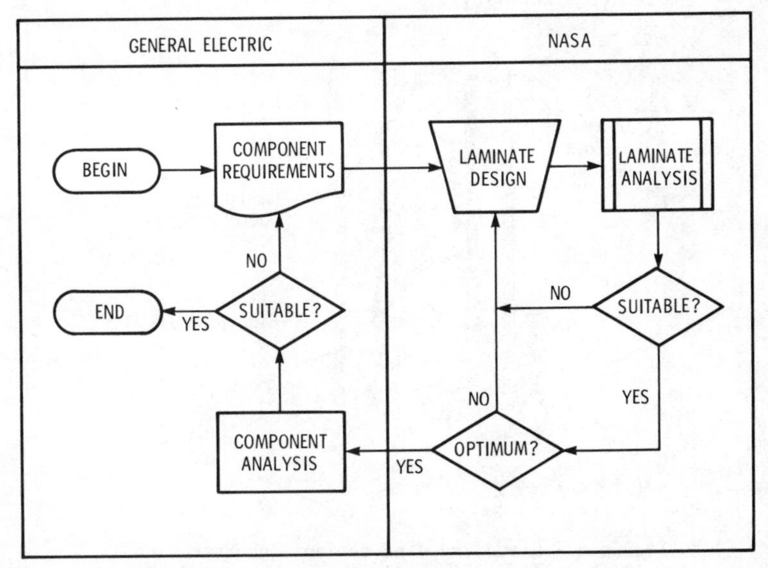

Figure 3 - "..Overview.." contract TFRS component design responsibility was shared by General Electric and NASA.

TFRS Airfoil Design Considerations

TFRS was used in only the airfoils of the blade, vane, and OGV preliminary designs. This section indicates some of the considerations that affect TFRS airfoil design.

Film cooling is currently the most efficient proven superalloy airfoil cooling technique. It has been used successfully for turbine blades and vanes. However, its utility with TFRS is questionable. The reason being that film cooling holes would probably have to cut through tungsten reinforcing fibers, and that could cause two problems. First, the TFRS airfoil would be weakened because discontinuous fibers provide less reinforcement. Second, unless the holes were subsequently coated, the airfoil's susceptability to oxidation damage would increase. Although these are not necessarily insurmountable problems, it would be best to avoid them by avoiding film cooling, if possible.

TFRS may be convection cooled and impingement cooled with or without a Thermal Barrier Coating (TBC). In fact, a TFRS airfoil with impingement cooling plus a TBC might be an attractive combination of technologies (Fig. 4). Previous calculations indicated that Impingement/TBC (ITBC) cooled TFRS airfoils could theoretically operate in near stoichiometric temperature inlet gas streams (Ref. 5). Furthermore, the tailorability of TFRS thermal expansion could allow increased durability of TBC on TFRS relative to its durability on superalloys. Moreover, ITBC cooled TFRS airfoils without film cooling holes should exhibit greater aerodynamic efficiency and lower LCF (low cycle fatigue) stresses than film cooled airfoils. The greater aerodynamic efficiency would be due to the lack of undesirable air turbulance normally associated with air jetting out of film cooling holes. The lower LCF stresses would be due to the lack of stress concentrations caused by film cooling holes cutting through the airfoil wall. Finally, as will be indicated below, ITBC cooled TFRS may have the same cooling efficiency as film cooled superalloys.

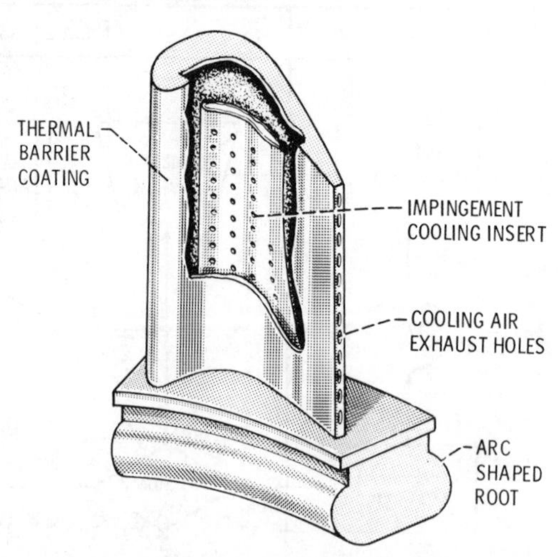

Figure 4 - Blade and Vane designs incorporating both impingement cooling and a Thermal Barrier Coating (TBC) may be optimum for TFRS.

Only moderately complex internal geometries may be feasible with hollow TFRS airfoils. The TFRS fabrication process indicated in figure 5 utilizes a leachable steel core to provide the hollow cavity (Ref. 2). This approach permits the easy incorporation of trailing edge cooling holes and simple ribs. However, complex serpentine rib schemes could be difficult to achieve.

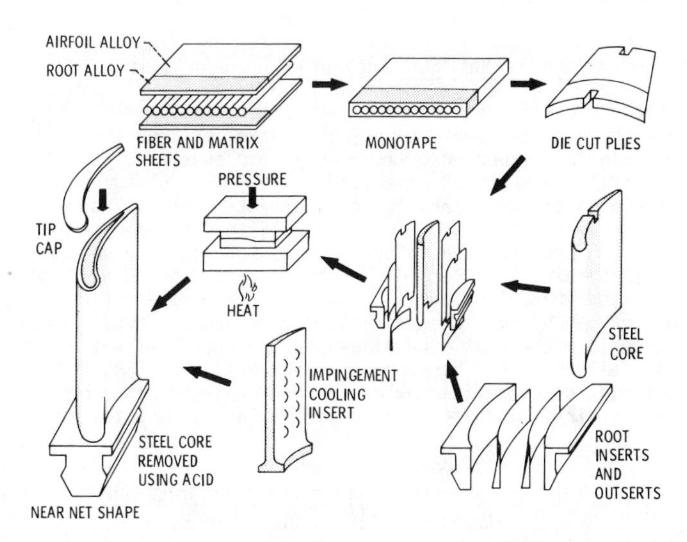

Figure 5 - The TFRS blade fabrication sequence is similar to that used for B/Al fan blades (note dual alloy plies) (Ref. 2).

The leading edge of hollow TFRS airfoils must be of wrap around construction. The exact construction shown in figure 5 leads to a seam at the leading edge of the airfoil (Fig. 1). Calculations indicate that failure could occur at the seam of hollow airfoils (not a problem with solid airfoils). Thus the plys must wrap around the leading edge of hollow airfoils to eliminate the seam. The feasibility of the wrap around technique has been demonstrated for the blade shown in Fig. 1 (Ref. 2).

The allowable tungsten fiber diameter is limited by several conflicting considerations. Fiber reaction (see below) requires that the largest possible fiber diameter be used. On the other hand, the minimal thickness of hollow airfoil walls combines with the need for laminate symmetry (i.e., several plys must be used) to keep allowable diameters small. Using a weak matrix (e.g., FeCrAlY) has the same effect because smaller diameter fibers are needed to decrease fiber critical length and, thereby, increase reinforcement efficiency (Ref. 6). Yet, if impact strength is needed, larger diameters should be used (Ref. 7). Typically, allowable fiber diameters range between 0.1mm and 0.2mm; although, for larger components, diameters up to 0.4mm are useful.

Special attachment schemes must be developed for TFRS blades and vanes. Blade dovetail and vane bands will probably have to be made from superalloys to save weight and provide maximum strength at the component/engine interface. However, the low thermal expansion of TFRS laminates could cause problems if simple brazed or diffusion bonded attachments were used. The attachment problem is being addressed by NASA research.

Use of TFRS blades as substitutes for currently used turbine blades will probably be seriously hampered by disc size and strength limitations. Current discs are optimised for superalloy blades. Typically, there is no extra physical space around the disc to allow its enlargement to accomodate slightly heavier TFRS blades.

Turbine Material Property Considerations

TFRS properties cannot be directly compared to superalloy properties. Current turbine material requirements were established during over three decades of design/use experience with monolithic superalloys (Table I). There is no similar history of experience with TFRS. Consequently, there is a natural tendency to use superalloy requirements when evaluating TFRS properties. However, this practice can result in misleading evaluations. For example, TFRS seem to have inferior LCF capability relative to the best superalloys (appendix A and Ref. 8). But, in fact, the modulii, thermal conductivity, and thermal expansion properties of TFRS drastically reduce LCF stresses caused by thermal gradients (see below). Thus, TFRS may actually outperform the best superalloys in some applications where LCF is the limiting failure mode. Therefore, the suitabilty of TFRS for specific applications must be determined by design and analysis using composite theory - not by direct comparison between TFRS and superalloy properties.

Table I. - Material Characteristics Important in Turbine Blade Applications

Property	Significance to Design
Creep and rupture	Limit allowable airfoil metal temperature and stress
High cycle fatigue	Vibration stresses at all locations on the blade must be less than the endurance limit of the material, as determined in smooth and notched bar tests
Low cycle fatigue	Determines design, life: smooth bar data important to airfoil leading and training edges; notched bar data important to dovetail and bleed holes in air cooled blades
Tensile	Limits dovetail/shank design
Combined steady state and vibratory	Vibratory stress endurance limit is reduced by presence of steady state stresses
Shear and torsion	Adequate in conventional superalloys, but could be limiting in anisotropic materials, particularly in the dovetail
Density	Affects blade and disk stresses
Thermal expansion	Affects blade expansion, important to gas leakage and tip rub
Incipient melting	Affects over-temperature capability of airfoil in the event of hot spots
Elastic constants	Affect blade material frequencies, and thermal stresses

Material properties must be well characterized to allow detailed component design and analysis. Table II indicates the minimum property characterization generally desired for initial design consideration of superalloys.

Table II. - Minimum Property Data Needed to Design Turbine Blades for Development Engines

Property	Temperature or temperature range, °K						
	RT	775	900	1025	1150	1275	1400
0.2 percent	X	X	X	X	X	X	X
UTS	X	X	X	X	X	X	X
Percent El.	X	X	X	X	X	X	X
R of A	X	X	X	X	X	X	X
100 hr SR				X	X	X	X
1000 hr SR				X	X	X	X
Plastic creep (0.2 percent)				X	X	X	X
Low cycle fatigue			X		X	X	X
High cycle fatigue			X		X	X	
Joint efficiency (if applicable)							
- UTS	X	X	X	X	X	X	X
- Percent El.	X	X	X	X	X	X	X
- 1000 hr SR			X		X	X	X
Lowest melting temp.							
Density	X						
Thermal exp.		E S T I M A T E D					
Thermal cond.		E S T I M A T E D					
Spec. heat		E S T I M A T E D					
Poisson's ratio		E S T I M A T E D					
Mod. of elasticity	X	X	X	X	X	X	X

Unfortunately, TFRS properties cannot be easily summarized as can superalloy properties. The reason being that TFRS are laminated structures, not simple materials. Literally thousands of valid permutations of fiber diameter, volume percent, and fiber angle versus ply sequence exist for even simple TFRS laminated structures. The overwhelming quantity of variations makes thorough characterization of each variation impossible. Moreover, the properties of TFRS laminates are highly geometry dependent; for example, merely changing the width of an angle plied test panel can drastically affect the strength properties. Therefore, unlike the situation for superalloys, tests conducted on simple TFRS laboratory specimens can give grossly misleading indications of component performance. Consequently, TFRS components must be designed and analyzed using composite theory; then, the TFRS laminates determined by the design process must be tested in a form as close to their component geometry as possible.

The TFRS laminate must be custom designed for each specific component application. Such custom design is not always required for the better known low temperature composites (e.g., Graphite/Epoxy, Boron/Aluminum). The reason is that components made of such better known composites tend to be very large relative to the fiber critical length; therefore, critical length is not a factor. Furthermore, their reinforcing fibers have low density which makes maximum volume percent reinforcment practical. Also, fiber reaction degradation during service is usually not a factor. Consequently, standard panels of these composites can be characterized and the resulting properties used for design (e.g., ±15°, 50 volume percent, 0.2mm diameter Boron fiber, 1100 Aluminum matrix $\overline{B}$/Al for fan blade airfoils). By comparison, the small size of typical TFRS airfoils combined with the density and reaction degradation of tungsten fibers require that all TFRS laminate parameters be optimized.

Ideally, when designing TFRS components, one would like to have the capability to completly determine the component properties and performance using only fiber and matrix properties combined with composite theory (Fig. 6). Essentially, that is what was attempted in the NASA design of the "... Overview..." laminates. But the methods used are still under developement; thus, significant errors were possible. Hence, conservatism and safety factors were employed to increase the probability of successful design.

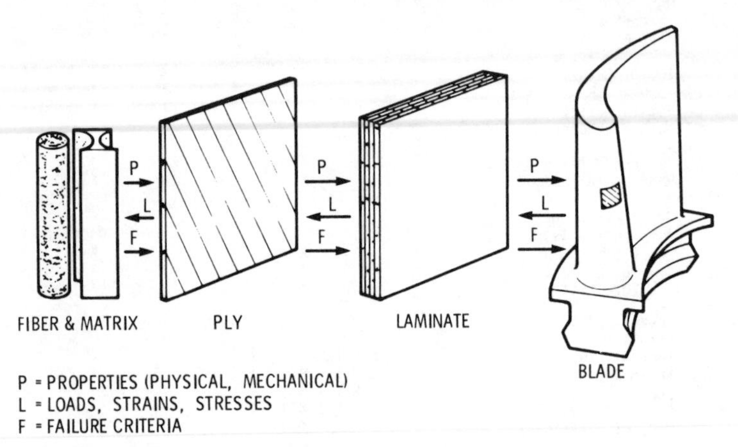

Figure 6 - Design of TFRS components requires the ability to predict component characteristics in terms of fiber and matrix characteristics.

TFRS Laminate Analysis Considerations

The NASA approach used to analyze the TFRS laminates used for the GE components is flow diagramed in figure 7. The rest of this section is an overview of principal considerations relevant to the TFRS laminate analysis methodology of figure 7. Some typical calculated laminate properties for W - 1.5ThO_2/FeCrAlY are given in the Appendix.

Fiber Degradation

Fiber degradation in the forms of diffusion induced recrystallization and partial dissolution is a chief cause of TFRS property degradation. The appearance of recrystallized fibers is illustrated in Fig. 8. The reaction penetration depth (P) is defined as the distance measured from the location of the original perimeter of the unreacted fiber to the perimeter of the unreacted core of the fiber.

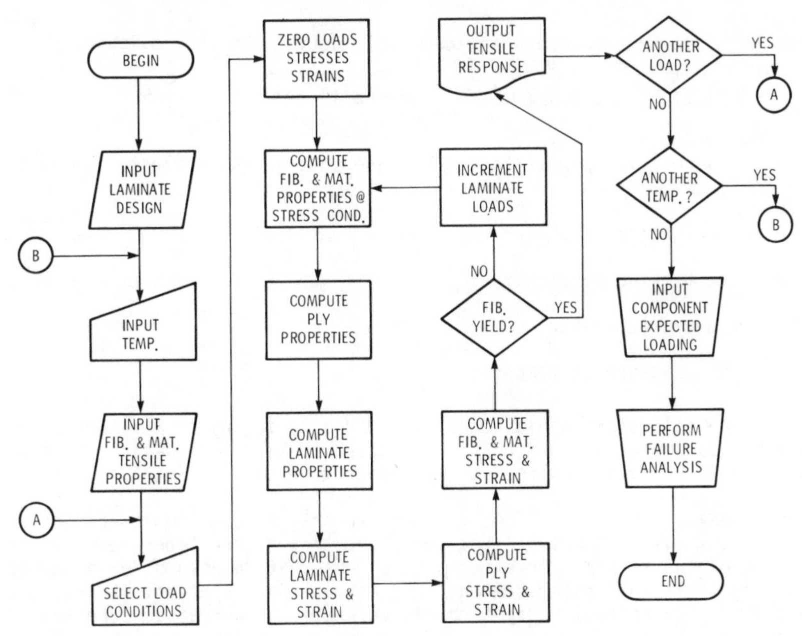

Figure 7 - TFRS laminate characteristics were computed for a range of load/temperature conditions using a piecewise linear approach to allow for plasticity.

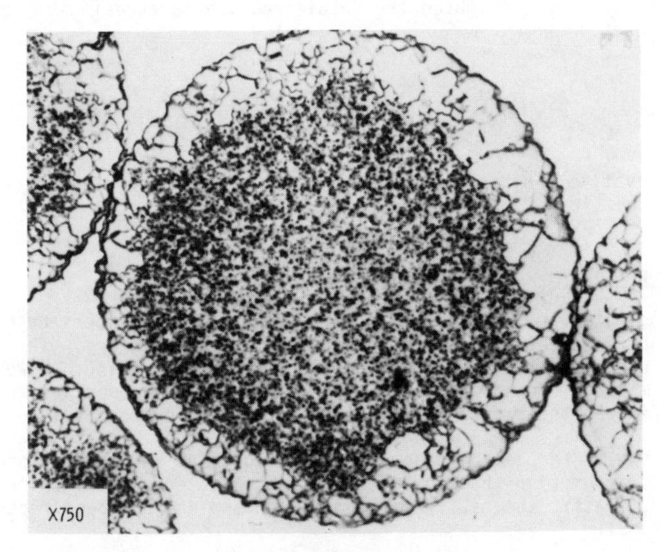

Figure 8 - Fiber recrystallization and dissolution due to elemental diffusion is the chief cause of TFRS property degradation with time.

Reaction penetration depth is adequately defined by the following equation.

$$P = (Po^2 + D \exp(-G/T)\, t)^{1/2}$$

where:

Po initial reaction penetration depth due to the TFRS laminate fabrication process

D a curve fitting parameter

G a curve fitting parameter

T temperature (absolute)

t time of exposure to temperature

This equation allows extrapolation of limited data to any temperature/time condition. But, because the equation is approximate, final designs must be based on reaction data acquired at the temperature/time of interest.

TFRS tensile, stress-rupture, and LCF property calculations were based on the conservative assumption that the properties of the fiber reacted zone were those of the matrix. In effect, this conservative assumption amounted to reducing the actual fiber content to a pseudo fiber content. The ratio of the pseudo content/actual content equals the ratio of unreacted core area/original fiber area.

Actual components usually have a complex temperature/time history. For example, a turbine blade might see hot spots of 1150 K for a total of 7000 hr, 1250 K for 1000 hr, 1275 K for 700 hr, and 1325 K for 300 hr during a 9000 hr lifetime. We estimated the total reaction penetration using the following equation.

$$P = \left(Po^2 + \int_0^t D \exp(-G/T)\, dt\right)^{1/2}$$

To be conservative, we assumed that the total reaction penetration (P) was present at the first instant of component operation.

Tensile Properties

TFRS laminate tensile properties were estimated from fiber and matrix tensile properties combined with appropriate composite theory. The estimates were used for preliminary component design optimization. However, before attempting to produce components, the estimated properties would have to be verified by tests of TFRS laminated panels and component-like shapes.

Derivation of the laminate constitutive equations we used was predicated on the assumption that the plys could be treated as homogenious, anisotropic, elastic sheets. Their derivation and use has been reported by others (Ref. 9).

These linear equations were used to generate TFRS laminate elastic-plastic stress versus strain curves by a piecewise linear, step loading approach. The stresses on the plys, fibers and matrix were calculated at each step of the laminate loading process (Fig. 7).

Two useful relationships resulting from the laminate piecewise linear analysis were laminate stress versus strain diagrams and maximum fiber stress versus laminate stress diagrams (Figs. 9 and 10). The plots shown in the figures are for a simple longitudinal loading case at 1225 K. Similar diagrams were generated for a variety of complex loading cases wherein longitudinal, transverse, and shear loads were applied in the proportions expected in the TFRS component. Moreover, behavior over the range of important temperatures was calculated. The final laminate designs optimized the response to complex loads over the operational temperature spectrum. In addition to the stress relationships, laminate "elastic" constants were generated for Stage I and Stage II deformation. These constants were used by GE during component analysis.

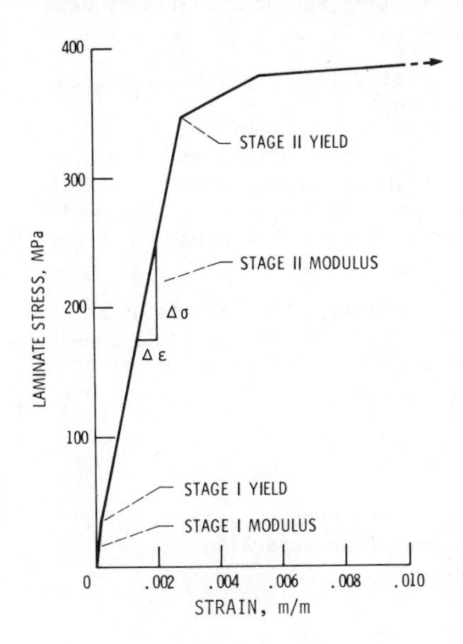

Figure 9 - Calculated TFRS laminate behavior is characterized by two stages of "elastic" deformation (W-1.5ThO$_2$/Fe CrAlY airfoil laminate in simple spanwise tension at 1225°K).

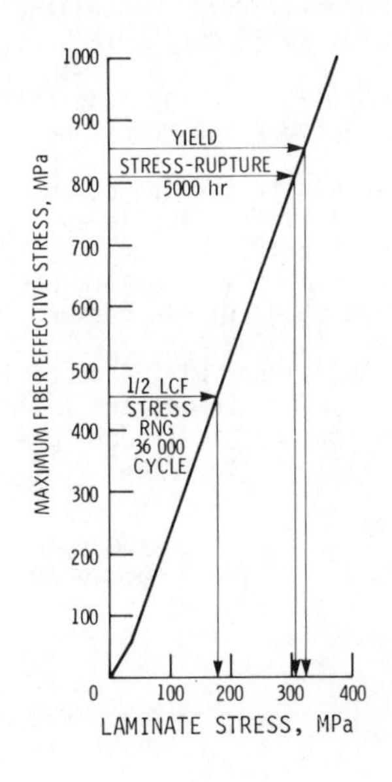

Figure 10 - Laminate Failure Calculation Approach.

Laminate Failure Prediction

Tensile, stress-rupture, and LCF failure criteria were used to determine the suitability of laminates used in the blade, vane, and OGV. A range of loading patterns and temperatures were assessed for each component.

Tensile Failure. Tensile "failure" was assumed when laminate strain exceeded 1 percent (Fig. 9) or fiber-effective-stress (von Mises stress) exceeded the fiber-yield-stress in any ply (Fig. 10). Actual fracture would normally occur at over 5 percent elongation.

Stress-Rupture Failure. The laminate stress-rupture strength was defined by the stress rupture strength of the unreacted fiber cores. Laminate stress-rupture strength for a given lifetime was defined as that laminate stress which produced a maximum fiber-effective-stress (in the core) equal to the fiber stress-rupture strength for the same lifetime (Fig. 10).

LCF Failure. The laminate LCF strength versus cycles was defined by the LCF strength versus cycles behavior of the unreacted fiber cores. As with stress-rupture, the laminate LCF strength was defined as that laminate stress which produced a maximum fiber-effective-stress (in the core) equal to the LCF strength of the fibers (Fig. 10).

Miscellaneous Properties

The following miscellaneous physical properties were calculated by NASA for use by GE during TFRS component analysis.

Density and specific heat were both calculated as the weighted average of the fiber and matrix values (Rule of Mixtures).

Laminate thermal expansion coefficients were calculated during the previous laminate tensile analysis. The instantaineous expansion coefficients are a function of the TFRS laminate matrix stress state. Hence, they vary from, typically, 10 μm/m per °K during stage I to about 5.5 μm/m per °K during stage II elongation.

Individual ply conductivities were calculated with the methods of Ref. 10. Typical values are indicated in Fig. 11. To calculate laminate conductivities, the individual ply conductivities were combined in series or in parallel as required.

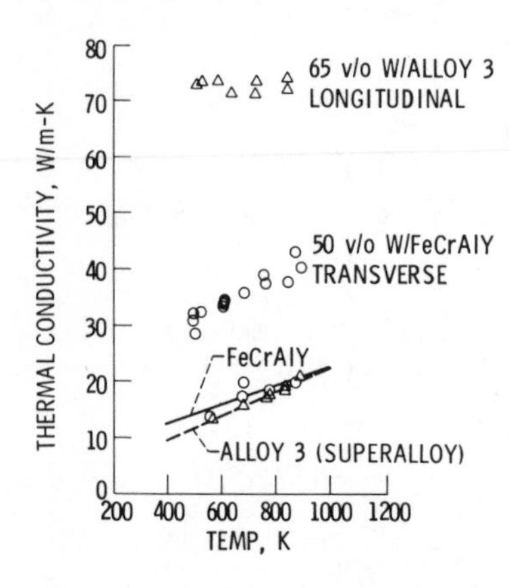

Figure 11. - TFRS conductivity is typically much higher than superalloy conductivity.

Components Considered in Design Overview

Three components received major attention during the GE TFRS design overview. As previously mentioned, these were a turbine blade, turbine vane, and an OGV. Only the key findings are alluded to below.

It is important to realize that all three components were designed and evaluated using W-1.5 ThO_2/FeCrAlY. This is a moderate strength TFRS being used primarily as a model FRS system but which may also have practical utility (Ref. 4). Much stronger fibers (e.g., W-Re-Hf-C, Ref. 1) have been tested, and stronger matrixes are under development. Consequently, the findings indicated below do not reflect the ultimate potential of TFRS.

Turbine Blade

An advanced Stage 1 turbine blade was redesigned and analyzed as a paper experiment to evaluate cooling techniques. The current superalloy blade operates at relatively high stress/low temperature; whereas, W/FeCrAlY is better suited to moderate stress/high temperature applications. Thus, in this analysis W/FeCrAlY performance was expected to be less desirable than the superlloy performance. Nonetheless, GE's design experience with this blade made it ideal as a vehicle to evaluate the efficency of ITBC cooling versus film cooling.

GE found that an ITBC cooled blade was the best design approach for TFRS, as expected (Fig. 12). TFRS film cooled and impingement cooled (without TBC) designs were also evaluated. The cooling efficiency of an ITBC cooled TFRS blade (0.190 mm TBC) equaled that of the advanced film cooled blade. However, the TFRS blade provided no significant use temperature advantage, as expected.

We infer from the NASA/GE results that a TFRS ITBC blade using stronger W-Re-Hf-C fibers could have significant potential. Aerodynamic efficiency should be higher than possible with film cooled superalloy blades. Moreover, the lack of cooling hole stress concentrations and the reduced temperature gradients due to high TFRS conductivity and the TBC might result in improved LCF capability (relative to superalloy blades). However, the target engine would have to be specifically designed to make optimum use of TFRS blades.

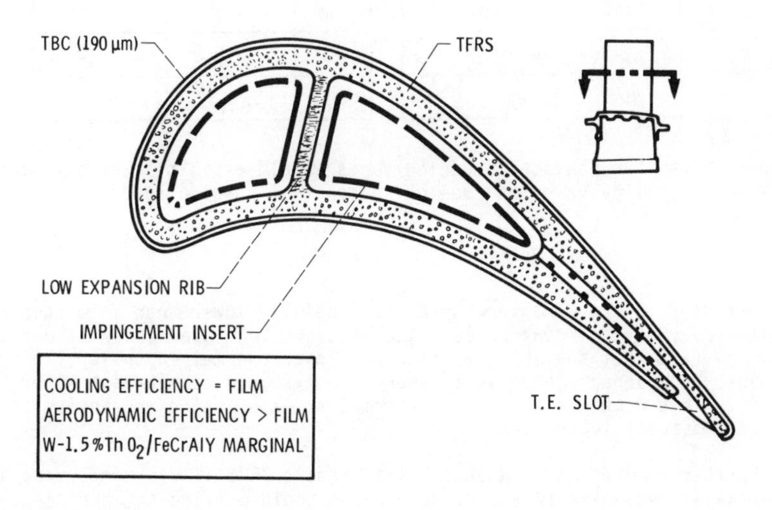

Figure 12. - TFRS blades of this general design could be viable if strong fibers (e.g., W-Re-Hf-C) were used.

Turbine Vane

A convection cooled vane was evaluated because it was considered to be an ideal application of current moderate strength TFRS technology (Fig. 13). The geometrically simple, convection cooled (no TBC) airfoil could be fabricated using previously developed techniques (Ref. 2). And the improved temperature capability of TFRS should permit reduced coolant flow for greater engine efficiency.

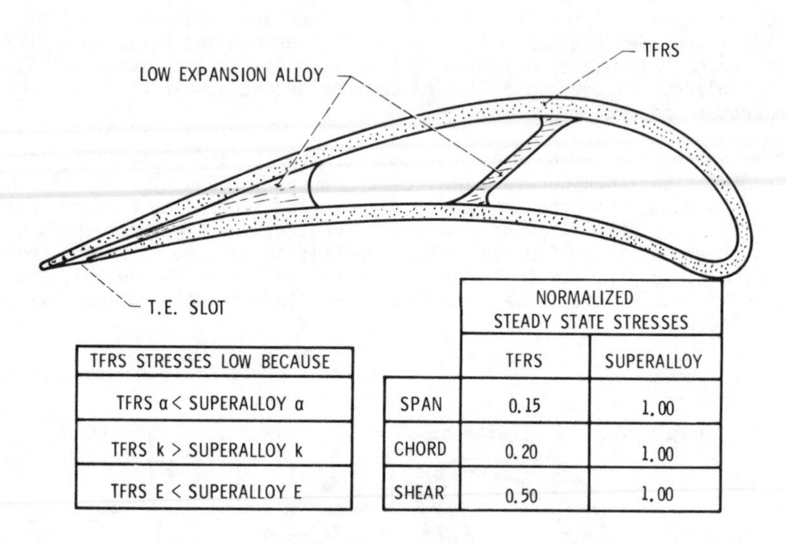

TFRS STRESSES LOW BECAUSE
TFRS α < SUPERALLOY α
TFRS k > SUPERALLOY k
TFRS E < SUPERALLOY E

	NORMALIZED STEADY STATE STRESSES	
	TFRS	SUPERALLOY
SPAN	0.15	1.00
CHORD	0.20	1.00
SHEAR	0.50	1.00

Figure 13. - TFRS convection cooled vanes should exhibit lower stresses than superalloy vanes because of substantial physical property differences.

Stresses in the TFRS vane were substantially lower than in a comparable superalloy vane according to GE calculations. The lower stresses derived from the fact that the TFRS laminate had lower thermal expansion, lower modulus, and higher conductivity than commonly used cobalt-base superalloys. This combination reduced thermal stresses which are the chief source of stress in the vane.

Further evaluation at NASA suggests that a TFRS ITBC Stage 1 vane is potentially very attractive. ITBC cooling could provide the efficiency of currently used film cooling without the associated aerodynamic penalty. Furthermore, cooling hole stress concentrations would be absent, and the TBC would lower temperature gradients in the TFRS. Those benefits combined with the inherently low thermal stresses in TFRS could significantly reduce TFRS vane susceptability to LCF failure (relative to superalloy vanes). Since LCF is a principal failure mode in Stage I vanes, the reduced LCF susceptibilty of TFRS would be highly advantageous.

Outlet Guide Vane

A TFRS OGV was considered because it offered a low risk application. Moreover, the relatively simple solid OGV airfoil was felt to be a good first candidate for rig tests (Fig. 14). The NASA/GE results indicate that a TFRS OGV may exhibit much longer life than a superalloy OGV.

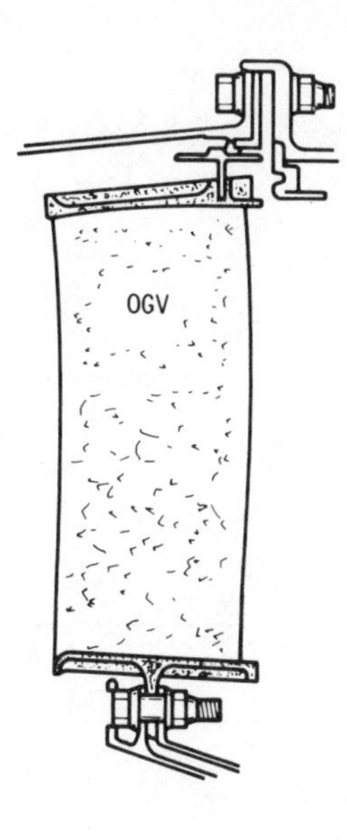

Figure 14. - A TFRS Outlet Guide Vane (OGV) could be used to gain design and rig test experience with first generation TFRS composites.

Concluding Remarks

TFRS has promise as an airfoil material in advanced aircraft turbines. The aerodynamic and cooling efficiencies of an ITBC cooled TFRS airfoil could exceed those of film cooled superalloy airfoils. Thermal stresses, which are a leading cause of failure in some applications, could be inherently lower in TFRS airfoils because TFRS has lower thermal expansion, lower modulii, and higher conductivity than most superalloys. Moreover, even moderate strength TFRS (W-1.5 ThO_2/FeCrAlY) seems adequately strong for some turbine vane and outlet guide vane applications.

None the less, more development and understanding is needed before TFRS will be ready for engine testing. For example, airfoil-to-engine attachment schemes must be developed and demonstrated. Stronger fibers (e.g., W-Re-Hf-C) must be developed to make TFRS blades more attractive. Component rig test experience must be acquired. And a larger TFRS laboratory specimen data base must be developed to expedite refinement of TFRS computational structural analysis and design methodologies. Current NASA programs are addressing these needs.

APPENDIX

Calculated TFRS Vane Airfoil Material Properties After Long Term Exposure to Vane Conditions Using Moderate Strength Tungsten Fibers

A. TFRS Composition

Fiber > W - 1.5 ThO_2 Matrix > Fe - 24 Cr - 6 Al - 1 Y
Average fiber volume fraction > 0.59
Density > 14000 kg/m^3

B. Thermal Properties

Thermal conductivity

Temp.	Span and chord	Through wall
920 K	56 W/mK	45 W/mK
1310 K	57 W/mk	52 W/mK

Thermal expansion (typical)
9.1 μm/m during Stage I elongation
7.5 μm/m during Stage II elongation

C. Mechanical Properties

Tensile properties at 1255 K
(Same in span and chord directions, semi-isotropic)

Ultimate strength	235 MPa
Ultimate elongation	> 5 percent

Property	Stage I elongation	Stage II elongation
Elastic mod.	210 GPa	52 GPa
Shear mod.	81 GPa	19 GPa
Poisson's rat.	0.301	0.334
Yield stress	21 MPa	165 MPa
Yield strain	98 μm/m	2700 μm/m

Low cycle fatigue (LCF) for 36000 cycles
(Same in span and chord directions)

Temp.	Alternating stress range
1255 K	185 MPa
1365 K	165 MPa

Stress-rupture at 1310 K for 500 hr = 145 MPa

References

1. D. W. Petrasek and R. A. Signorelli: NASA TM 82590, 1981.
2. P. Melnyk and J. N. Fleck: Report No. TRW-ER-8101, TRW, Inc., Cleveland, O., Dec. 1979 (NASA CR-159788).
3. C. F. Barth, D. W. Blake and T. S. Stelson: Report No. TRW-1ER-7930, TRW, Inc., Cleveland, O., Oct. 1977 (NASA CR-135203).
4. D. W. Petrasek, E. A. Winsa, L. J. Westfall and R. A. Signorelli: NASA TM-79094, 1979.
5. E. A. Winsa, L. J. Westfall, and D. W. Petrasek: NASA TM-73842, 1978.
6. R. W. Jech: NASA TN D-5735, 1970
7. E. A. Winsa and D. W. Petrasek: NASA TN D-7393, 1973.
8. G. I. Friedman and J. N. Fleck: Report No. TRW-ER-8135, TRW, Inc., Cleveland, O., Oct. 1979 (NASA CR-159720).
9. J. E. Ashton, J. C. Halpin and P. H. Petit: Primer on Composite Materials Analysis. Progress in Materials Science Series, Vol. 3, Technomic Publishing Co., Inc., Westport, Conn., 1969.
10. L. J. Westfall and E. A. Winsa: NASA TP 1445, 1979.

SiC-REINFORCED-ALUMINUM ALLOYS FOR AEROSPACE APPLICATIONS

B.J. Maclean and M.S.Misra
Martin Marietta Aerospace
Denver, Colorado 80201

Discontinuous SiC-reinforced aluminum alloys are newly emerging advanced materials for aerospace applications requiring high performance, isotropic mechanical properties. The aluminum alloys 6061 and 2024, reinforced with SiC whiskers or particulates, were tested for tensile, fatigue, impact toughness, and thermal expansion properties. Substantial improvements in modulus, strength, and fatigue resistance were observed when compared to the metal-matrix composite's wrought alloy counterpart. Depending on the degree of hot-working, elastic moduli on the order of 18 x 10^6 psi (124 GPa) are possible with tensile strengths of greater than 70,000 psi (480 MPa). Enhanced strength and stiffness evolve at the expense of elongation and impact toughness. Microstructure and fractography reveal the relation between reinforcement/matrix homogeneity and isotropy of properties. The coefficient of thermal expansion is seen to decrease from a nominal value of 13 x 10^{-6} in./in.^oF to 8 x 10^{-6} in./in.^oF.

Introduction

High performance space structures and antenna systems must meet increasingly stringent requirements for weight savings, dimensional stability, extended service life, and survivability. These requirements are most pronounced in the development of large space structures (LSS) where diameters may range from 20 meters to 200 meters or more. One such generic design prototype of a 1/15th scale 12-bay deployable box-truss is shown in Figure 1 in its final deployed configuration. The structural technology for LSS has progressed to the fabrication of full scale single cubes and is demonstrated in the 15 foot cube deployed in Figure 2. This cube was produced from state-of-the-art graphite/epoxy (Gr/E) composites. It is dimensionally stable, exhibits a high degree of stiffness and weighs less than sixty nine pounds.

Figure 1 -- Scale model of a 12-bay deployable box truss antenna system measuring 20 meters in diameter.

Organic-matrix composites such as graphite/epoxy offer high specific stiffness (elastic modulus divided by density) and near-zero coefficient of thermal expansion (CTE). However, they suffer from limited temperature capability, outgassing in space vacuum, low resistance to radiation damage, and problems with dimensional stability due to moisture absorption, creep, and micro-cracking and yielding. Other problems stem from graphite/epoxy's low thermal and electrical conductivity which results in overheating and space charging (1).

Figure 2 -- This full-scale single cube, made of graphite/epoxy, is dimensionally stable and weighs less than 69 lbs (31 kg).

It was less than ten years ago that beta-silicon carbide whiskers became available for use in metal-matrix composites. Utilizing the inexpensive process of pyrolizing rice hulls to produce fine whiskers (0.2 to 1.0 micron diameter), formable composites can be consolidated from a mixture of fine whiskers and standard metal alloy powders. Unlike continuous-filament reinforced metals, such as boron/aluminum and graphite/aluminum which are highly unidirectional in their mechanical behavior, SiC/Al composites are nearly isotropic. These materials can be forged, extruded, rolled, and pressed into a variety of shapes, yielding substantial improvements in elastic modulus and strength over the base alloy used as their matrix. With the introduction of SiC/Al composites which utilize silicon carbide particulate by-products from the abrasive industry, this whole class of discontinuous SiC-reinforced metals has provided low cost alternatives to designers for applications which require high specific stiffness and strength (2, 3).

One application being studied is the use of SiC/Al composites for spacecraft truss fittings. One such fitting, shown in Figure 3, joins the tubes and diagonal stays of the cube in Figure 2. It is presently made of compression molded chopped fiber-graphite/epoxy and exhibits high specific stiffness and low thermal expansion. However, its relatively low strength is a disadvantage. A comparison of metal-matrix and organic-matrix composites with more conventional metals and alloys is made in Figure 4. Utilizing SiC/Al composites for such fittings could generate improved environmental stability and considerable weight savings by offering twice the specific stiffness of conventional materials and with improved strength. The graphite/epoxy used in the corner fitting of Figure 3 is shown below and to the left of the SiC/Al group. In addition, the continuous-filament-reinforced metal matrix composites are indicated (for mechanical properties parallel to the reinforcement) for comparison but it is the need for isotropic mechanical properties in truss fittings that rules out the use of these materials.

The objective of this ongoing study is to characterize the mechanical and thermal behavior of SiC/Al composites and evaluate their application as a structural material for spacecraft and large space structures.

Figure 3 -- A spacecraft truss fitting presently made by compression molding chopped-fiber graphite/epoxy.

Table I -- SiC/Al Product Forms Evaluated

AS-PRESSED SHEET	25 v/o SiC_P/2024 (0.020" THICKNESS)
AS-PRESSED BILLET	20 w/o SiC_W/6061 (6" dia. x 6")
EXTRUDED TUBE	20 w/o SiC_W/6061-T6 (1.20" O.D. x 0.10" WALL)
ROLLED SHEET	20 w/o SiC_W/6061-T6 (0.100" THICKNESS)
CROSS-ROLLED SHEET	20 w/o SiC_W/6061-T6 (0.050" THICKNESS)

P = Particulate-reinforced

W = Whisker-reinforced

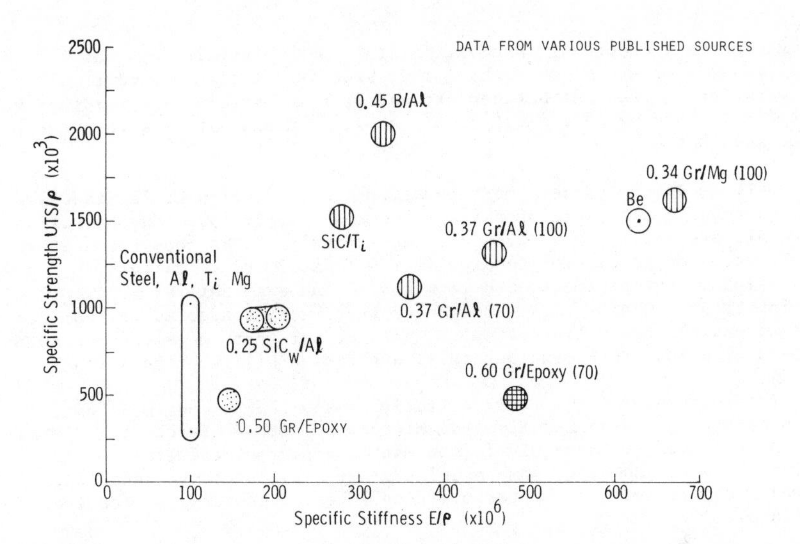

Figure 4 -- Specific strength/specific stiffness comparison of metal-matrix and organic-matrix composites with more conventional alloys.

Experimental Procedure

Material Selection

The materials evaluated in this study are listed in Table I. It should be mentioned that all of these materials were fabricated during 1980 except the whisker-reinforced 0.050 inch (1.30 mm) sheet fabricated in early 1982. The 2024 aluminum sheet, reinforced with 25 volume percent (v/o) SiC particulate, was received in the 'as-pressed' condition (no heat treatment). A slight longitudinal taper in thickness was noticed; probably the result of using non-parallel plattens during consolidation. The 6061 aluminum billet, sheets, and tube were reinforced with 20 weight percent (w/o) SiC whiskers (Exxon grade F-9). The billet was also received in the 'as-pressed' condition (no heat-treatment), while the tube was extruded from a similar billet. The sheet materials were formed by rolling plate, also extruded from consolidated billets. The 0.100 inch (2.54 mm) sheet was rolled parallel to the extrusion direction. However, presently, the common practice is to cross-roll the plate (rolling transverse to the extrusion direction), as was performed on the 0.050 inch (1.30 mm) sheet. The sheets and tube were subsequently heat treated to the T-6 condition.

Test Procedures

Evaluation of the SiC/Al product forms included microstructural investigation, tensile, fatigue, Charpy impact toughness testing, associated fractography, and coefficient of thermal expansion measurements.

Mechanical testing was performed on all five of the SiC/Al product forms. Specimens were machined per ASTM methods E8-69 and room temperature tensile tests were run (per ASTM B557-73) at 0.005 in./in./minute (1.27×10^{-2}cm/cm/minute) on a 100 kip MTS machine while simultaneously recording load vs. strain. Both a strain gage and extensometer were used on each of four specimens tested per material orientation. Agreement between extensometer and strain gage readings was excellent and gave no indication of non-uniform elongation within the gage sections.

Fatigue specimens were machined from the SiC/6061-0 billet the same as the tensile test specimens (gage section: 1.250 inch long by 0.250 inch diameter; 3.18 cm by 0.64 cm, respectively). The room temperature tests were run in tension-tension (R = 0.1) at 20 Hz to failure, utilizing seven specimens each from radial and axial billet orientations. Charpy impact toughness tests were conducted on the whisker-reinforced billet as well. Notched bar tests were run per ASTM E23-73 with four specimens per axial and radial billet orientations.

Fractography for the tensile, fatigue, and Charpy specimens was conducted on the Scanning Electron Microscope (SEM). Fracture surfaces were ion sputtered with gold to minimize charging of the SiC reinforcement. However, some specimens were left unplated, enabling Kevex X-ray analysis to be performed on selected features of the fracture surface.

Coefficient of thermal expansion (CTE) measurements were made on a laser ballistic dilatometer. Approximately 6 inches (15 cm) in length, specimens were mounted with thermocouples and stood upright against a quartz standard. A tilt mirror was then placed between the standard and specimen allowing thermal expansion to deflect the incident laser beam. The tests were conducted in vacuum as the temperature was changed by quartz element heating or liquid nitrogen cooling. Readings were taken at the end of a 22 foot (6.6m) optical path which generated a resolution of 0.05 μin/in $°F^{-1}$ (0.09 μin/in $°C^{-1}$). The temperature range of measurement for the apparatus was approximately -80° to +200°F (-62° to +93°C) depending on the emissivity and absorptivity of the specimens tested.

Results and Discussion

Material Characterization

The microstructure of the particulate sheet material is shown in Figure 5. Although this material is described as sheet, it is not a rolled mill product but, rather, a hot-pressed shape. Accordingly, the microstructure does not exhibit the banded appearance typical of rolled material but appears uniform with direction within the sheet. Some areas (light) of the microstructure are nearly depleted of SiC particles; however, the distribution of SiC appears to be relatively uniform throughout the remaining regions of the matrix.

Scanning electron microscopy results on the as-pressed SiC/6061-0 billet material are presented in Figure 6. The billet, unlike the particulate-reinforced aluminum sheet, utilizes SiC whiskers which are seen either fully exposed or partially imbedded in the aluminum matrix. The orientation of the whiskers is random although some non-uniformity in whisker distribution is present.

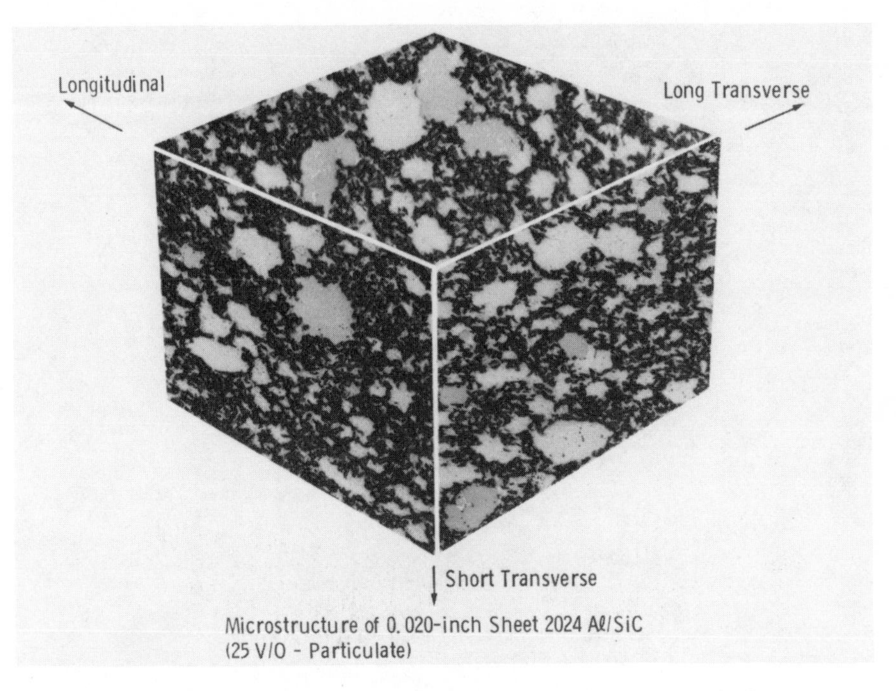

Figure 5 -- Three-dimensional microstructure for the 25-volume percent particulate SiC/Al 0.020-inch sheet.

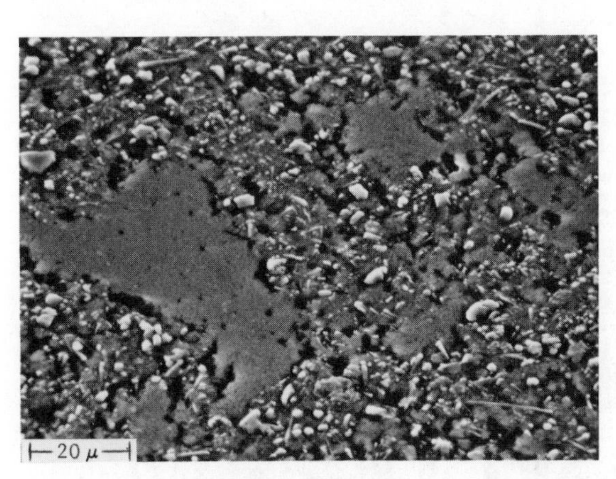

Figure 6 -- SEM micrograph of the 20-weight percent whisker-reinforced SiC/6061-O Al as-pressed billet.

The 0.100 inch thick whisker-reinforced sheet is rolled parallel to the extrusion direction from the same powder metallurgy billet material and exhibits a microstructure (Figure 7) more typical of rolled aluminum alloys. The microstructure of the extruded tube is shown in Figure 8. Again, as with the sheet material, the tubing exhibits a worked or banded microstructure with some alignment of the SiC whiskers in the major direction of the working.

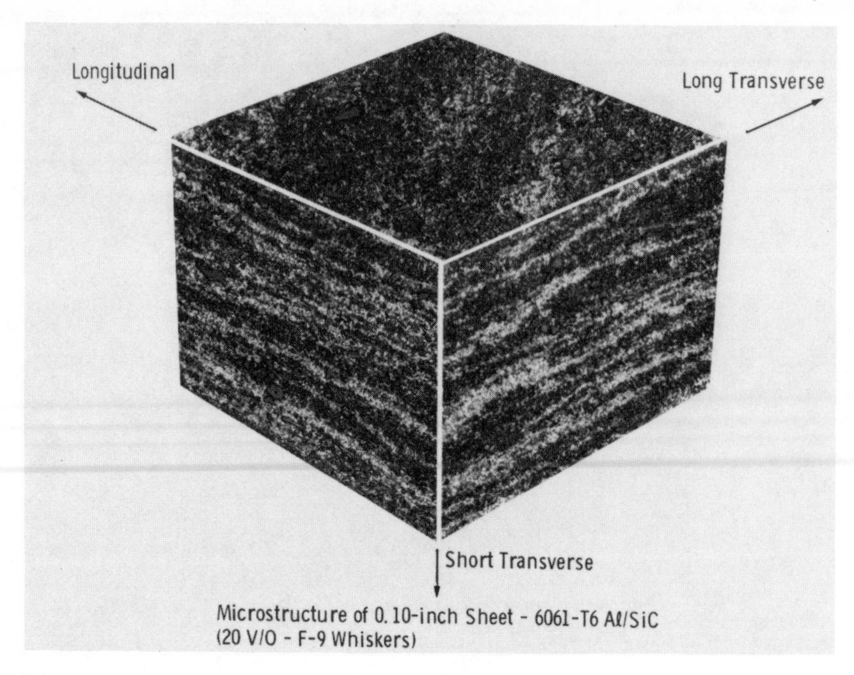

Figure 7 -- Three dimensional microstructure for the 20 weight percent whisker-reinforced/6061-T6 sheet 0.100 inch thick. Extruding and rolling were in the same direction.

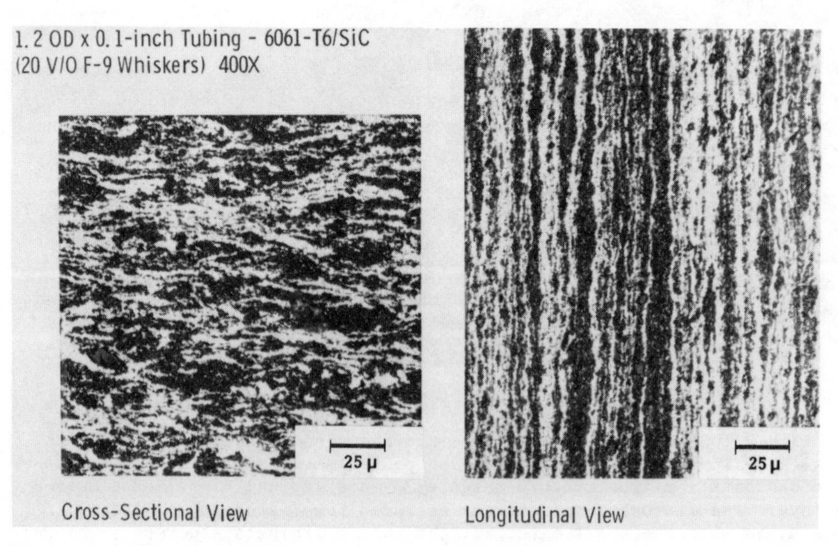

Figure 8 -- Microstructures of the extruded tube.

Tensile Properties

Results of the tensile tests on the SiC/Al product forms are shown in Table II. Tensile properties of the matrix alloys are also presented. Density values were ascertained by water immersion to determine the effect of SiC additions on the composite's weight.

Table II -- Tensile Properties of SiC/Al Product Forms and Related Wrought Aluminum Alloys

MATERIAL	PROPERTY			
	UTS ksi (MPa)	0.2% Y.S ksi (MPa)	Elongation %	E Msi (GPa)
25 v/o SiC_p/2024 [b]				
As-Pressed Sheet 0.020"				
- Longitudinal	____[a]	____[a]	____[a]	18.0 (124)
- Transverse	50 (345)	48 (331)	2.5	17.4 (120)
Typical 2024 Aluminum				
- Annealed	27 (186)	11 (76)	21	10.6 (73)
- T3	69 (476)	50 (345)	18	
20 w/o SiC_W/6061-T6 [b]				
Rolled Sheet 0.100"				
- Longitudinal [c]	63 (434)	54 (372)	2.4	18.0 (124)
- Transverse	59 (407)	49 (338)	7.0	13.5 (93)
Cross-Rolled sheet 0.050"				
- Longitudinal [c]	73 (503)	56 (386)	2.1	14.7 (101)
- Transverse	71 (490)	56 (386)	3.8	14.2 (98)
Extruded Tube 1.20" O.D.				
- Longitudinal [c]	67 (462)	56 (386)	1.9	14.5 (100)
Typical 6061-T6 Al	45 (310)	40 (276)	15	10.0 (69)
20 w/o SiC_W/6061-0 [b]				
As-Pressed Billet 6" dia. x 6"				
- Axial	33 (228)	17 (117)	4.1	12.6 (87)
- Radial	39 (269)	18 (124)	2.7	12.8 (88)
Typical 6061-0 Al	18 (124)	8 (55)	30	10.0 (69)

a Longitudinal sheet taper caused failure within the thinner of the two grip sections

b Four specimens tested per orientation

c Longitudinal to the extrusion direction

It is clear that the mechanical properties, in general, are less attractive for the 6061/SiC as-pressed billet. This material has not been hot-worked and, as indicated during the metallography, the orientation of the reinforcing SiC whiskers is random. There is, however, some preferred orientation in the radial plane of the billet (induced during consolidation) resulting in slight increases in modulus and strength compared to the axial direction. A corresponding decrease in radial strain to failure is likewise seen. The nearly isotropic behavior of the billet yields only a 25% improvement in specific stiffness (elastic modulus divided by density) compared to wrought 6061 aluminum, while 50 to 75% is more typical for the sheet products. Also, there is strength improvement over the matrix alloy in the annealed condition, but in the T-6 condition, 6061 aluminum is still slightly stronger than the as-pressed billet material.

The whisker-reinforced 6061-T6 sheet material further demonstrates the effects of whisker alignment on mechanical properties. The 0.100 inch thick sheet, rolled parallel to the extrusion direction, retains considerable alignment of the whiskers in this direction. The properties are far more anisotropic and, even though the SiC loading is the same as the billet material (20 wt.% SiC whiskers), the elastic modulus has been greatly enhanced from 13 x 10^6 psi (90 GPa) for the billet to 18 x 10^6 psi (125 GPa) for the sheet. Transverse to the extrusion direction, however, the elastic modulus remains low, about 13.5 x 10^6 psi (93 GPa), but a substantial ductility of 7% is still available in this direction. Although strength appears to be only slightly affected by whisker alignment, as it was in the billet, the elongation to failure is quite sensitive to the orientation of whisker reinforcement. Because of the very low strains to failure in the extrusion direction, rolling in this direction is both difficult and causes considerable whisker breakage and degraded mechanical properties (a 6061 aluminum matrix appears to be the most forgiving in this operation compared to the alloys 2024, 7075, and so on). For this reason, most sheet material is cross-rolled, transverse to the extrusion direction, as was the 0.050 inch thick sheet material.

The cross-rolled whisker-reinforced sheet possessed far more isotropic mechanical properties. Some preferred whisker orientation, aligned in the extrusion direction, resulted in very slight increase in modulus and strength (and a corresponding decrease in available strain to failure). Despite the overall drop in modulus, from 18 x 10^6 psi (125 GPa) in the extrusion direction for the 0.100 inch thick sheet to about 14.5 x 10^6 psi (100 GPa) for the cross-rolled sheet, the strength values are higher.

The whisker-reinforced 6061 tube properties provide additional information. It is interesting to note that while the yield strengths of the sheets and tube are nearly constant, about 55 x 10^6 psi (380 MPa), the ultimate tensile strength of these product forms appear to improve with increasing amounts of hot-working. Except in the case of the 0.100 inch thick sheet, where rolling parallel to the whisker alignment no doubt damaged much of the reinforcement, it can be assumed that additional amounts of hot-working of this material further eliminates such anomolies as unmixed regions of whiskers and matrix. Likewise, improvements in the whisker/matrix bond and reduced numbers and sizes of anomolies increases ultimate tensile strength and plasticity before fracture.

For the particulate SiC 2024 sheet material, the properties also remain nearly isotropic. With an elastic modulus of 18 x 10^6 psi (125 GPa) and tensile strength of 50,000 psi (345 MPa), the sheet shows a marked improvement over the properties of annealed 2024 aluminum, though not as high as the 69 x 10^6 psi (476 MPa) obtained in the T-3 condition. Elongation to failure is comparable with the whisker-reinforced materials at 2.5%. With additional heat-treatment, the tensile strength should improve further. The isotropy of this material is the result of the as-pressed condition of the sheet (no rolling or hot-working was performed) and the low aspect ratios of the particulates. It would seem that the particulate-reinforced SiC/Al materials lend themselves well to further hot-working, remaining less sensitive to reinforcement alignment and retaining a higher degree of isotropy. In any case, the substantial improvements in modulus and strength in these composites come at the expense of ductility. Whether the strain to failure can be improved by eliminating internal anomolies through improved fabrication and processing techniques, will remain to be seen.

Tensile Fractography

Fractography, conducted on the scanning electron microscope, revealed several major characteristics of SiC/Al failure modes. In Figure 9 the fracture surface of a whisker-reinforced billet specimen shows the presence of cavities within an otherwise normal, dimple tensile overload surface. Closer examination of these pits reveals clusters of unmixed whiskers (Figure 10) which can act as fracture initiation sites. Secondary cracking can be seen connecting several of these sites in Figure 9. Despite the low elongation measured during tensile testing, the fracture surfaces indicate a high degree of localized ductility. For a 0.100 inch thick sheet specimen oriented parallel to the extrusion and rolling directions, whiskers can be seen imbedded normal to the fracture surface at the bottoms of the dimples (Figure 11). In Figure 12, however, the specimen was oriented transverse to the hot-working and the whiskers are found predominantly on their sides at the bottoms of much larger dimples. From the mechanical property data, elongation in this direction was greater than the longitudinal direction.

Concern has been expressed regarding the strength of the interfacial bond between whiskers or particulates and the aluminum matrix. In Figure 13 a large whisker is seen at the fracture surface of a sheet specimen. Although an apparent disbond and separation of whisker and matrix has occurred, a close examination of the whisker surface shows aluminum matrix still adhering to the SiC. The separation, in fact, leads through the aluminum matrix and X-ray analysis confirmed that the material adhering to the silicon carbide is aluminum.

An examination of the particulate-reinforced sheet material fracture surface reveals what appears to be inclusions (Figure 14). Upon further examination (Figure 14a) these inclusions were found to be partially consolidated, oxidized aluminum powder devoid of SiC particulates. The fact that they cleaved and shattered so easily is an indication of brittleness and potential fracture initiation. The appearance of the surrounding regions of particulate-reinforced matrix is more ductile (Figure 15) with SiC particles found, typically, in the bottoms of many dimples. The presence of shallow shear-type dimples in this material may be due more to the 2024 matrix than the effect of particulates (compared with the whisker-reinforced 6061 discussed above).

The problem of unmixed regions of oxidized aluminum powder was not restricted to the as-pressed particulate-reinforced sheet material. A fatigue specimen machined from the whisker-reinforced billet failed on the first cycle at less than half of the tensile yield strength of this material. The fracture shown in Figure 16 is seen radiating from a large particle at the bottom. A closer view and X-ray analysis determined that this was unmixed, oxidized aluminum powder. There is a need to minimize

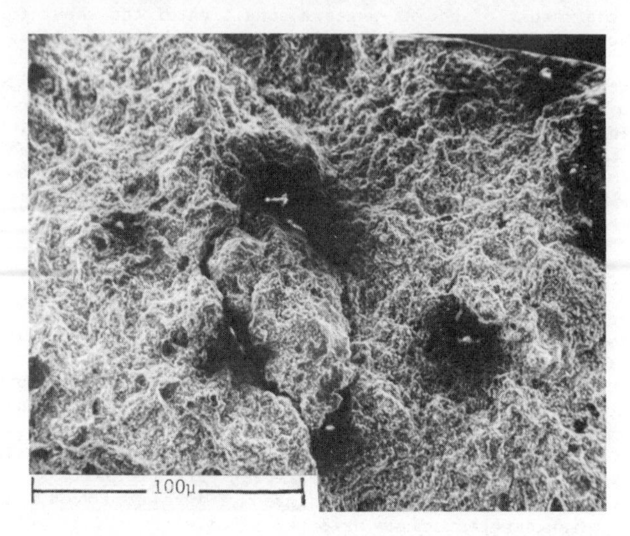

Figure 9 -- Fracture surface of a whisker-reinforced billet tensile specimen.

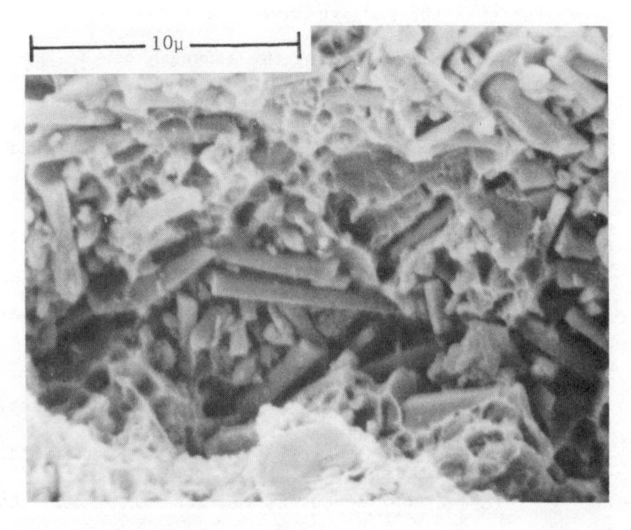

Figure 10 -- Unmixed whisker clusters in the pits shown in Figure 9.

oxidation of powders during atomization and handling to prevent this problem. Hot-working should improve strength by eliminating these regions of unmixed reinforcement where clusters of whiskers or particulates act like voids or fracture initiation sites.

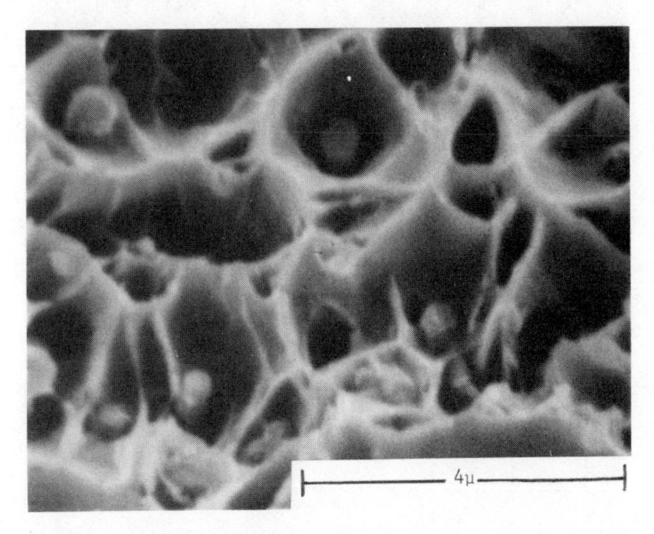

Figure 11 -- Fracture surface of the 0.100-inch thick sheet specimen with the tensile axis parallel to the extrusion and rolling directions showing whiskers embeded normal to the surface at the bottoms of dimples.

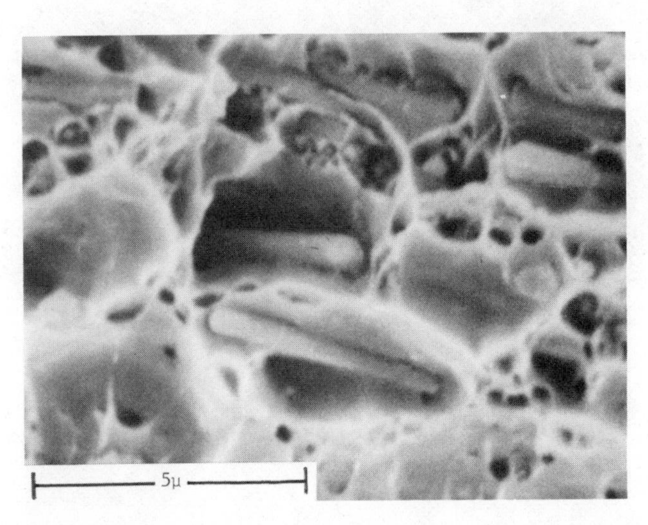

Figure 12 -- Fracture surface of a tensile specimen perpindicular to the rolling direction.

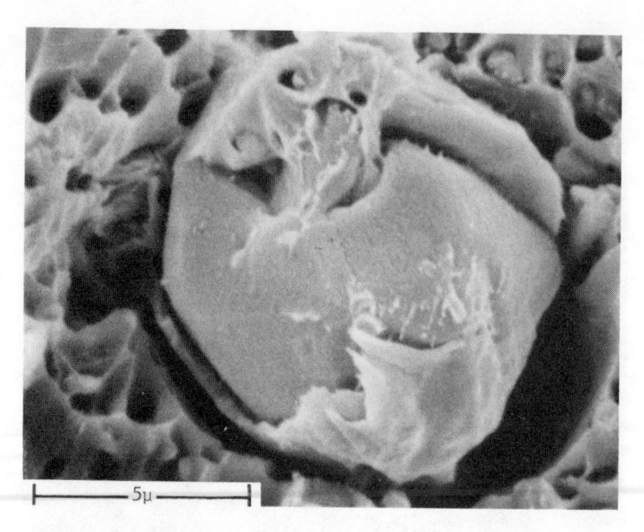

Figure 13 -- A large whisker at the fracture surface of a sheet specimen showing the aluminum matrix still adhering to the SiC.

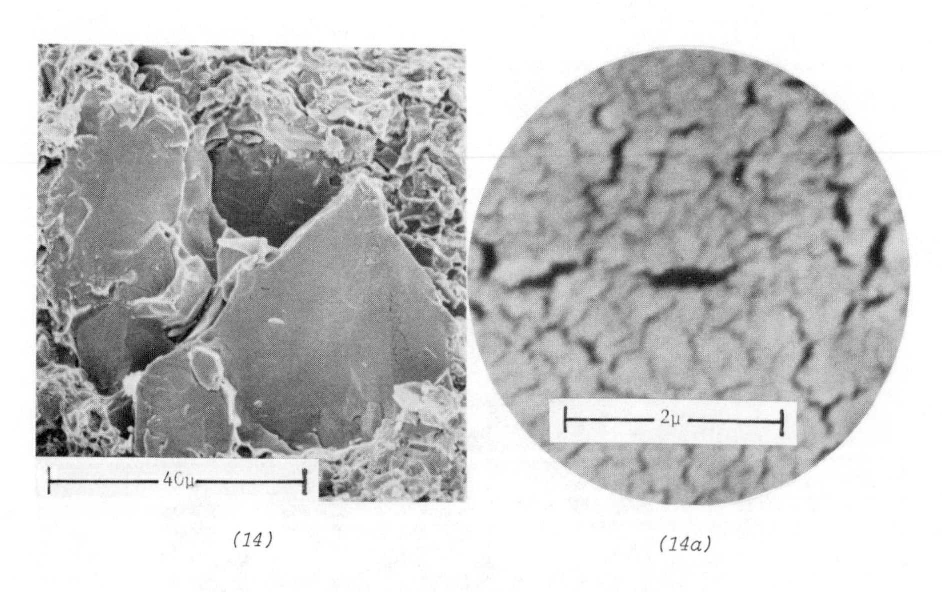

(14) *(14a)*

Figure 14 -- Fracture surface of particulate-reinforced sheet material (0.020" thick). Inclusions (14a) are partially consolidated oxidized aluminum powder devoid of SiC particulates.

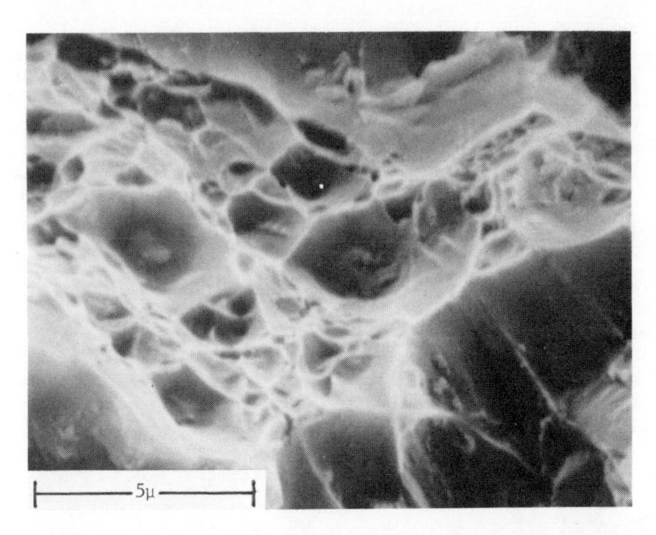

Figure 15 -- Fracture surface of particulate-reinforced 2024 sheet showing matrix ductility and SiC particles at the bottom of many dimples.

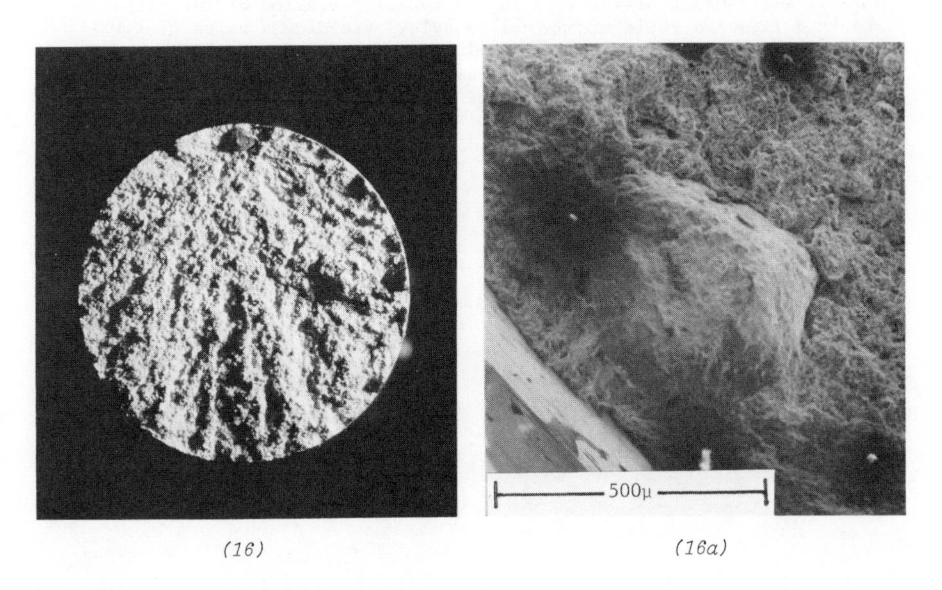

(16) (16a)

Figure 16 -- Fracture surface of the whisker-reinforced billet fatigue specimen which failed on the first cycle (16). The fracture radiates from a large region of unmixed, oxidized aluminum powder (16a).

Fatigue Properties

Results of fatigue tests run in tension-tension (R = 0.1) at 20 Hz on the billet material are presented in Figure 17. Although no reference data was available for this type of fatigue for 6061 aluminum in the 0-condition, data for the T-6 heat-treated condition is shown. The distinct difference in fatigue resistance between the axial and radial directions can be correlated with the respective differences in whisker orientation and tensile strengths. The radially oriented tensile specimens demonstrated greater strength (and lower ductility) than those axially oriented (from Table II), showing that fatigue resistance is improved in the direction of whisker alignment.

The fracture surface of fatigue specimens shows dimpling and localized ductility similar to that of tensile overloading. Fatigue striations could not be found, although as seen in Figure 18, fracture crescents can usually be traced back to such subsurface defects as oxidized powder inclusions, whisker clusters, or pores. In several cases, multiple damage sites were also found. Fatigue failure in the SiC/Al billet further illustrates the need for improvements in powder quality, mixing, and consolidation, and the advantages of hot-working in reducing the amount and size of internal defects.

Impact Toughness

Charpy impact toughness data was obtained for both orientations of the as-pressed SiC/6061-0 billet material. Once again, the more extensive alignment of whiskers in the radial direction of the billet resulted in anisotropic mechanical behavior, yielding a value of 1.10 ft-lb of energy compared to 0.75 ft-lb of energy for the axial orientation. Although these values are far from the typical 3 to 7 ft-lb of most aluminum alloys, it should be emphasized this material has neither been heat-treated nor hot-worked.

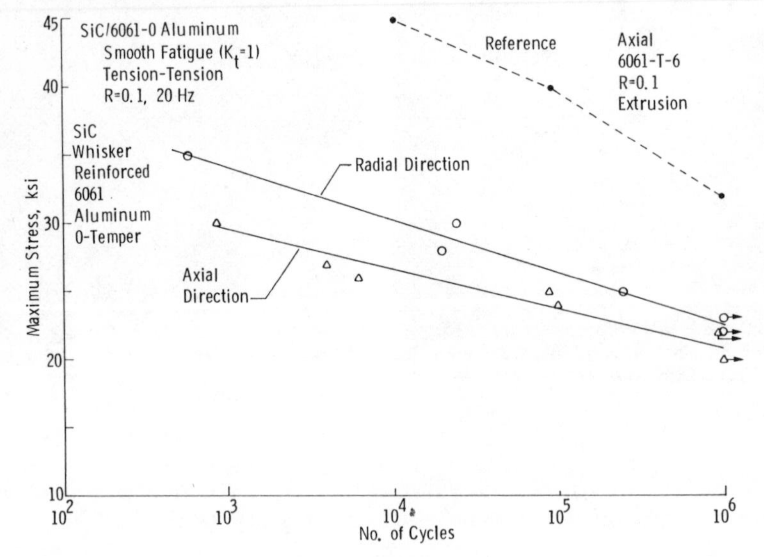

Figure 17 -- Results of fatige tests run in tension-tension (R=0.1) at 20 Hz on the as-pressed billet material.

(18)

(18a) (18b)

Figure 18 -- Fracture surface of the whisker-reinforced billet fatigue specimen (18) depicting fracture crescents (18a) tracing back to subsurface defects (18b).

Fractography on the Charpy impact toughness specimens also shows the extensive localized ductility of the SiC/Al, as seen in Figure 19. The primarily dimpled structure near the notch of the specimens begins to transform from tensile overload to shear as the far edge is reached.

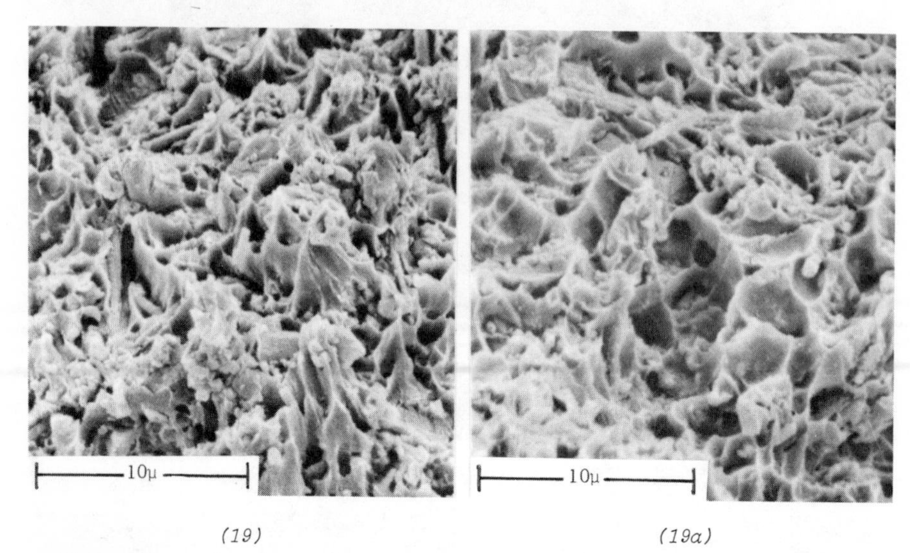

(19) (19a)

Figure 19 -- The charpy impact fracture surface of the as-pressed SiC/6061-O Al billet material showing areas near the notch (19) and near the back of the specimen (19a).

Coefficient of Thermal Expansion (CTE)

The CTE of SiC is approximately 2.0 to 2.4 x 10^{-6} inch/inch°F (3.6 to 4.3 x 10^{-6} cm/cm°C), depending on the crystalline phase present. When SiC particulates or whiskers are imbedded in an aluminum matrix, the resulting composite's CTE will be reduced compared to the matrix. The CTE of 0.020 inch thick particulate-reinforced and 0.050 inch thick whisker-reinforced sheet materials, and the whisker-reinforced extruded tube, was measured and the data is presented in Table III. It is encouraging to see the CTE improve from a nominal value of 13 x 10^{-6} inch/inch°F (23 x 10^{-6} cm/cm°C) for most aluminum alloys to around 8 x 10^{-6} inch/inch°F (14 x 10^{-6} cm/cm°C) for the SiC/Al specimens tested.

A closer look at the anisotropy of thermal expansion in the sheet materials provides a good correlation between SiC orientation and CTE. As seen from the microstructure and mechanical data, if alignment of whiskers or particulates occurs, improvement in the elastic modulus for that direction is realized. In the particulate sheet material, for example, an increase in modulus from 17.4 to 18.0 x 10^6 psi resulted in an improvement in CTE from 8.38 to 8.03 x 10^{-6} inch/inch°F (15.1 to 14.4 x 10^{-6} cm/cm°C). The correlation continues with the cross-rolled 0.050 inch thick whisker-reinforced sheet where a change in modulus from 14.2 to 14.7 x 10^6 psi resulted in an even greater improvement in CTE from 8.70 to 7.34 x 10^{-6} inch/inch°F (15.7 to 13.2 x 10^{-6} cm/cm°C) for the rolling and extrusion directions, respectively. With a modulus of 14.5 x 10^6psi (100 GPa), the extruded tube's CTE was between the two values for the whisker-reinforced sheet. It is expected, in any case, that further improvements in CTE will result in higher SiC loading levels, and enhanced SiC alignment.

Table III -- Average Coefficient of Thermal Expansion for SiC/Al Product Forms*

MATERIAL	Average CTE ppm°F^{-1} (ppm°C^{-1})	Temp. Range	Elastic Modulus x10^6psi (GPa)
25 v/o SiC_p/2024			
0.020" THICK SHEET			
- Longitudinal	8.03 (14.4)	-25 to +215°F	18.0 (124)
- Transverse	8.38 (15.1)	(-32 to 102°C)	17.4 (120)
20 w/o SiC_W/6061-T6			
1.20" O.D. TUBE			
- Longitudinal	8.34 (15.0)	-70 to +170°F (-57 to 77°C)	14.5 (100)
20 w/o SiC_W/6061-T6			
0.050" THICK SHEET			
- Longitudinal	7.34 (13.2)	-70 to +170°F	14.7 (101)
- Transverse	8.70 (15.7)	(-57 to 77°C)	14.2 (98)
Aluminum Alloys (Typical)	13.0 (23.4)	————	10.0 (69)

* Laser Ballistic Dilatometer Method

P - Particulate

W - Whisker

ppm = 10^{-6} in/in (10^{-6} cm/cm)

Summary/Conclusions

SiC/Al composite product forms were evaluated in this study, including 25 v/o particulate in 2024 aluminum (as-pressed sheet) and 20 w/o whiskers in 6061 aluminum (rolled and cross-rolled sheet, extruded tube, and as-pressed billet materials). Mechanical and thermal expansion testing results were correlated with microstructure and associated fractography to determine composite failure modes and the relation between SiC reinforcement/matrix homogeneity and isotropy of properties.

The mechanical properties of the SiC/Al product forms were found to be highly dependent on the degree of hot-working (rolling, extruding, etc.) and subsequent heat treatment. For the whisker-reinforced 6061 aluminum, the greatest degree of whisker alignment resulted in the highest elastic modulus in that direction. Likewise, the lowest strain-to-failures were recorded. The yield strength of the 6061-based materials appeared to be unaffected by hot-working. However, the greater the degree of hot-working, the higher the tensile ultimate values became. Microstructure and fractography revealed that hot-working improves SiC distribution, breaking up unmixed clusters of whiskers and unconsolidated regions of powder. As fabrication and processing of SiC/Al composites improves, these internal anomolies will be minimized and fatigue resistance, ultimate tensile strength, elongation to failure, and fracture toughness should improve. Furthermore, the fact that the elastic modulus and tensile strengths of SiC/Al represent nearly a 75% improvement over the base alloys, is great encouragement toward the ultimate optimization of these composites.

The thermal expansion of the SiC/Al products was measured to be 40% lower than typical aluminum alloys, and with increased loading levels of reinforcement (as well as orientation) further improvements should be expected. Because of the high specific stiffness and strength of SiC/Al composites, isotropy of mechanical properties, and low thermal expansion characteristics, many aerospace applications can take advantage of the performance and cost advantages of their use.

References

1) R. Stedfeld, R. H. Wehrenberg II, W. K. Kinner, "Advanced Composites," Materials Engineering, January 1980, pp. 24-66.

2) A. P. Divecha and S. G. Fishman, "Progress in the Development of SiC/Al Alloys," SAMPE Quarterly, April 1981, pp. 40-42.

3) A. P. Divecha, S. G. Fishman, and S. D. Karmarkar," Silicon Carbide Reinforced Aluminum - A Formable Composite," Journal of Metals, September 1981, pp. 12-17.

THE OXIDATION AND HOT CORROSION OF DIRECTIONALLY-SOLIDIFIED, MOLYBDENUM-STRENGTHENED EUTECTIC ALLOYS

G. L. Leatherman and S. R. Shatynski

Department of Materials Engineering
Rensselaer Polytechnic Institute
Troy, New York 12181

Certain directionally solidified eutectic alloys are being considered as future possibilities for components which must simultaneously withstand high temperatures and high stresses. This paper examines the hot corrosion of γ/γ' alloys. These alloys should display catastrophic attack when exposed to molten Na_2SO_4 at $900^\circ C$ in .21 atm. P_{O_2}. The corrosion mechanism will be discussed. The results of this study are then related to current theoretical treatments. The results indicate the α-Mo fibers are somewhat more resistant than the γ/γ' matrix thus questioning the use of acid fluxing in this system.

Introduction

Directionally solidified eutectic (D.S.) alloys because of their extremely anisotropic properties, have been proposed for gas turbine blades and vanes. These components are subjected to combined conditions of high temperature and large axial stresses. Although D. S. alloys have exceeded the mechanical requirements necessary for such components, the composition of these proposed alloys indicates extremely poor corrosion resistance. Prior studies[1-4] on a variety of D. S. alloys have shown that coatings are necessary if adequate corrosion resistance is to be obtained.

One alloy series of particular interest is the directionally solidified α-Mo strengthened γ/γ′ type. An initial investigation by Whelan[5] has shown that the Mo rich phase is more corrosion resistant than the matrix. The conclusion of this work was that Mo is not detrimental to the corrosion resistance of such D. S. alloys. These hot corrosion and oxidation morphological studies have shown the Mo-rich phase to extend into the corrosion product.

Hot corrosion is an accelerated form of oxidation commonly associated with the presence of a condensed molten salt on the alloy, generally Na_2SO_4. One current theory[6] is based on the fluxing of the oxide product in the salt thereby rendering the metal surface to the oxidizing medium. The fluxing can occur in one of two ways: a) if dissolution of the oxide into the fused salt makes the salt more basic then an accelerated form of attack will occur. b) if dissolution of the oxide into the fused salt makes the salt more acidic then an even greater catastrophic attack results. The former case is commonly referred to as basic fluxing while the later is called acid fluxing. Acid fluxing is a self-perpetuating form of attack which requires a salt coating to initiate the attack and is independent of the amount of salt present on the alloy. Basic fluxing however is totally dependent on the amount of salt present and will only occur if salt is continually available. Hot corrosion due to acid fluxing will be particularly pronounced if significant amounts of Mo or W are present in the alloy.

A second theory by Rapp and Goto[7] is based upon a negative solubility gradient. This theory requires the generation of electrochemical half cell reactions in which the local concentration of salt is modified hence resulting in dissolution at one location and reprecipitation at another. The fluxing model above requires the formation of a sulfide beneath the oxide while a sulfide precipitation is not required for the model of Rapp and Goto.

The results of Whelan's work[5] on the hot corrosion of γ/γ′ α-Mo strengthened alloys showed that the Mo rich phase was more resistant to attack than the matrix. If this is indeed the case, then a question can be raised as to the validity of acid fluxing for such complex alloys. Because of the important consequences of these conclusions, the authors have decided to re-evaluate α-Mo strengthened D. S. alloys.

Experimental

The composition the alloys used in this study are shown in Table I (These alloys were supplied by United Technologies Research Laboratory and Climax Molybdenum Corp.).

The as-received materials were in the form of random lengths of 9.5mm diameter rod. These alloys were directionally solidified at the rate of 3cm/hr. All alloys showed an exceptionally well defined directionally solidified microstructure (see Fig. 1)

Table I. Composition of Alloys Used in This Study

Alloy	Ni	Mo	Al	Cr	Ta	W
A75-660	62	31	7	-	-	-
A75-842	57.3	27.5	6.2	9	-	-
A75-716	62.3	31.5	6.2	-	-	-
A76-505	58.5	30	5.5	3	3	-
A75-738	60.8	31	6.2	2	-	-
A76-509	57.5	31	5.5	3	-	3
A75-831	58.8	29	6.2	6	-	-
A75-795	60.7	34.6	4.7	-	-	-

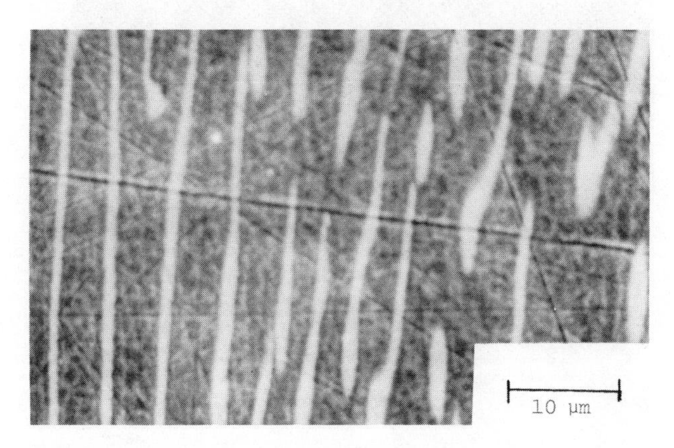

Figure 1 - Secondary electron image of as received microstructure of Ni-31Mo-5.5Al-3Cr-3W alloy.

The samples were cut into lengths of approximately 3 mm and sectioned into half-rounds to conserve material. They were polished down to 3μm diamond paste and ultrasonically cleaned in acetone and methanol. Crucible tests were conducted on all alloys for 1000 min. at 900°C in air using uncovered alumina crucibles. Approximately 17 g of Na_2SO_4 were put in each crucible. Samples of each alloy were also coated with Na_2SO_4 laid in alumina boats, and exposed to 1atm O_2 at a flow rate of $10cm^3$/min. The temperature was held at 900±2°C for intervals of 10, 100, and 1000 minutes. The coating was ∿0.005g/cm^2 and was applied by heating the specimen to 150°C and spraying with a saturated Na_2SO_4 solution.

Both crucible and coated samples were examined using conventional metallographic techniques. Scanning electron microscopy and energy dispersive x-ray analysis were used in phase determination.

Results and Discussion

The specimens subjected to the 1000 min. crucible tests were attacked to such a degree that none of the original alloy remained. A typical corrosion microstructure is shown in Fig. 2.

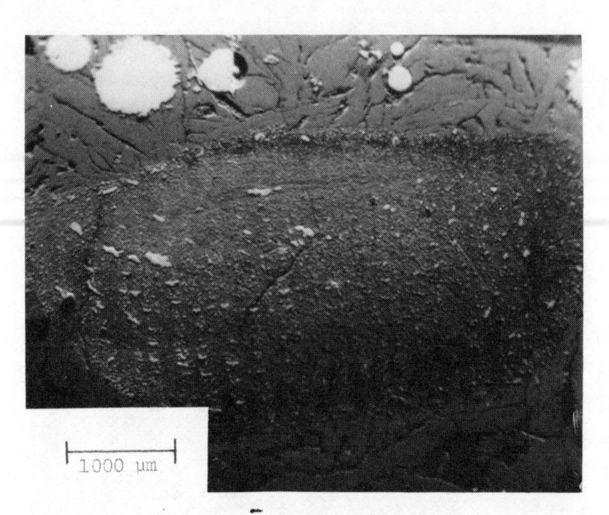

Figure 2 - Backscattered electron image of microstructure from a crucible test of Ni-31Mo-7Al in molten Na_2SO_4 for 1000 min at 900°C in air.

It consisted of a core of light green $NiMoO_4$ containing small round metal inclusions surrounded by a thin layer of dark green NiO. The solidified Na_2SO_4 surrounding the sample contained small spheres of alloy composition. These spheres showed none of the directionally solidified microstructure of the original alloy. Stewart[8] also observed such spheres in crucible tests of B-1900. The spherical nature of these particles and the lack of D. S. microstructure suggests localized melting. This melting may be due to large heat of oxidation and/or rapid corrosion kinetics.

In addition to crucible tests, samples were also subjected to salt coated hot corrosion tests. A thin film of salt on the alloy surface resulted in a less severe but still extensive hot corrosion. Optical microscopy performed on such samples revealed a number of interesting features. Fig. 3 is a typical optical micrograph of Ni-34.6mo-4.7Al alloy coated with $0.005 g/cm^2 Na_2SO_4$ and exposed at 900°C for 100 minutes in 1 atm. O_2.

Figure 3 - Photomicrograph of Ni-34.6Mo-4.7Al alloy located with 0.005 g/cm^2 Na_2SO_4 exposed at 900^0C for 100 min. in 1 atm. O_2.

The corrosion product can be divided into three regions: a porous inner layer consisting mainly of $NiMoO_4$ and two outer regions consisting of a mixture of alloy particles and oxide. $NiMoO_4$ undergoes a phase transformation from B to γ resulting in a powdery product. This product was extensively examined by Whelan.[6] The outer region consisted of mainly NiO with large elongated particles of alloy with little composition change. These particles again had no remnants of the aligned α-Mo fibers. Fig. 4 is an electron micrograph of a typical metallic particle in the outer zone.

Note the absence of any second phases in these particles. These metallic particles are generated during the initial stages of the reaction as they are definitely present after only 10 minutes exposure as seen in Fig. 5.

A second region beneath the outer zone (see Fig. 3) consists of mixed oxide laced with a high density of alloy particles. This region is also highly porous.

The interface between the alloy and the corrosion product appear on the macroscopic level to be somewhat uniform. However, higher magnification reveals the existence of α-Mo fibers protruding into the oxide. A photomicrograph of the corrosion interface of a Ni-29Mo-6.21Al-6Cr alloy coated with 0.005g/cm^2 Na_2SO_4 after exposure to 900^oC for 100 minutes at 1 atm. P_{O_2} is shown in Fig. 6.

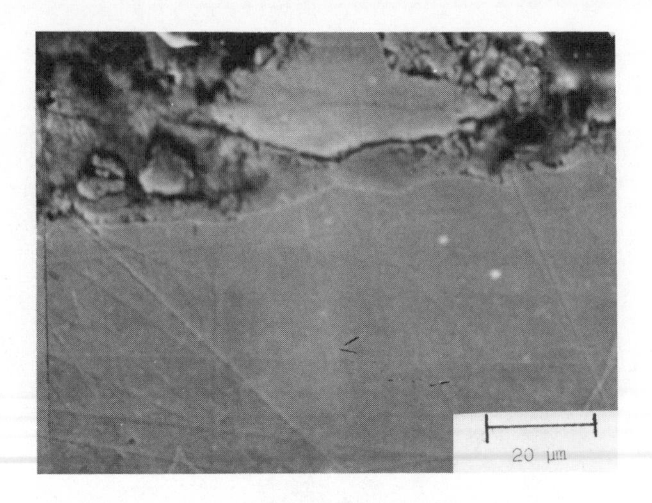

Figure 4 - High magnification secondary electron image of Ni-31Mo-5.5Ai-3Cr-3W alloy coated with 0.005 g/cm^2 Na_2SO_4 exposed at 900°C for 10 min. in 1 atm. O_2.

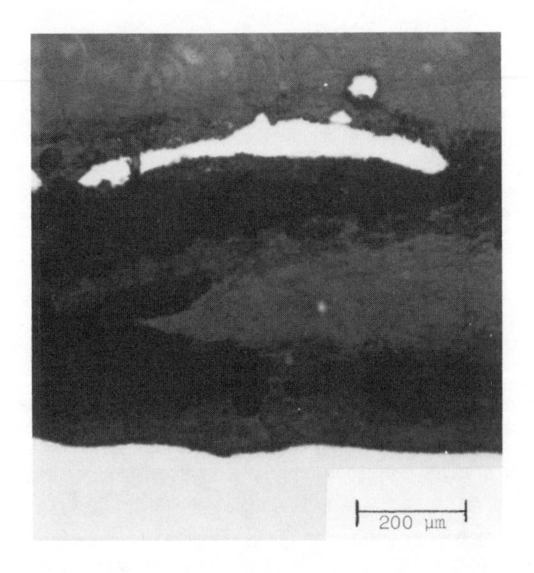

Figure 5 - Photomicrograph of Ni-31Mo-5.5Al-3Cr-3W alloy coated with 0.005 g/cm^2 Na_2SO_4 exposed at 900°C for 10 min. in 1 atm.

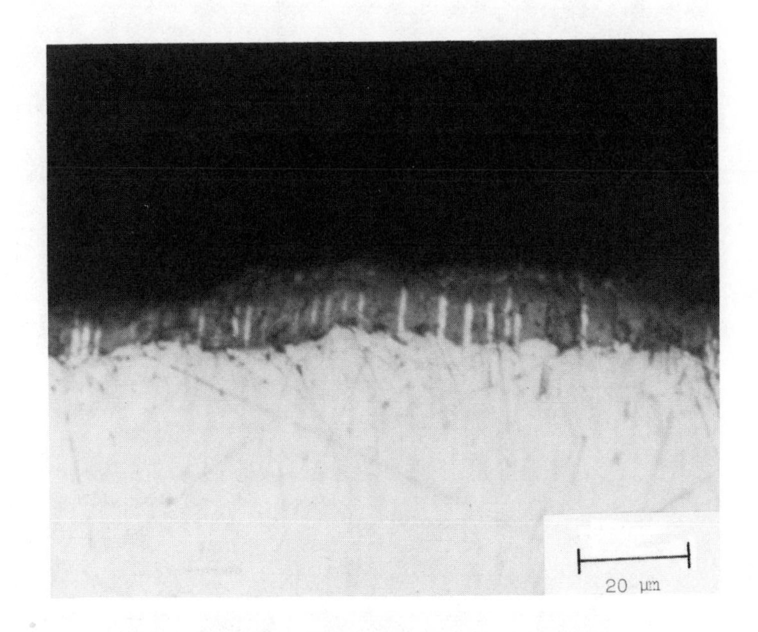

Figure 6 - Photomicrograph of Ni-29Mo-6.2Al-6Cr alloy coated with 0.005 g/cm^2 Na_2SO_4 exposed at 900°C for 100 min. in 1 atm. O_2.

Notice that some α-Mo pegs extend well into the oxide. This indicates greater stability of the α-Mo phase. A similar alloy with a composition of Ni-31Mo-5.5Al-3Cr-3W under similar conditions but exposed for only 10 minutes also exhibited preferential oxidation of the matrix as seen in Fig. 7.

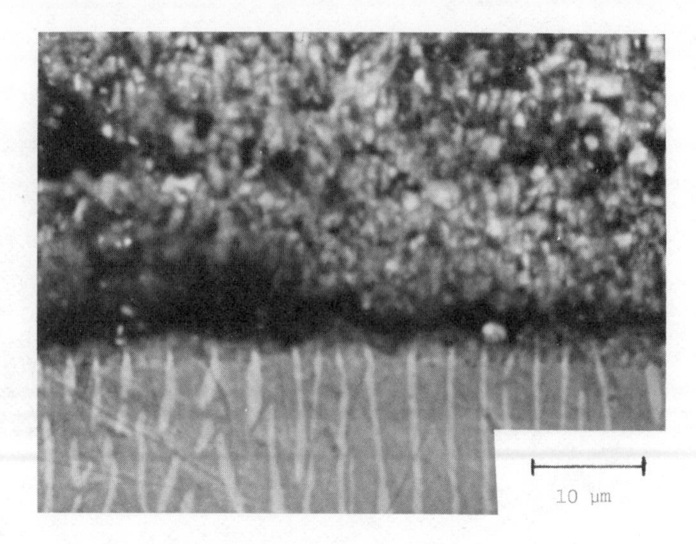

Figure 7a - High magnification secondary electron image of Ni-31Mo-S. 5Al-3Cr-3W alloy, coated with 0.005 g/cm^2 $Na_2 SO_4$ exposed at 900^oC for 10 min. in 1 atm. O_2.

Figure 7b - The corresponding Al map.

Figure 7c - The corresponding Ni map.

Figure 7d - The corresponding Mo-S map.

Figure 7e - The corresponding Cr map.

Notice the oxidation is uniform with no noticeable attack along the interphase interfaces. The associated x-ray maps indicate that little gross segration of the alloying elements occur in the oxide. Furthermore there is no Cr or Al depletion zone beneath the oxide. The lack of segregation in the oxide is further illustrated by examining the microstructure of a corroded Ni-34.6Mo-4.7Al alloy after 100 min. exposure to 900°C to 1 atm. O_2 with a Na_2SO_4 coating of 0.005g/cm^2 as shown in Figure 8.

Figure 8a - High magnification secondary electron image.

Figure 8b - The corresponding Al map.

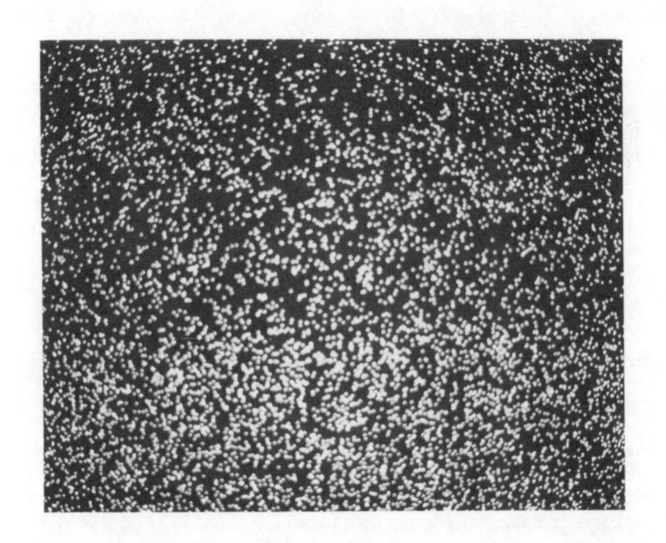

Figure 8c - The corresponding Ni map.

Figure 8d - The corresponding Mo-S map.

A very porous oxide is clearly shown above the interface. The α-Mo fibers are present in the oxide yet the matrix is beginning to break down. A large amount of subsurface porosity is noticed in the matrix adjacent to the interface. Rapid degradation of the matrix will soon follow. The associated x-ray maps again indicate little segragation of the elements beyond the D. S. nature of the alloy.

The role of α-Mo is unclear in such experiments. Previous studies by Goebel et al[6] indicated catastrophic attack of numerous conventional nickel base alloys. Catastrophic attack also occurred in alloys used in this study. However, one would expect preferential attack of the α-Mo rich phase. This certainly was not observed by Whelan[5], and in this expanded study was also not observed. Perhaps the dissolution of MO_3 in Na_2SO_4 results in modifying the salt composition by increasing the concentration of MoO_4 species by the following reaction

$$MoO_3 + SO_4^{-2} \rightarrow MoO_4^{-2} + SO_3$$

The now acidic salt would preferentially attack the Al-Cr rich mixture. However, if this were true then substantial sulfidation due to the build-up of SO_3 would result. This was not observed.

The theory of Rapp and Goto[7] when applied to such alloys also is not straight forward. The porous oxide is the result of a phase transformation. This transformation masks the degree of porosity generated during the corrosion reaction. Using a series of localized half cell electrochemical reactions, Rapp and Goto[7] have demonstrated the possibility of producing catastrophic hot corrosion without sulfidation. Certainly extensive sulfadation was not seen using optical microscopy. (x-ray mapping cannot distinguish between S and Mo). In any case neither theory gives a complete satisfactory explanation of the observed phenomenam, and it appears that the acid fluxing rule does not apply to this alloy group.

Conclusion

This work upholds a recent study by Whelan[5] which indicated that for α-Mo-γ/γ′ type D. S. alloys the matrix is selectively attacked during hot corrosion. Extensive corrosion was noted in all alloys studied with the bulk of the reaction product being $NiMoO_4$. Particles of alloy which contained all metallic elements were found on the outer surface of the corrosion product. These particles did not have a directionally solidified microstructure and showed some indications of melting. Additions of Ta and W did not dramatically influence the corrosion resistance. The role of Mo during the corrosion of these alloys does not agree with current theories but still is detrimental to adequate corrosion resistance.

Acknowledgement

The authors would like to acknowledge E. P. Whelan of Climax Molybdenum Corporation for supplying the alloys. Also, we would like to acknowledge the technical services of M. Doody. Finally the authors are grateful to acknowledge the financial assistance of the Division of Basic Sciences, U. S. Department of Energy under contract No. DE-AC02-79 Er10428.

References

1. S. R. Shatynski, Rev. of High Temp. Materials, in press.

2. S. R. Shatynski and K. A. Dannermann, Oxid. of Metals 14 (1980) pp. 531-548.

3. J. J. Aliprando and S. R. Shatynski, Oxid of Metals, 15 (1981) pp. 455-469.

4. J. J. Aliprando and S. R. Shatynski, "Thermal Stability of Directionally Solidified Co-W Eutectic Alloys" in New Developments and Applications in Composites ed. D. Huhlmann-Wilsdorf, and W. C. Harrigan, AIME 1978 pp. 31-39.

5. E. P. Whelan, Climax Molybdenum Co., Report J-4297, February 16, 1978.

6. J. A. Goebel, F. S. Pettit and G. W. Goward, Met. Trans. 4 (1973) pp. 261-278.

7. R. A. Rapp and K. S. Goto, "The Hot Corrosion of Metals by Molten Salts" Symposium on Fused Salts, Pittsburgh, Pennsylvania, The Electrochemical Society, Princeton, New Jersey, 1979.

8. S. F. C. Stewart, M. S. Thesis, Rensselaer Polytechnic Institute, Troy, New York, August 1981

MARINE CORROSION OF GRAPHITE/ALUMINUM COMPOSITES

M.G. Vassilaros+, D.A. Davis+, G.L. Steckel*, J.P. Gudas+

+David W. Taylor Naval Ship R and D Center
Annapolis, Maryland 21402

*Aerospace Corporation
El Segundo, California

The marine corrosion and mechanical properties of two types of VSB-32/Al 6061 graphite aluminum composite materials were characterized. Corrosion tests were performed in natural flowing seawater, tidal immersion and atmospheric exposure. The residual mechanical properties of the composites were evaluated after exposure. Results of environmental exposures showed that the galvanic driving force dominated the corrosion of the composite materials, and the overall performance of the composites was related to both the corrosion of the surface foil and the substrate. Residual mechanical properties did not show latent effects of the environment where corrosion was not visible, but were substantially reduced in response to visible corrosion damage.

Introduction

Metal matrix composite materials are being considered for structural applications because of the range of mechanical properties attainable. The graphite/aluminum system is particularly promising because of the very high specific strength and modulus levels attainable over other alloys and composites materials. As mechanical properties of graphite/aluminum composites have been improved, it became necessary to evaluate the environmental sensitivity of the composite system.

The objective of this investigation was to characterize the marine corrosion performance of two similar types of graphite/aluminum composites materials. The approach included the production of VSB-32/Al-6061 uniaxially reinforced composite plates which were exposed in marine environments including natural flowing seawater, tidal immersion, and atmospheric exposure. Mechanical property tests of the composite material were performed prior to and after marine environment exposure. Analysis of test results were performed to correlate type and degree of corrosion attack with the residual mechanical properties of the composite.

Background

A recent review of results of corrosion tests of graphite/aluminum has been prepared by Pfeifer[1]. This review detailed the results of corrosion exposures of T-50/Al 6061 in 3.5% NaCl and distilled water, T-50/Al 201 and T-50/Al 202 panels exposed to the marine atmosphere, Gr/Al 202 with Al 1100 interlayer foils exposed to the marine atmosphere and alternating immersion in laboratory seawater, and T-50/Al 201 with 6061 with various combinations of alloy foils including 1100, 2024, 3003, 5056, and 6061 exposed in the marine atmosphere, alternate seawater immersion, salt spray and relative humidity cabinets. The summary of the corrosion exposures as discussed by Pfeifer[1] suggested that the mode of corrosion observed with graphite/aluminum composites was predominantly pitting and severe exfoiliation. Metal/matrix interfaces were found to limit corrosion resistance and both chemical and mechanical factors contributed to corrosion. As expected, the aluminum alloy composition affected corrosion resistance, particularly when comparing the Al 6061 matrix and Al 201 matrix alloys. Finally, it was found that corrosion behavior of graphite aluminum panels were quite sensitive to fabrication process and efficiency. To date there have been no controlled experiments to evaluate the residual mechanical properties of graphite/aluminum composites.

The results of the corrosion evaluations performed to date were employed in the design of this experimental program. Specifically, Al 6061 was chosen as both a matrix and foil cladding material in order to provide some inherent corrosion resistance. Extensive NDE was performed to eliminate (as much as possible) material produced with consolidation defects. Finally, mechanical property tests were performed before and after marine exposure to determine extent of latent environmental effects, as well as quantify the degree of degradation caused by the environment.

Material

The metal matrix composite plates used in this investigation were produced from pitch-based VSB-32 fibers and a matrix of 6061 aluminum alloy. The VSB-32 fibers were supplied by Union Carbide Corporation and displayed typical properties as follows:

Tensile Strength	1720 MPa
Young's Modulus	$3.8 \times .10^5$ MPa
Fiber Diameter	7-11μm
Numbers Fiber/Tow	2000

The fiber tows were coated with a TiB layer which was used to promote wetting during the subsequent liquid metal infiltration process. The infiltrated fiber tows appeared as aluminum wires which typically has 45 volume percent fiber as supplied by Materials Concepts Inc. DWA Composites Specialties Inc. then diffusion bonded the infiltrated wires between surface foil claddings to produce the metal matrix plates.

The graphite/aluminum plates used in this investigation were produced in two configurations. The standard plates consisted of three uniaxial layers of wires between 0.15 mm thick Al 6061 surface foils which resulted in plate thicknesses ranging from 1.8 to 1.9 mm. Figure 1 is a photomicrograph of a typical cross section of the standard material. Encapsulated plates were also produced with three uniaxial layers of wire between 0.15 mm thick surface foils. However, additonal foils were wrapped around the wires to reduce the fiber volume and increase the transverse strength of the composite. Figure 2 is a photomicrograph of the encapsulated composite material, which was produced in thicknesses of 2.0 to 2.1 mm. A total of 6 panels of each type of composite plates were produced with planar dimensions of 216 mm x 216 mm.

Experimental Procedures

The marine corrosion exposures were accomplished by removing panels from the composite plates and exposing them to three types of environments for varying lengths of time. Separate specimens were exposed in each environment for corrosion characterization and residual mechanical property measurements. The planar dimensions of the two types of specimens were as follows:

Corrosion Panels	101 x 67 mm.
Residual Mechanical Property Panels	101 x 101 mm.

The graphite/aluminum composites were exposed with and without edge protection. Edge protection consisted of a continuous bead of RTV compound applied to the edges of selected panels.

All panels were exposed to one of three marine environments at the LaQue Center for Corrosion Technology, Wrightsville Beach, North

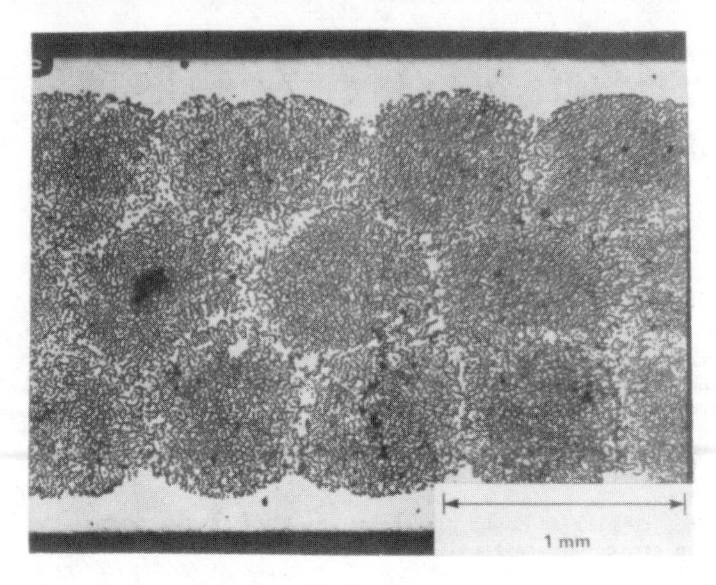

Figure 1- Photomicrograph of Cross Section of Standard VSB-32/Al 6061 Plate

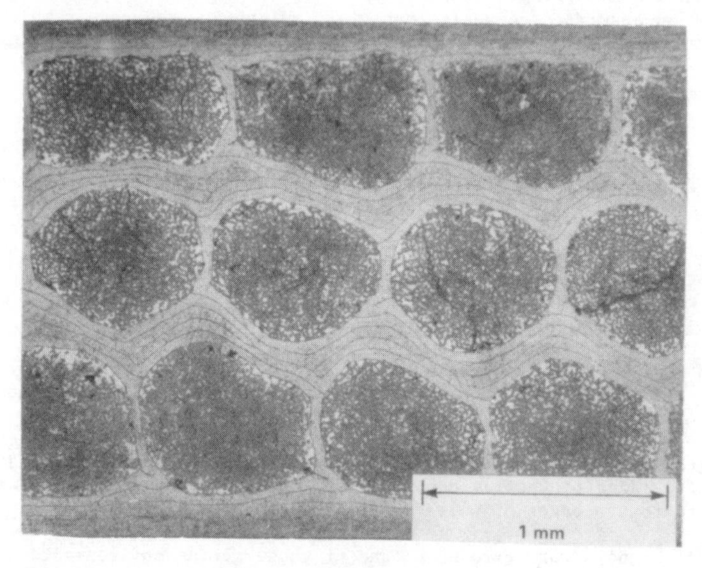

Figure 2- Photomicrograph of Cross Section of Encapsulated VSB-32/Al 6061 Plate

Carolina, USA. The environments included:

(1) Complete submergence in natural, flowing seawater (0.6-0.9 m/S);
(2) Alternate immersion in tidal zone;
(3) Marine atmospheric exposure, 25 meters from the ocean.

Separate panels of both types of composite were removed for corrosion evaluation and residual mechanical property tests after 6 and 12 week exposures.

Residual Mechanical Properties Testing

Longitudinal and transverse tensile strength and Young's modulus were determined for the baseline and exposed plates using a standard Instron Universal testing machine. The exposed plates were nominally 100 mm square and the tensile samples were prepared from these panels according to the geometry shown in Figure 3. The longitudinal samples were 9.5 mm wide by the full length of the plates and the transverse samples were typically 38-75 mm. long by 12 mm wide. Both types of samples were the full thickness of the composite plates. In some instances, particularly when the plate edges were left unprotected, swelling at the edges required removal of a small amount of material. This is indicated by "edge corrosion" in Figure 3. Unless an excessive amount of material had to be removed from the edges, four longitudinal and five transverse samples were prepared from each plate. The samples were usually prepared by shear cutting. However, corrosion of the standard composite after the 12 week flowing seawater exposure was too severe to allow shearing and samples from these plates were prepared by hand sawing and carefully filing the edges smooth.

One mm thick aluminum tabs were bonded to each side of the sample ends in order to minimize stress concentrations at the testing machine grips. Despite the tabs, many of the longitudinal samples fractured near or within the grips. However, no correlation could be made between the fracture location and tensile stength.

Strain measurements for the Young's modulus calculations were made with a 13mm gauge length clip-on electrical extensometer. Residual stresses which result from the composite processing were removed by loading the samples to 50% of their anticipated maximum load, unloading to 5% of maximum, and then reloading to failure. The load-extension curve obtained during reloading was used to determine the material's modulus in the absence of residual stresses. One longitudinal and one transverse sample from each set was tested without an extensometer in order to determine the maximum load.

Results and Discussion

Corrosion of Graphite/Aluminum Composites

All of the panels which were exposed to marine environments were subjected to corrosion analysis. This included the panels employed in

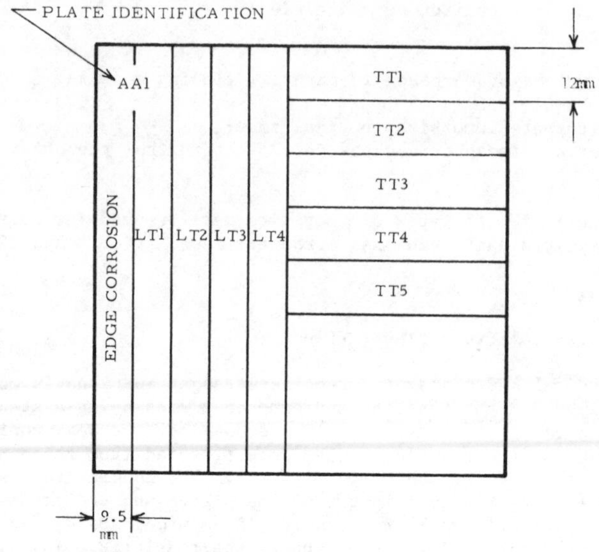

Figure 3- Typical Sample Geometry for Exposed Plates

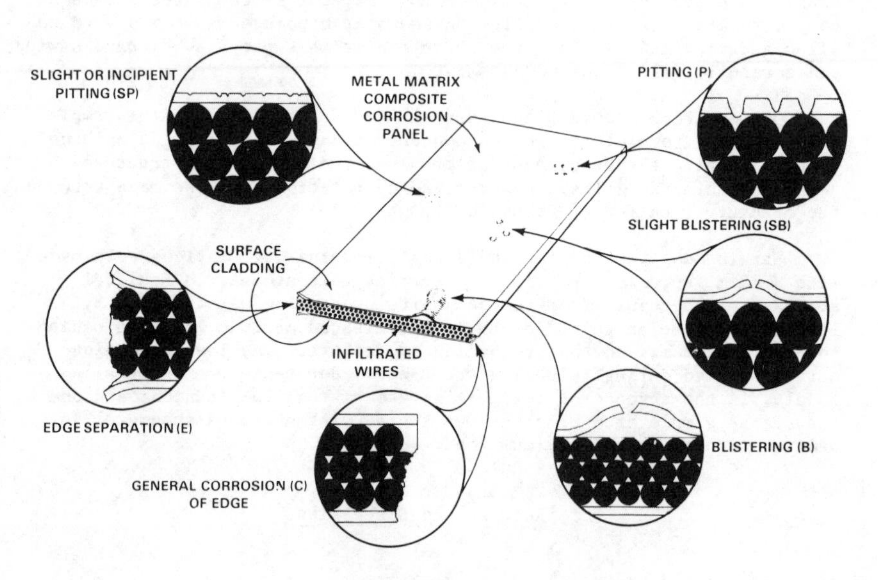

Figure 4- Schematic Illustration of Types of Corrosion Attack

Table I

Corrosion Exposure Results for VSB-32/6061 Aluminum in Flowing Seawater

Specimen Number	Length of Exposure (weeks)	Edge Protection (yes or no)	Visual Observations surface	edge
		STANDARD		
AB-1	6	yes	SP	N
AB-2	6	no	P, B	E, corr
AC-1	6	yes	B	E
AC-2	6	no	B	E, corr
AE-3	12	yes	P, B	N
AE-4	12	no	P, B	E, corr
AF-1	12	yes	P, B	E, corr
AF-2	12	no	P, B	E, corr
		ENCAPSULATED		
BA-1	6	yes	SP	N
BA-2	6	no	P, B	E, corr
BB-1	6	yes	SP	N
BB-2	6	no	P, B	E, corr
BA-3	12	yes	P, B	N
BA-4	12	no	P, B	E, corr
BF-1	12	yes	P, B	E
BF-2	12	no	P, B	E, corr

N - No visible attack
SB - Slight Blistering; B - Blistering
SP - Slight or incipient Pitting; P - Pitting
corr - general corrosion; E - edge separation

the residual mecahnical properties analysis as well as the corrosion test panels. The usual practice of describing corrosion behavior is to report weight loss, thickness reduction, and depth of pitting attack. Due to the nature of the attack observed with these composites, such descriptions were inapplicable. For example, most of the graphite/aluminum composites experienced weight gain from the oxide formed and trapped during the corrosion process. The analysis used herein is in the form of qualitative descriptions of the corrosion and pitting behavior.

The corrosion behavior exhibited by the exposed panels is described according to five different types of attack observed in these tests. These include:

(a) Slight and incipient pitting;
(b) Pitting of surface foils;
(c) Slight blistering of surface foils;
(d) Blistering of surface foils;
(e) Edge separation of matrix, fiber and foils;
(f) Uniform edge corrosion.

All types of observed corrosion attack are illustrated schematically in Figure 4, and these descriptions are included in the tabulations of corrosion test results to be presented in the following sections.

Corrosion of Graphite/Aluminum in Flowing Seawater

The results of the corrosion tests of both types VSB-32/Al 6061 composites in flowing seawater are shown in Table 1. The extent of corrosion attack experienced by the standard and encapsulated composites after six weeks exposure to flowing seawater was related to the edge protection provided each panel. When a good sealant was maintained as shown in Figure 5, no edge corrosion or blistering occurred, and the panels experienced only slight surface pitting. This pitting is typical of 6061 aluminum alloys. This behavior was related to all six week exposure panels with edge protection with one exception (AC-1). In this case, a small failure of the sealant compound resulted in intrusion of seawater to the edge which caused edge attack and blistering. Panels exposed without edge protection experienced edge attack and blistering along the fiber path, Figure 5.

Both types of panels exposed for twelve weeks in flowing seawater displayed substantially different behavior than the six-week exposures. The twelve-week exposure panels displayed pitting attack as shown in Figure 6 which pierced the surface foils and allowed seawater to come in contact with the graphite/aluminum interfaces. When this situation occurred, the aluminum corrosion product blistered the composite and exposed more surface area for corrosion attack as the corrosion process progressed from the pit. Figure 7 is a photomicrograph of a typical blister. This form of corrosion attack was not related to the edges of the panels as both the protected and unprotected panels experienced similar corrosion attack.

Corrosion of Graphite/Aluminum under Alternate Tidal Immersion

Table 2 presents results of the corrosion analysis of VSB-32/Al 6061

Figure 5- Corrosion Panels of VSB-32/Al 6061 After 6 Weeks Exposure to Flowing Seawater

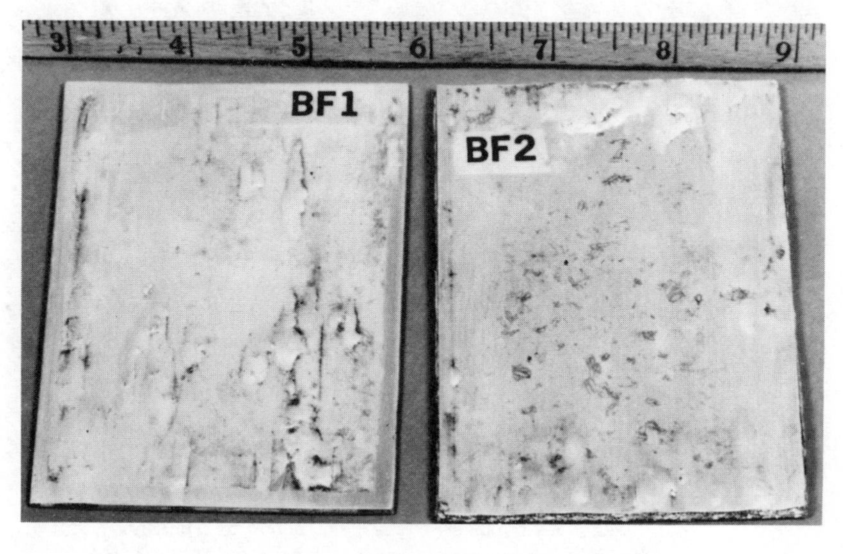

Figure 6- Corrosion Panels of VSB-32/Al 6061 After 12 Weeks Exposure to Flowing Seawater

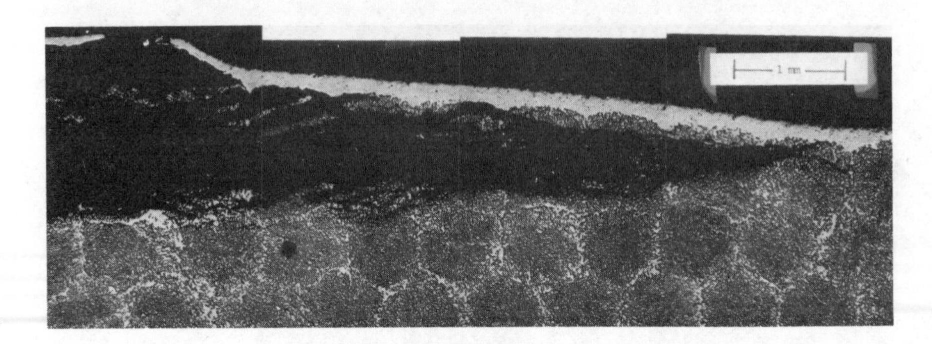

Figure 7- Photomicrograph of Blistered VSB-32/Al 6061 Composite After 12 Weeks Exposure to Flowing Seawater

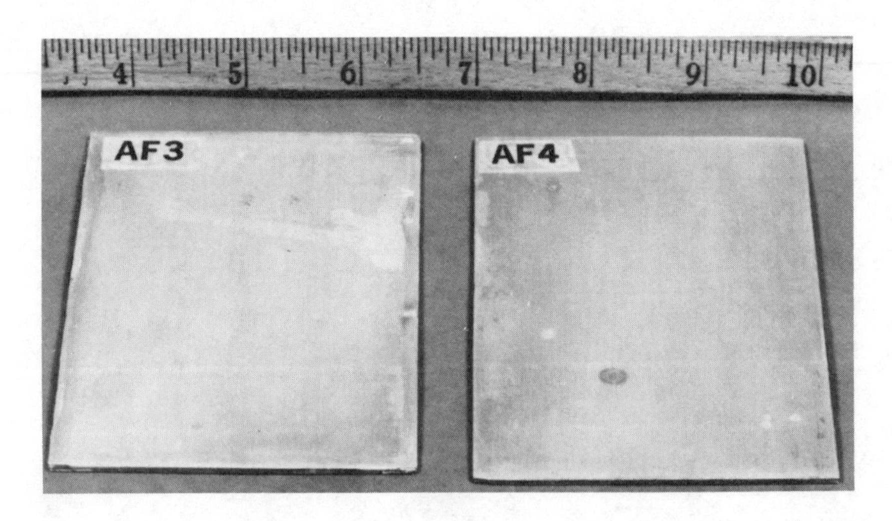

Figure 8- Corrosion Panels of VSB-32/Al 6061 After 12 Weeks Exposure to Tidal Zone Environment

composites exposed to the tidal zone environmment. All of the standard and encapsulated specimens experienced similar corrosion except for the effects of edge protection. The surface of the panels had incipient to light pitting as shown in Figure 8, with some pits causing small blisters as seen in the flowing seawater exposures. The edges of the corrosion panels which were protected did not experience any corrosion where the protective compound remained intact. Two panels (AB-3 and AF-3) did experience some edge attack due to bond failure of the sealant. The panels without any edge protection suffered slight to moderate edge corrosion. The encapsulated panels appeared to have slightly greater resistance to edge corrosion which is likely the result of the greater volume of aluminum in the matrix. Interestingly, the tidal environment did not show a clear exposure time dependence, and proved to be the least aggressive marine environment included in this test program.

Corrosion of Graphite/Aluminum in the Marine Atmosphere

Results of the corrosion exposures of the VSB-32/Al 6061 composites in the marine atmosphere are shown in Table 3. The surfaces of the standard and encapsulated panels experienced incipient to light pitting with some slight blistering around the pits as shown in Figure 9. This is not unusual for the 6061 aluminum alloy. The panels with edge protection were free of edge corrosion. The panels exposed to the marine atmosphere without edge protection experienced edge corrosion which was usually severe enough to cause separation.

The twelve-week exposure panels with edge protection did not appear significantly different from similar six week exposures as shown in Figure 10. The twelve week exposures without edge protection displayed advanced edge attack. This attack appeared to accelerate with time, indicating that the process would most probably continue until all of the aluminum matrix was oxidized.

Summary of Corrosion Exposures

The results of the corrosion exposures performed in this program show that the overall performance of the composite reflects both the performance of the surface foils and the performance of the graphite/ aluminum substrate. The dominant factor in the corrosion of VSB-32/Al 6061 appears to be th galvanic cell between the graphite /aluminum with a driving force of 1.0 to 1.2 volts.[2] As long as the surface foils of the composites prevent matrix invasion, the corrosion performance was equivalent to that of the surface foils. After matrix invasion occurred, the galvanic couple was activated, accelerated corrosion took place. The corrosion attacked both the matrix material and the matrix/foil interface, and was seen to be assisted by the production of corrosion products. There was a slight difference in the behavior of the two variations of graphite/aluminum tested due to the difference in fiber loading of the composites. However, there was no difference in inherent corrosion mechanism when comparing the standard and encapsulated materials.

Three different marine environments were included in this program. Regarding performance of the surface foils, the flowing seawater was the most aggressive, while the tidal immersion and atmospheric exposures were similar in pitting performance, but less aggressive than the flowing seawater. Where matrix invasion occurred, and free edges were exposed,

Table II

Corrosion Exposure Results for VSB-32/6061 Aluminum in the Tidal Zone

STANDARD

Specimen I.D.	Length of Exposure (weeks)	Edge Protection (yes or no)	Visual Observations surface	edge
AA-1	6	yes	SP	N
AA-2	6	no	SP	corr
AB-3	6	yes	P	corr
AB-4	6	no	P	corr
AC-3	12	yes	SP	N
AC-4	12	no	SP	E, corr
AF-3	12	yes	P, SB	corr
AF-4	12	no	SP	corr
ENCAPSULATED				
BB-3	6	yes	SP	N
BB-4	6	no	SP	corr
BC-1	6	yes	SP	N
BC-2	6	no	SP	corr
BE-1	12	yes	SP	N
BE-2	12	no	SP	N
BF-3	12	yes	P, SB	N
BF-4	12	no	P, SB	N

N - No visible attack
SB - Slight Blistering; B - Blistering
SP - Slight or incipient Pitting; P - Pitting
corr - general corrosion; E - edge separation

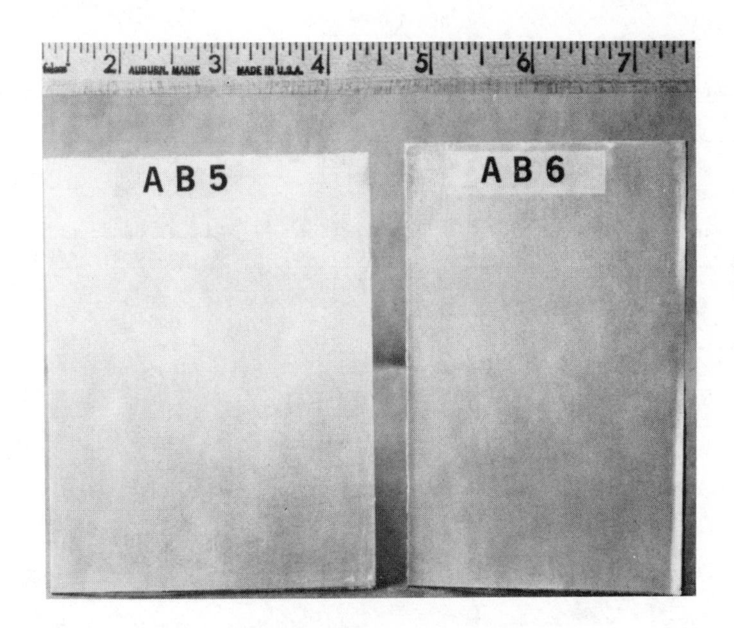

Figure 9- Corrosion Panels of VSB-32/Al 6061 After 6 Weeks Exposure to Marine Atmosphere

Figure 10- Corrosion Panels of VSB-32/Al 6061 After 12 Weeks Exposure to Marine Atmosphere

Table III

Corrosion Exposure Results for VSB-32/6061 Aluminum in Atmosphere

Specimen I.D.	Length of Exposure (weeks)	Edge Protection (yes or no)	Visual Observation surface	edge
		STANDARD		
AA-3	6	yes	SP	N
AA-4	6	no	P	E, corr
AB-5	6	yes	P	N
AB-6	6	no	P	E, corr
AE-1	12	yes	SP	N
AE-2	12	no	SP	E, corr
AF-5	12	yes	P	N
AF-6	12	no	P	E, corr
		ENCAPSULATED		
BB-5	6	yes	SP	N
BB-5	6	no	SP	corr
BC-3	6	yes	P, SB	N
BC-4	6	no	P, SB	corr
BE-3	12	yes	SP	N
BE-4	12	no	SP	E, corr
BF-5	12	yes	P, SB	N
BF-6	12	no	P, SB	E, corr

N - No visible attack
SB - Slight Blistering; B - Blistering
SP - Slight or incipient Pitting; P - Pitting
corr - general corrosion; E - edge separation

the marine atmosphere and the flowing seawater were the most severe environments, and the tidal exposure resulted in substantially decreased level of corrosion.

Residual Mechanical Properties

The purpose of the residual mechanical property tests was to evaluate possible latent effects of the marine environments on the mechanical properties of the VSB-32/Al 6061 composites. Due to the variation of fiber volume, and consolidation differences inherent in the batch processing of composite wire and plate, the mechanical properties of the unexposed plates varied not only from plate to plate, but also within each plate. Therefore, small changes in residual mechanical properties could be related in inherent material scatter, as well as effects of environmental exposures. In this context, it should be noted that a total of 24 individual panels were exposed for various durations in the three marine environments and subsequently used for post-exposure mechanical property tests. Only one panel of each type of VSB-32/Al 6061 composite was employed in evaluating mechanical properties prior to exposure.

Table 4 is a summary of the mechanical property test results from both types of graphite/aluminum composites where tests were performed in the as-received, unexposed condition. These tests show that the standard

Table IV

Mechanical Properties of VSB-32/6061 Aluminum Composites

	STANDARD	ENCAPSULATED
Longitudinal Tensile Strength	650 MPa	510 MPa
Longitudinal Modulus	177 GPa	140 GPa
Transverse Tensile Strength	25 MPa	35 MPa
Transverse Modulus	32 GPa	32 GPa
Fiber Content of Plate	40.1%	30.6%

Reported values are the average of five tests

composite displays superior longitudinal strength and modulus when compared to the encapsulated composite. This clearly reflects the higher fiber volume loading of the standard material. On the other hand, the transverse tensile strength of encapsulated composite is superior as expected, again because of the difference in fiber loading.

The residual tensile strength test results for all corrosion exposures of the standard VSB-32/Al 6061 composite are plotted in Figure 11, while similar data is presented for the encapsulated composite in Figure 12. In most cases, the data points presented in Figure 11 and 12 are the average of five tensile tests with specimens removed in such a way as to exclude material with obvious corrosion attack, except for surface foil pitting.

The results of residual mechanical property tests of the standard graphite/aluminum composite shown in Figure 11 suggests that there was no significant deleterious effect of any of the marine environments provided no evidence of corrosion attack was observed. The average tensile strength data for all exposures, (except twelve week tests in flowing seawter) fell in the range of 550-670 MPa. This range compared favorably with the data from the unexposed panel, and was consistent with the scatter inherent in the mechanical properties of these materials. In general, a tensile strength degradation on the order of 10-15% occurred in all environments for which specimens without visible corrosion attack could be tested. This is a reflection of the effects of the incipient/light pitting which occurred in all test environments.

Figure 11 also includes two other data points which were the average of tests performed on tensile specimens which had visible corrosion damage such as blistering or foil delamination. These specimens experienced twelve weeks of exposure in flowing seawater. These data show that after the matrix was invaded and corrosion occurred, there was a substantial decrease in average tensile strength. The order of this decrease reflects the severity of corrosion attack in the matrix of the standard composite material.

The residual mechanical property test results for the VSB-32/Al 6061 encapsulated composite are shown in Figure 12. All data points are average values for tensile tests performed with specimens displaying no visible corrosion damage. These data agree with the trends developed with the standard composite in that there was only slight degradation of tensile properties resulting from marine environment exposures. All data points are lower than for the standard composite because of the lower fiber loading of the encapsulated plates. The three lowest points in Figure 12, which occurrred in the twelve week tidal and marine atmosphere exposures, were all developed from the same plate of material. This suggests that inherently lower mechanical properties were responsible for the results rather than effects of the environmental exposures.

The results of the residual mechanical property tests with the VSB-32/Al-6061 composites indicate that there are no latent effects with any of the marine environments which result in severe degradation of tensile properties in the case where no visible corrosion attack is evident. In the case where corrosion attack has occurred as evidenced by severe pitting, blistering and foil delamination, the composites displayed substantial reduction in tensile strength. This reduction was in response to the vigorous attack on the 6061 aluminum matrix and the matrix foil interface separation.

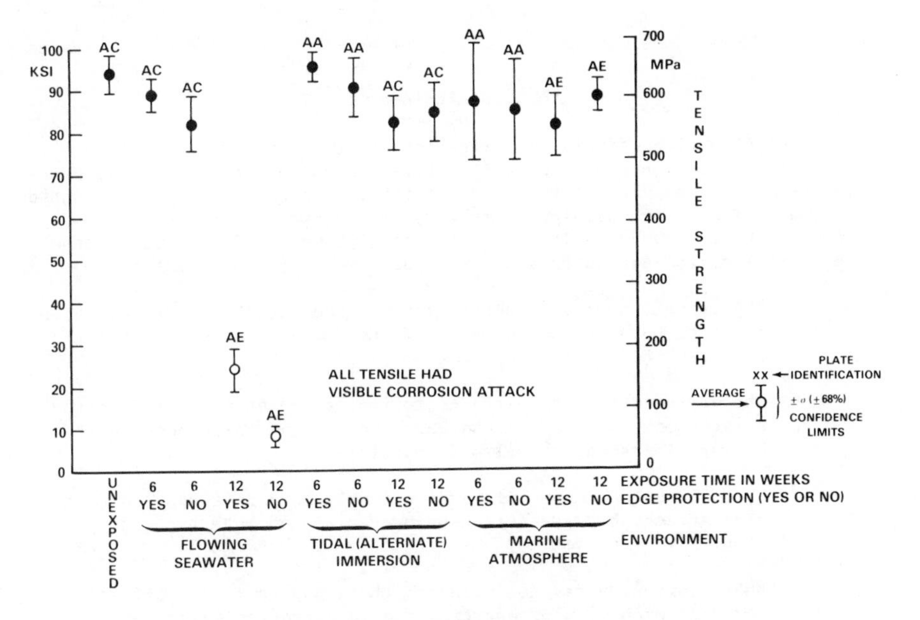

Figure 11- Residual Tensile Mechanical Properties of Standard Gr/Al Composite vs Corrosion Exposures

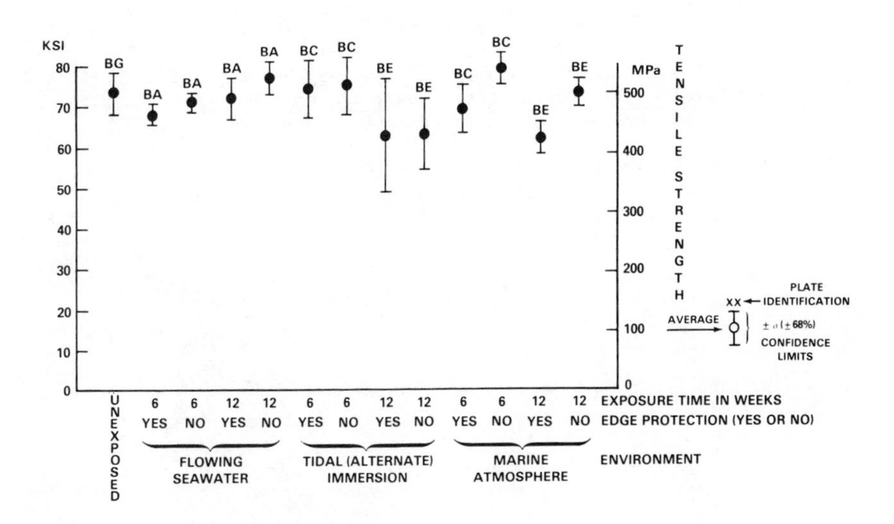

Figure 12- Residual Tensile Mechanical Properties of Encapsulated Gr/Al Composite vs Corrosion Exposures

Conclusions

The objective of this investigation was to characterize marine corrosion performance of two similar types of graphite/aluminum composite materials. A key element in the corrosion investigation was an evaluation of the residual mechanical properties of the composite in response to environmental exposure. The following conclusions are suggested by the results of marine exposures and tests performed in this investigation:

- The overall corrosion performance of the VSB-32/Al-6061 composite reflects both the performance of the surface foils and that of the graphite/aluminum substrate;

- The dominant factor in the corrosion of the standard and encapsulated compossites was the galvanic driving force between the VSB-32 fibers and the Al 6061 matrix;

- The encapsulated composite was slightly more resistant to corrosion attack, but the corrosion mechanisms of both types of composite were similar;

- Where corrosion was not evident, there was no latent effects on residual mechanical properties due to corrosion exposure;

- Visible corrosion damage resulted in significant degradation of residual mechanical properties.

References

1. Pfeifer, Will, "Graphite/Aluminum Technology Development", Hybrid and Select Metal-Matrix Composites: State of the Art Review, American Institute of Aeronautics and Astronautics, New York, NY 1977.

2. Tuthill, A.W. and C.M. Schillwaller, "Guidelines for Selection of Marine Materials" Proceedings of Ocean Science Engineering Conference, Marine Technology Society, Washington, DC, June 14-17, 1965.

Subject Index

Author Index